Emerging Trends and Innovations in Industries of the Developing World: A Multidisciplinary Approach

T0074876

CRC Press
Taylor & Francis Group
Boca Raton London New York

CRC Press is an imprint of the
Taylor & Francis Group, an **informa** business

First edition published 2024
by CRC Press
4 Park Square, Milton Park, Abingdon, Oxon, OX14 4RN

and by CRC Press
2385 NW Executive Center Drive, Suite 320, Boca Raton FL 33431

CRC Press is an imprint of Informa UK Limited

ISBN: 978-1-032-60103-8 (pbk)
ISBN: 978-1-003-45760-2 (ebk)

DOI: 10.4324/9781003457602

Typeset in Sabon LT Std
by Ozone Publishing Services

About

ISC 2022 is dedicated to the Niti Aayog policies to promote sustainability through exchange of ideas emerging out of the academia. The ISC is an annual conference that would be held in virtual mode until COVID restrictions on travel exist.

Vision

The vision of the conference is to capacitate Academia with the necessary ideas that will provide insights of the grassroot level development to various stakeholders of the Niti-Aayog policies. Towards this goal, the conference will create a conjunction of various stakeholders of Niti-Aayog policies that include- academic institutions, government bodies, policy makers and industry.

Mission

The ISC organizers will make concerted efforts to promote academic research that would provide technological, scientific, management & business practices, and insights into policy merits & disruptions. The framework of exchange of ideas will be geared towards adoption of deep technologies, fundamental sciences & engineering, energy research, energy policies, advances in medicine & related case studies. This framework will enable the round table discussions between the academia, industry and policy makers through its range of plenary and keynote speakers.

Chairs
Ravindra Pratap Singh, PhD
Indira Gandhi Tribal University, India.

Dimitrios A Karras, PhD
NKUA Greece.

Amarendra Pratap Singh, PhD
Indira Gandhi Tribal University, India.

K.K. JACOB, PhD
Faculty of Applied Sciences and Technology, Universiti Tun Hussein Onn Malaysia.

Organizing Commitee
Mengistu Tulu Balcha, PhD
Ambo University Ambo, Ethiopia.

Kannaiya Raja, PhD
Ambo University Ambo, Ethiopia.

Karthikeyan Kaliyaperumal, PhD
Ambo University Ambo, Ethiopia.

Publishing Committee:
Dimitrios A Karras, PhD
University of Athens (NKUA), Greece.

Karthikeyan Kaliyaperumal, PhD
Ambo University Ambo, Ethiopia.

Sudenshna Ray,PhD
RNTU-AIISECT University Bhopal, India.

Contents

Editor's Biography

Proceedings of the 1st International Sustainability Conference (ISC 2022)

Edited by

Dimitrios A Karras

Bio: Dimitrios A. Karras received his Diploma and M.Sc. Degree in Electrical and Electronic Engineering from the National Technical University of Athens (NTUA), Greece in 1985 and the Ph. Degree in Electrical Engineering, from the NTUA, Greece in 1995, with honors. From 1990 and up to 2004 he collaborated as visiting professor and researcher with several universities and research institutes in Greece. Since 2004, after his election, he has been with the Sterea Hellas Institute of Technology, Automation Dept., Greece as associate professor in Intelligent Systems-Decision Making Systems, Digital Systems, Signal Processing, till 12/2018, as well as with the Hellenic Open University, Dept. Informatics as a visiting professor in Communication Systems (the latter since 2002 and up to 2010). Since 1/2019 is Associate Prof. in Intelligent Systems-Decision Making Systems, Digital Systems and Signal Processing, in National & Kapodistrian University of Athens, Greece, School of Science, Dept. General as well as Adjunct Assoc. Prof. Dr. with the EPOKA university, Computer Engineering Dept., Tirana (1/10/2018–25/9/2020). He has published more than 70 research refereed journal papers in various areas of intelligent and distributed/multiagent systems, Decision Making, pattern recognition, image/signal processing and neural networks as well as in bioinformatics and more than 185 research papers in International refereed scientific Conferences. His research interests span the fields of intelligent and distributed systems, Decision Making Systems, multiagent systems, pattern recognition and computational intelligence, image and signal processing and systems, biomedical systems, communications and networking as well as security. He has served as program committee member as well as program chair and general chair in several international workshops and conferences in the fields of intelligent Systems-Decision Making Systems, signal, image, communication and automation systems. He is, also, former editor in chief (2008–2016) of the International Journal in Signal and Imaging Systems Engineering (IJSISE), academic editor in the TWSJ, ISRN Communications and the Applied Mathematics Hindawi journals as well as associate editor in various scientific journals, including CAAI, IET. He has been cited in more than 2220 research papers (https://scholar.google.com/citations?user=IxQurTMAAAAJ&hl=en), (Google Scholar) and 1626 (ResearchGate Index, after recent recounting based only on ResearchGate database, with more than 9105 ResearchGate Index Reads metric) citations, as well as in more than 1101 citations in Scopus peer reviewed research database, with H/G indexes 20/48 (Google Scholar), Scopus-H index 15, RG-index 30.96 and RG-Research Interest index 886.1 (https://www.researchgate.net/profile/Dimitrios_Karras2/).

Sai Kiran Oruganti

Indian Institute of Technology Patna India.

ORCID: https://orcid.org/0000-0003-4601-2907

Areas: Wireless Power Transfer, Wireless technologies, IoT, Radio Science, Electromagnetics & applications.

<u>Profile Summary</u>

Prof. Dr. Sai Kiran Oruganti is with the School of Electrical and Automation Engineering, Jiangxi University of Science and Technology, Ganzhou, People's Republic of China as a full Professor since October 2019. He is responsible for establishing an advanced wireless power transfer technology laboratory as a part of the international specialists team for the Center for Advanced Wirless Technologies. Between 2018–2019, he served as a senior researcher/Research Professor at Ulsan National Institute of Science and Technology. Previously, his PhD thesis at Ulsan National Institute of Science and Technology, South korea, led to the launch of an University incubated enterprise, for which he served as a Principal Engineer and Chief Designer in 2017–2018. After his PhD in 2016, he served Indian Institute of Technology, Tirupati in the capacity of Assistant Professor (Electrical Engineering) between 2016–2017.

<u>Research</u>

Prof. Dr. Oruganti, prime research focus is in the development of Wireless Power Transfer(WPT) for applications- Internet of Things (IoT) device charging, Agriculture, Electric Vehicle Charging, Biomedical device charging, Electromagnetically induced transparancy techniques for military and defence applications, Secured shipping containers, Nano Energy Generators.

<u>Achievements</u>

Prof. Dr. Oruganti has more than 21 patents pending on his credit and with several of those patent applications passing the NoC stage. As of 2021, 16 of 21 patents have been granted. He is credited with the pioneering work in the field of Zenneck Waves based Wireless Power Transfer system. Most notably, he has been regarded as one of the only few researchers in the field of WPT to be able to conduct power and signal transmission across partial Faraday shields. His recent paper accepted by Nature Scientific Reports has generated a lot of interest and excitement in the field. International Union of Radio Science(URSI) recognized his research efforts and awarded him Young Scientist Award in 2016. He is also recipient of IEEE sensors council letters of appreciation.

Sudeshna Ray

Dr. Ray's research at the interface of Chemistry and Material Science is focused on the development of Novel Inorganic based Luminescent materials for the application in white Light Emitting Diodes (LEDs) and as Spectral Converters in Solar Cell. The novel materials can be a single composition multi-centred phosphor or near UV/blue excitable blue, green, yellow and red emitting phosphor. The utilization of 'Green' solution based Synthesis Methodology for the precise control of the composition of the phosphors and achievement of a homogenous distribution of small amounts of activators in the host compounds is the main paradigm of my Research. In addition, to my previous focus on Synthesis, Characterization and Optical studies of size and shape tuned nanocrystalline Y2O3, YVO4 and YPO4 based phosphors, I am extensively involved into the research for the development of Advanced Luminescence Materials so called Quantum Cutters for the application in Solar Cell. A unifying theme of my research is the compositional tuning of the properties of extended solids through solid solution; sometimes referred to as the game of x and y, as, for example, in Sr2(1-x-y/2)Eu2xLaySi1-yAlyO4. Design of New phosphor for LEDs and fabrication of LEDs using the phosphors is an integral part of my Research. Currently, I am involved in the synthesis of Persistent Phosphors for the fabrication of Glow Bullet for Defense Application. RESEARCH INTERESTS • Exploration of New Phosphors by Mineral Inspired Methodology • Development of water soluble silicon compound by alkoxy group exchange reaction • Synthesis of Eu2+ and Ce3+ doped silicate phosphors using water soluble silicon compound • Solution Synthesis of 'Size' and 'shape' tuned Nanomaterials • Characterization of Nanocrystalline phosphors by XRD, TEM, FE-SEM and Raman spectroscopic measurement • Study of 'Up-conversion', 'Down-conversion', 'Down-shifting' phenomena by steady state photoluminescence and lifetime measurement and analysis • Measurement of 'Quantum Efficiency' and Thermal stability of phosphors • Development and optical study of 'Quantum Cutting' Materials as Spectral Converter for Solar cell.

RESEARCH EXPERIENCE

- Postdoctoral Fellow Phosphor Research Laboratory, Department of Applied Chem. September (2012) – July (2013) National Chiao Tung University, Taiwan Advisor: Prof. Teng Ming Chen
- Study of Energy Transfer from sensitizer to activator by lifetime analysis.
- Measurement of Quantum Efficiency

Foreword

I had the privilege to serve as the convenor for the first ISC2022 (formerly ICTSGA-1) which is dedicated to the realization of Niti-Aayog policies of the government of India. As a chairman for the Technology Innovation Hub IIT Patna whose National Mission on Interdisciplinary Cyber-Physical Systems is dedicated to Analytics, I have always felt the need to make an outreach to the Indian as well as international academia. ISC2022 is a prelude to the larger vision to capacitate Academia with the necessary ideas that will provide insights of the grassroot level development to various stakeholders of the Niti-Aayog policies. Towards this goal, the conference has aimed to create a conjunction of various stakeholders of Niti-Aayog policies that include- academic institutions, government bodies, policy makers and industry.

I hope that the contents of this series will serve as a guide for young researchers and policy makers for their future endeavours towards a holistic growth of the human society.

Trilok Nath Singh, PhD
Professor & Director, Indian Institute of Technology Patna.
Chairman and Board of Directors, Technology Innovation Hub,
Indian Institute of Technology, Patna-INDIA.

Preface

ISC 2022 (Formerly ICTSGA-1) is dedicated to the Niti Aayog policies to promote sustainability through exchange of ideas emerging out of the academia. The ISC is an annual conference that would be held in virtual mode until COVID restrictions on travel exist. The conference featured several plenary and keynote speeches from UN, African Environmental Sustainability, Mahatma Gandhi University, Universidad Politécnica de Valencia, Technical University of San Luis Potosí Mexico, Delhi Technological University, University at Johannesburg. The sessions were divided into two major sections: (a) Sciences & Engineering (b) Management & Humanities. In addition, a dedicated session on women in sciences, and academia was held. The sessions included ~50% representation from Women in academia, and research.

Statistics

The ISC 2022 received 816 abstract submissions of these only 340 were selected.

- Total Plenary Talks: 4
- Total Keynote Talks: 8
- Total Invited Talks: 7
- Total Oral Talks spread across two days: 35
- Total Women in SDG talks: 26

Conference Chairs

Ravindra Pratap Singh,
IGNTU Amarkantak, India

Dimitrios A Karras
National and Kapodistrian
University of Athens, Greece

Amarendra Pratap Singh
IGNTU Amarkantak, India

Introduction

Vision

The vision of the conference is to capacitate Academia with the necessary ideas that will provide insights of the grassroot level development to various stakeholders of the Niti-Aayog policies. Towards this goal, the conference will create a conjunction of various stakeholders of Niti-Aayog policies that include- academic institutions, government bodies, policy makers and industry.

Mission

The ISC organizers will make concerted efforts to promote academic research that would provide technological, scientific, management & business practices, and insights into policy merits & disruptions. The framework of exchange of ideas will be geared towards adoption of deep technologies, fundamental sciences & engineering, energy research, energy policies, advances in medicine & related case studies. This framework will enable the round table discussions between the academia, industry and policy makers through its range of plenary and keynote speakers.

Details of Programme Committee

Organizing Committee:

T. Sunder Selwyn, PhD
Prince Dr. K. Vasudevan College of Engineering and Technology, Chennai India.

K.K. JACOB, PhD
Faculty of Applied Sciences and Technology, Universiti Tun Hussein Onn Malaysia.

Mengistu Tulu Balcha, PhD
Ambo University Ambo, Ethiopia.

Kannaiya Raja, PhD
Ambo University Ambo, Ethiopia.

Karthikeyan Kaliyaperumal, PhD
Ambo University Ambo, Ethiopia.

Publishing Committee:

Dimitrios A Karras, PhD
University of Athens (NKUA), Greece.

Karthikeyan Kaliyaperumal, PhD
Ambo University Ambo, Ethiopia.

Sudenshna Ray, PhD
RNTU-AIISECT University Bhopal, India.

Insuretech: Saviour of Insurance Sector in India

M. Umasankar,[*a] Kavitha Desai,[a] and S. Padmavathy[b]

[a]Christ University, Bangalore, India
[b]Kongu Engineering College, Erode, India
E-mail: [*]umasankar.m@christuniversity.in, kavitha.d@christuniversity.in, padmavathy.mba@kongu.edu

Abstract

Technology in finance has propelled financial literacy and inclusiveness and may give the insurance sector an edge to reach its potential consumers. The current study aimed to identify the role of Fintech in transforming the insurance sector and improving the penetration rate in India. With the descriptive research design, the study collects the primary data through a survey technique targeting the general public and personnel in the insurance sector as a study population. A conceptual model is proposed to understand the interlink between consumer attitude towards Insurance, factors influencing their decision, and the role of Fintech in bridging the gap in insurance penetration. This study focuses on three areas, namely health insurance, life insurance, and vehicle insurance. The study's findings reveal that the insurtech will significantly improve the efficiency of the insurance sector which will result in significant financial performance.

Keywords: Fintech, Health Insurance, Insurance Sector, Structural Equation Model, Consumer Attitude, InsurTech

Introduction

Technology is ruling today's world, and there are no stones unturned by it. It becomes part and parcel of human lives. The word technology means techniques used to reduce or ease the human work burden. With this meaning, if we get insight into technological development, it includes the evolution of human lives. With the invention of the wheel and the proper utilization of fire, everything can be called technology. We are in the highest growing phases of the technological life cycle. It is pretty impossible to think of a life without technology. The technology becomes a suffix word, which can be added along with any noun. The way the experts' coin finTech is that finance is suffixed by technology that becomes FinTech. Fintech is nothing but technological adoption in the financial sector, which eases financial transactions and activities. It is because of Artificial Intelligence and the Internet of Technology in the financial industry. Finance is a broader term that included many aspects that involve money. One of the components in the financial sector is the Insurance sector.

As human life becomes so fast and insecure, insurance is a vital element. Insurance is devised to compensate for the losses and damages in human life. It could be the loss of a life or articles we use. It ensures compensation and assures backup plans. Insurance is considered a vital element and also a lifesaving investment among people in developed countries. At the same time, this is the opposite scenario in countries like India. The penetration of the insurance sector in India is slower than the global average. Many factors affect insurance sector growth in India. The factors include accessibility, affordability, social norms, the psychological character of consumers, etc. The arrival of technology has brought up developments in many sectors, and its inclusion in the financial sector has proved a significant contribution. Keeps this in mind, the technocrats come out with a concept of technology in the insurance sector. This study expects to comprehend the impact of different InsurTech results on changing the protection area in India. For this, five distinct elements were recognized as the result of InsurTech after the careful writing survey and conversation with the business specialists. The dimensions are named as Efficiency Improvement, Improved Risk assessment, Greater Consumer Satisfaction, Cost Reduction, and Financial Inclusion. Through technology boosts development, it also has some downgrades in terms of security, and the successful implementation of the technology in any sector strictly depends on infrastructure capabilities. Both the issues tended to in this exploration and are important for the examination system as mediating factors. The change of the protection area is the reliant variable in this review, and it is estimated through the monetary execution of insurance agency over the most recent five years.

Efficiency Improvement

Efficiency has been identified as bottleneck for insurance sector growth in India as the settlement of finances after the claim is taking too much time. It involves an unidentifiable problem in the settlement process, and it results in detention in the concurrence of claims. The main cause behind this was the homemade operation of the claim agreement process. The introduction of technology in the insurance sector might significantly reduce the time taken for settlement and other functions involved. Hence, InsurTech leads to improve efficiency in the insurance service sector, which is considered one of the variables in this study.

Cost Reduction

As mooted in the former point, the insurance sector consists of multitudinous manual processes and mortal intervention, leading to high costs in delivering Insurance-related services to the end-user. The insurance sector also has multitudinous intercessors between the insurance companies and the customers. Hopefully, InsurTech will especially remove the intercessors, and the ultimate of the repetitive processes will be automated with the help of artificial intelligence. Hence the service cost can be reduced to range that is affordable for the consumer. Cost Reduction is considered one of the outcomes of InsurTech in this study.

Improved Risk Assessment

Even the insurance companies also have some constraints in providing the service to some consumers based on few reasons. One of the reasons could be many times, the consumer making false claims. This also slows down the settlement of claims due to thorough verification and assessment of claims. Hence assessment of the risk of claims is a crucial element in the insurance service delivery. The involvement of technology could ease the process of assessing the risk with the help of Artificial Intelligence. Since InsurTech can improve risk assessment, it is added as one factor in this study.

Greater Customer Satisfaction

Customers neither know the background process nor are interested in understanding them, but they need exemplary service at an affordable cost. As denoted, technology addition could ameliorate efficiency, which will result in cost reduction and accuracy of claim assessment. With this, the insurance service providers can offer a good service and can achieve greater client satisfaction. Hence customer satisfaction was included in the study as an outcome of the InsurTech.

Huge Financial Inclusion

Though India is a highly populated country globally, the penetration of the insurance sector is least on the global average. This could be because the insurance sector reached only a few segments of the population because of its complex process. InsurTech can make the process simple and reach the untapped consumer segment.

Literature

Practitioners and incumbents are looking for InsurTech and new technologies, led by improved client satisfaction and efficiency (Greineder et al., 2018). Aside from the main task of providing Protection, Insurance companies are looking with interest also to providing Prevention to create value. This possibility is considered to have strong market potential and probability of success by practitioners (Oliver Wymann, 2019). The theme has high importance at the policy level as well: InsurTech can bring both opportunities, such as higher Insurance inclusiveness (Altamirano & van Beers, 2018) and public health improvement (Yamasaki & Hosoya, 2018), as well as threats, such as privacy concerns.

From the current literature review, InsurTech and new technologies can directly affect how insurance and Prevention are interconnected. Due to the emergence of new technologies allowing to collect and analyze huge amounts of data from individuals and objects (Gidaris, 2019), Geste-predicated the models of Insurance are displaying, eventuality in order to enable a highly integrated relationship between prevention and Insurance, given, for case, the effect that geste-predicated Insurance pricing has on individual behavior (Wiegard & Breitner, 2019). New technologies and InsurTech could play a momentous part also in enabling and backing Insurance companies in furnishing Prevention services (Stoeckli et al., 2018).

The importance of Insurance is evident also looking at the numbers of the Insurance sector, for instance, in Europe. In 2017, the total gross written premia were equal to 1231 billion euros. Total claims and benefits paid equal to 1014 billion euros, and the average spending per capita on Insurance was similar to 2030 euros. There is two main kinds of Insurance: Property and Casualties (P&C) Insurance, which protect the insured against possible losses coming from damages to person, objects or capital, and Life Insurance, where the insurer recognizes a certain amount of money to the insured in case of a life event, such as death. However, the current panorama for Insurance companies is changing (Stoeckli et al., 2018), with the increasing dimension and disruptive nature of digitalization and the emergence of a new phenomenon labeled with the term "InsurTech."

Yet, there is a search for a well-known definition of InsurTech. For someone, InsurTech is "a marvel involving advancements of at least one customary or non-conventional market players misusing data innovation to convey arrangements explicit to the Insurance business" (Stoeckli et al., 2018). For someone else, "InsurTechs" is, more specifically, "innovation drove organizations that enter the Insurance area, exploiting new advances to give inclusion to an all the more carefully shrewd client base.

For instance, the above-portrayed change in clients' conduct to utilize wearable innovation for personal growth and self-checking sets out new open doors for well-being and extra security. Nowadays, in the data-driven world, most of the data related to a product were not utilized by the producer, so there is considerable potential that needs to be exploited to help InsurTech within the insurance industry. Albeit the collection of literature on FinTech and InsurTech is scarce, earlier examination contains endeavors to conceptualize FinTech (Puschmann, 2017). Here the study tries to match the meaning of InsurTech to the insurance sector rather than it commonly mentioned with banking operations (Puschmann, 2017).

A detailed review of FinTech reported by (Zavolokina et al., 2016) delivers a complete groundwork. Initially, the strategies for insurance companies were mainly regarded as InsurTech (Chuang et al., 2016), and InsurTech the involvement of technology in the financial sector is called FinTech. Moreover, this study emphasizes that InsurTech development can have its starting point in customary monetary specialist co-ops and non-conventional organizations, like new businesses and organizations from different areas. Puschmann (2017).

Objectives

The study's objective is to identify the role of Fintech in transforming the insurance sector in Inia. For this, five variables were identified as dimensions of Fintech that contribute to the transformation of the insurance sector.

Methodology

Since very little literature available on Fintech and comparing its role in transforming the insurance sector, this study adopted the descriptive research design to identify the dimensions of Fintech in Insurance and its role in transforming the insurance sector. There are two types of people selected as the target population to collect the research data. They are common public who have taken any one or more insurance and insurance, sector employees. The probability of sampling is ensured by adopting a simple random sampling technique to choose the target respondents.

Theoretical System

Given the goal and factors remembered for the review, the accompanying reasonable system is created with autonomous factors, subordinate variables, and mediating factors.

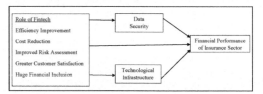

Fig. 1. Conceptual Framework

Result and Discussion

Table 1. The relationship between the hypothesized Paths

Hypothesized Path Relations	Unstandardized Estimate	Standardized Estimate	S.E.	C.R.	P	R^2
Efficiency Improvement → Technological Infrastructure	.379	.329	.080	4.738	***	
Huge Financial Inclusion → Technological Infrastructure	1.000	.597	.127	7.846	***	0.43
Improved Risk Assessment → Technological Infrastructure	.312	.262	.088	3.535	***	
Cost Reduction → Technological Infrastructure	.338	.512	.050	6.724	***	
Cost Reduction → Data Security	.094	.148	.039	2.425	.015	
Improved Risk Assessment → Data Security	.429	.376	.080	5.354	***	
Greater Customer Satisfaction → Data Security	.254	.245	.064	3.977	***	0.54
Technological Infrastructure → Data Security	.232	.241	.059	3.926	***	
Efficiency Improvement → Data Security	.220	.198	.073	2.999	.003	
Data Security → Financial Performance	.172	.170	.049	3.507	***	
Efficiency Improvement → Financial Performance	.822	.733	.048	17.263	***	
Cost Reduction → Financial Performance	.167	.259	.033	5.086	***	
Technological Infrastructure → Financial Performance	.194	.199	.047	4.128	***	0.81
Huge Financial Inclusion → Financial Performance	.220	.135	.086	2.568	.010	
Greater Customer Satisfaction → Financial Performance	.147	.140	.045	3.287	.001	

According to the conceptual model, five variables related to InsurTech are termed exogenous variables influencing the fiscal performance of the Insurance firms. In the meantime, it again deals with two intermediating variables, Data Security and Technological structure, which are anticipated to intervene between the exogenous and endogenous variables. The theoretical framework is tested with structural equation model. The result of the same is presented in Table no. 2. It's found that technological structure is pivotal for the effective perpetration and success of InsurTech. Huge Financial Inclusion and Cost Reduction are the two factors that significantly influence performance of insurance sector, which can be linked from the measured value of 0.597 and 0.512, independently.

Hence, efficient technological infrastructure will reduce the cost involved in Insurance and lead to substantial financial inclusion. The second intervening variable is Data Security, an essential factor when technology remains engaged in financial transactions. Risk assessment becomes the vital factor that catalyst the data security elements in InsurTech.

The study results revealed that the fiscal performance of the insurance sector had been largely told by four Fintech confines and two intermediating variables. effectiveness enhancement embedded by Fintech was set up to be the loftiest impacting factor with a measured value of 0.733. Cost reduction is also another factor that appreciatively influences the fiscal performance of the insurance sector.

Conversation/Conclusion and Recommendations

With the outcome examined in the review, it is distinguished that the consideration of innovation in protection innovation will raise numerous potential changes and lead to change also. Productivity improvement will take special care of higher monetary execution with the assistance of mechanical foundation and information security. Creating an appropriate framework is the way to the effective execution of protection. Effectiveness will empower the insurance agency to arrive at additional customers by simplifying the cycle and direct.

The study identified that consumers do not get any return on investment which is an important reason for not opting for Insurance. It is pretty standard in countries like India, where most of the population belongs to the middle-class and lower-middle-class categories. They will be more cautious about every penny they spent and need more awareness to believe that the amount they spend on Insurance is an investment. Thus the insurance agency needs to exercise systems to plan the expense element of Insurance to draw in additional purchasers. While managing the expense element of Insurance the organizations likewise need to deal with information security issues, which consume a critical part of the expense engaged with Insurance in the mechanical climate. The review discoveries additionally substantiate that price reduction variables are not straightforwardly connected with information security in the end.

Whereas cost of a service is a major worry meant for consumers, risk evaluation is a critical worry meant for organizations that offer protection administrations. Insurance agencies invest more energy evaluating the safety net provider's gamble due to multi-overlap misrepresentation claims rising consistently. InsurTech is a shelter for an insurance agency that deals with risk evaluation with the assistance of man-made brainpower. According to this review, risk evaluation isn't straightforwardly connected with neither innovative foundations nor the monetary exhibition of the insurance agency. Interestingly, it is a pivotal variable for information security.

Consumer loyalty is a definitive point of any business activity. The review distinguished that there is no immediate connection between consumer loyalty and innovative foundation. The respondents are worried about information security, which makes them happy with the cycle and drive the presentation of the insurance agency. Information security goes about as an interceding variable between consumer loyalty and monetary execution. Subsequently, the insurance agency holding back nothing fulfillment can give a high information security climate without compromising the innovative office.

Technology has a tremendous advantage over any business activity due to its reachability and affordability. Today's technological advancement doesn't require the user to be well-qualified to use it. With the help of artificial intelligence and the internet of things, communication between companies and consumers is so quick and easy. With the arrival of 4G and smartphones, everyone has access to technology at an affordable cost. This gives more space for the segment of the population deprived of insurance awareness and services in the past. The study predicts that with the help of InsurTech, more consumers will be attracted to an untapped potential segment.

Conclusion

Though India's insurance sector is not penetrating high, compared with the other countries, there is a massive positive change in its performance after the inclusion of technology. Wherever the technology is involved, it is always coupled with other issues as well. Hence the study has identified Data security and Technological Infrastructure that curbs the financial performance of the Insurance sector in the presence of InsurTech. This study brings up a new dimension in predicting the financial performance of insurance sector through the InsurTech dimension. The result will recommend the insurance industry stakeholders to develop infrastructure facilities to launch the upcoming changes in the industry and ensure data security elements to create trust among the insurers.

References

Altamirano, M. A., and van Beers, C. P. (2018). Frugal innovations in technological and institutional infrastructure: Impact of mobile phone technology on productivity, public service provision and inclusiveness. The European Journal of Development Research, 30(1), 84–107.

Chuang, L.-M., Liu, C.-C., and Kao, H.-K. (2016). The adoption of fintech service: TAM perspective. International Journal of Management and Administrative Sciences, 3, 1–15.

Gidaris, C. (2019). Surveillance Capitalism, Datafication, and Unwaged Labour: The Rise of Wearable Fitness Devices and Interactive Life Insurance. Surveillance & Society, 17(1), 132–138.

Greineder, M., Riasanow, T., Böhm, M., and Krcmar, H. (2018). The Generic InsurTech Ecosystem and its Strategic Implications for the Digital Transformation of the Insurance Industry. Gi Emisa, May, 1–14.

Oliver Wymann, P. (2019). The Future Of Insurtech In Germany PricewaterhouseCoopers. (2014). Early days. In. https://www.pwc.com/us/en/health-industries/health-research-institute/publications/health-wearables-early-days.html

Puschmann, T. (2017). Fintech. Business & Information Systems Engineering, 59, 1.

Stoeckli, E., Dremel, C., and Uebernickel, F. (2018). Exploring characteristics and transformational capabilities of InsurTech innovations to understand Insurance value creation in a digital world. Electronic Markets, 28(3), 287–305.

Wiegard, R. B., and Breitner, M. H. (2019). Smart services in healthcare: A risk benefit analysis of pay-as-you-live services from customer perspective in Germany. Electronic Markets, 29(1), 107–123.

Yamasaki, K., and Hosoya, R. (2018, August). In I. International (Ed.), Resolving Asymmetry of Medical Information by using AI: Japanese People's Change Behavior by Technology- Driven Innovation for Japanese Health Insurance (pp. 1–5). Conference on Management of Engineering and Technology (PICMET). IEEE.

Zavolokina, L., Dolata, M., and Schwabe, G. (2016). The {FinTech} phenomenon: antecedents of financial innovation perceived by the popular press. Financial Innovation, 2(1). https://doi.org/10.1186/s40854-016-0036-7

Efficiency Study of Coconut Producer Companies in India – A DEA Approach

Shine Raju Kappil,[*,a] *Ajeet Kumar Sahoo,*[b] *and Alwin Joseph*[c]

[a,b]Central University of Haryana, India
[c]CHRIST (Deemed to be University), India
E-mail: [*]shineraju15@gmail.com

Abstract

The concept of the Farmer Producer Company (FPC) model has been a hot issue, especially during the 2020–21 Indian farmers' protest. Considering the pioneering initiatives of the Coconut Development Board (CDB) in setting up CPCs, we compare the technical efficiencies of CPCs that focus on coconut and its byproducts in Rural India for two consecutive financial years (2018–19 and 2019–20). Coconut Producer Companies' efficiency scores are estimated using Data Envelopment Analysis (DEA), a mathematical technique to assess technical efficiencies across homogeneous units. The results reveal that 35.11 percent of the sampled CPCs for FY 2018–19 are overall technical efficient, and approximately 76 percent are purely technical efficient. It is found that technical inefficiency is reported for a few CPCs due to scale inefficiency. The overall technical and pure technical efficiency have improved in FY 2019–20 compared to the previous period.

Keywords: FPC Model, Technical Efficiency, EBITDA, DMU

Introduction

Coconut is considered as a fruit, seed, or whole palm tree that has been widely grown in the southern regions of the Indian subcontinent, especially in the coastal areas. Small and marginal farmers dominate the Indian coconut sector as most farmers engage in small-scale agricultural activities by operating at landholdings below two hectares. Usually, these coconut growers are exploited by the mediators who deny fair prices for coconut products. In addition, coconut farmers face problems in getting subsidized financial assistance, storage facilities, access to quality seeds and other inputs, etc. This adversely affects the earnings of coconut growers and puts strain on their family life. The case of other crop producers is the same across the region.

Realizing these facts, an agricultural committee chaired by YK Alagh recommended in 2003 to set up farmer producer companies (FPCs), a blended form of company and cooperative features. Until then, many states had promoted Farmer Producer Organisations (FPOs) registered under Cooperative Societies Act, Multi-State Cooperative Societies Act, Society Registration Act, etc. But over time, these collectives were hijacked by local traders, mediators, and political intruders, and finally, most of these collectives walked away from their objectives. This prompted the agricultural experts to develop a new institutional arrangement that accommodates the welfare features of cooperatives by discarding their weaknesses and working under a professional company format.

Considering the recommendation of agricultural exports, the Government of India has amended the Indian Companies act 1956 by incorporating section 581(C) in 2013, allowing the registration of producer companies in Indian territory. Thereby, FPCs have turned out to be a form of farmer collectives under the umbrella of the FPO category.

In this paper, the authors attempt to evaluate the performances of these Coconut Producer Companies (CPCs) promoted by the Coconut Development Board (CDB) across the country. CDB is among the pioneers among various stakeholders in implementing the FPC model.

Since its inception, the Central government has pumped huge funds and resources to promote CPCs. It is high time to assess the performance and decide whether CDB needs to focus on either establishing new CPCs, strengthening the existing CPCs, or both. There are numerous initiatives by the governments, both directly and indirectly, necessitating institutional arrangements such as FPOs at the village level to revive the Indian agricultural sector. This again prompted us to examine the performance of those existing FPOs and suggest policy recommendations if required.

Table 1. Inclusion and Exclusion criteria for efficiency analysis of CPCs in India (2018-2020)

		Inclusion Criteria	
S.No.	Criteria	Selected No. of CPCs After Each Criterion	Description
1	CPCs promoted by CDB	67	Inclusion of all CPCs registered under the Indian Companies Act 2013. It excludes cooperatives and FPCs to fulfil the homogeneity assumption
2	CPCs formed before 31st March 2017	65	Inclusion of mature CPCs, i.e., completed at least one financial year after registration.
3	Active CPCs	55	Inclusion of CPCs which files financial documents since inception and are not stricken off by MCA.
		Exclusion Criteria	
S.No.	Criteria	Selected No. of CPCs After Each Criterion (2018–19, 2019–20)	Description
1	Lack of Financial documents	43, 41	Exclusion of CPCs whose financial documents are not available due to technical reasons
2	Lack of Certain variable data	43, 34	The exclusion of CPCs with certain variable data is unavailable.
3	Presence of Zero values	27, 26	Excluding those CPCs with variables with zero values is part of the positivity requirement for DEA analysis.

While reviewing the previous literature, we come across several conceptual and case study reports by government organizations regarding CPCs. But the number of empirical works related to CPCs and their role in India was limited in the public domain. We have used advanced DEA models like SBM models and validated them by conducting stability tests suggested by Avkiran (2007) to check the robustness.

Material and Methods

There are various approaches to evaluating firms' performance, including methods such as ratio indicators, and parametric and non-parametric strategies in the literature. Data envelopment analysis (DEA) is a non-parametric approach that utilizes optimization techniques to estimate the relative efficiencies of entities under consideration. DEA models are also widely used in assessing efficiencies of banks, non-profit organizations, regions, etc., in the efficiency literature. Compared to other efficiency approaches, DEA models have great applicability in complex situations as it opens the scope to accommodate multiple inputs and outputs in the analysis. In addition, these models need not assume any functional forms to justify the relationships found in econometric approaches. Among the DEA models, we choose the Slack-based measures of efficiency model (SBM), a non-radial model, over the radial and traditional models of DEA as they suffer from few shortcomings. We calculate the efficiency scores using non-oriented SBM models, which accommodate the slacks while estimating the relative efficiencies of DMUs and providing the best results (Tone, 2001).

We estimate the technical efficiency that highlights the capacity of a CPC to produce the maximum outputs from the available inputs. We consider overall technical efficiency (OTE), pure technical efficiency (PTE), and scale efficiency scores of CPCs in India. PTE measures the actual levels of technical efficiency that contributed to the production process. SE refers to the role of scale size on the OTE scores.

Selection of FPCs

Homogeneity assumption plays a crucial role when choosing the DMUs as the DEA models assume that all the selected DMUs are homogeneous. There are 67 CPCs established and promoted by CDB across India, and these CPCs are located in major coconut-growing states such as Kerala, Tamil Nadu, Karnataka, and Andra Pradesh. We adopt an inclusion and exclusion approach to shortlist the targeted CPCs and create a DMU set to estimate their relative efficiencies. Inclusion criteria are the specific features of the population that the researcher chooses to find solutions to research questions. On the other hand, exclusion criteria are those features of the eligible people under inclusion criteria but have certain additional features that may hinder the study's outcome (Patino and Ferreira, 2018).

Three inclusion and three exclusion criteria are used to shortlist the DMUs, and these conditions are framed by considering the assumptions of the DEA technique and data availability (Table 1). As per inclusion criteria, we have shortlisted active CPCs promoted by CDB and formed before 31st March 2017. Eventually, the number of qualified CPCs has come down to 55 CPCs. Under exclusion criteria, we have excluded the CPCs that lack financial documents, variable data, and the presence of zero values for the targeted years; therefore, the number of CPCs has reduced from 67 to 27 and 26 for FY 2018–19 and FY 2019–20 respectively. Our DMU sets consist of CPCs hailing from all major coconut-growing states except Andra Pradesh for the FY 2018–19.

Selection of Variables

In DEA analysis, the right decision on choice and the number of variables, including input and output variables and the number of DMUs, play an inevitable role in determining the ability of the model to discriminate the DMUs into efficient and inefficient units. DEA analysis has rules regarding the required number of inputs, outputs, and DMUs. The researcher has the freedom to increase the number of DMUs to improve the model's discriminatory power without compromising the assumption of homogeneity. Boussofiane et al. (1991) said that the number of DMUs must be greater than the product of the number of inputs and outputs. Golany and Roll (1989) put forward a rule of thumb where the number of DMUs must be at least twice the summation of the number of inputs and outputs. On the other hand, Dyson et al. (2001) stated another rule of thumb where the number of DMUs is at least twice the product of the number of inputs and outputs. We have three inputs and two output models with several DMUs greater than the minimums recommended by the rule of thumbs.

Following the earlier studies on firm-level efficiency analyses, we choose three input and two output model frameworks. We also consider the availability of variable data and commonly used variables in the efficiency literature while selecting the input and output variables. The input variables are materials &

purchases, operating expenses, and fixed assets. Total revenue and EBITDA are chosen as output variables in the study. EBITDA of some CPCs is found to be harmful for the financial year and it hinders the non-negativity assumption of DEA analysis. Therefore, following Bowlin (1998), we have added a positive constant to all the output variable (EBITDA) values. This positive constant must be large enough to convert all the negative values of the out variable into positives. We also checked the descriptive statistics of variables as it helps provide a basic summary of the variables, and correlation helps to know the potential relationships between the variables. It is evident from the descriptive statistics that there exist huge variations in the variables across the CPCs. The correlation matrix results assure a positive relationship between input and output variables.

Selection of DEA Model

To develop the non-oriented SBM model, we choose a set of N DMUs $\{DMU_j : j = 1, 2,....N\}$, manufacturing o outputs Y_{rj} ($r = 1, 2,.. o$) by utilizing i inputs X_{kj} ($k = 1, 2... i$) for each CPC. Based upon the constant returns to scale, the SBM models is shown as follows. It holds non-negativity assumption for all variables i.e., $X_{kj} > 0$ ($k = 1, 2...i$ $j = 1, 2, 3.... N$) and $Y_{rj} > 0$ ($r = 1, 2...o$ $j = 1, 2, 3....N$).

Model 1: Non-Oriented SBM (FP)

$$\rho_{NO} = Min\left(\lambda, Z^-, Z^+\right) = \frac{1 - (1/i)\sum_{k=1}^{i}(z_k^- / X_{ks})}{1 + (1/o)\sum_{r=1}^{o}(z_r^+ / Y_{rs})}$$

$$s.t.: X_{ks} = \sum_{j=1}^{N}\lambda_j X_{kj} + z_k^- \quad \left(k = 1,2,3,....i\right)$$

$$Y_{rs} = \sum_{j=1}^{N}\lambda_j Y_{rj} - z_r^+ \quad \left(r = 1,2,3,......o\right)$$

$$\lambda_j \geq 0, z_k^- \geq 0, z_r^+ \geq 0$$

The above model is called the SBM-CRS model and we use it to calculate the Overall Technical Efficiency (OTE) for all the CPCs (Tone, 2001). We can also convert this fractional program into the equivalent linear program using the Charnes-Cooper transformation.

Model 2: Non-oriented SBM (LP)

$$\tau = \min(t, \delta, Z^- Z^+) = t - \frac{1}{i}\sum_{k=1}^{i}Z_k^- / X_{ks}$$

$$s.t.: \quad 1 = t + \frac{1}{o}\sum_{r=1}^{o}Z_k^+ / Y_{rs}$$

$$tX_{ks} = \sum_{j=1}^{n}X_{kj}\delta_j + Z_k^- \quad \left(k = 1,2,3,....i\right)$$

$$tY_{rs} = \sum_{j=1}^{n}Y_{rj}\delta_j - Z_k^+ \quad \left(r = 1,2,3,.....o\right)$$

$$\delta_j \geq 0, \ Z_k^- \geq 0, Z_k^+ \geq 0, t > 0$$

An optimal solution of model 2 be $(\tau^{*}, t^{*}, Z^{(-*)}, Z^{(+*)}, \delta^{*})$. Now we define the optimal solution of model 1 i.e.,

$$\rho^* = \tau^*, \ \lambda^* = \frac{\delta^*}{t^*}, \ z^{-*} = \frac{Z^{-*}}{t^*}, \ z^{+*} = \frac{Z^{+*}}{t^*}$$

The sth CPC is considered as pareto efficient when all the slacks are equal to zero i.e., for all inputs and outputs. The efficiency score of the sth CPC will be optimum (. If the efficiency score is less than one and presence of input or output slacks (non-zero) shows the level and reasons for inefficiency prevailing in the sth CPC. SBM-DEA model also suggests targets for inputs and outputs for inefficient CPCs to improve their performances.

We also calculate Pure technical efficiency (PTE) and Scale efficiency (SE) to understand the performance of CPCs under variable returns to scale. For this, we include the below convexity constraint in model 2.

Table 2. Efficiency Scores of CPCs in India (2018–2020)

CPC	FY 2018–19			CPC	FY 2019–20		
	CRS	VRS			CRS	VRS	
	OTE	PTE	SE		OTE	PTE	SE
CPC Tirur	0.187	1.000	0.187	CPC Tirur	0.121	0.276	0.436
CPC Nilambur	0.159	0.474	0.335	CPC Nilambur	0.162	0.269	0.602
CPC Perambra	0.325	0.417	0.780	CPC Perambra	0.060	0.366	0.163
CPC Vatakara	0.049	1.000	0.049	CPC Vatakara	0.058	1.000	0.058
CPC Thirukochi	0.021	0.050	0.423	CPC Onattukara	0.055	1.000	0.055
CPC Onattukara	0.032	0.398	0.079	CPC Ananthapuri	1.000	1.000	1.000
CPC Kozhikode	0.154	0.341	0.452	CPC Kozhikode	0.274	0.409	0.671
CPC Kodungallur	0.113	0.625	0.180	CPC Kodungallur	0.352	0.353	0.997
CPC Palakkad	0.182	1.000	0.182	CPC Thejaswini	1.000	NA	NA
CPC Thejaswini	0.171	1.000	0.171	CPC Iritty	0.090	1.000	0.090
CPC Ponnani	1.000	1.000	1.000	CPC Kottayam	0.225	0.599	0.376
CPC Iritty	0.110	1.000	0.110	CPC Valluvanad	0.072	0.262	0.274
CPC Kottayam	0.160	0.466	0.343	CPC Kalpatharu	1.000	1.000	1.000
CPC Valluvanad	NA	NA	NA	CPC Econut	0.141	1.000	0.141
CPC Kalpatharu	0.564	0.742	0.760	CPC Kalpatheertha	NA	NA	NA
CPC Econut	0.152	0.806	0.189	CPC Pudukkottai	0.033	1.000	0.033
CPC Kalpatheertha	1.000	NA	NA	CPC Pollachi	0.190	1.000	0.190
CPC Pudukkottai	1.000	NA	NA	CPC Dindigal	0.626	0.902	0.694
CPC Pollachi	0.130	1.000	0.130	CPC Karpagavirkshm	1.000	1.000	1.000
CPC tirupur	1.000	NA	NA	CPC Coimbatore	1.000	1.000	1.000
CPC Dindigal	0.230	0.793	0.290	CPC Vinayaga	0.302	1.000	0.302
CPC Karpagavirkshm	0.324	0.503	0.644	CPC Udumelpet	1.000	1.000	1.000
CPC Coimbatore	0.139	1.000	0.139	CPC Madathukulam	1.000	1.000	1.000
CPC Vinayaga	0.149	1.000	0.149	CPC Kanyakumari	1.000	1.000	1.000
CPC Udumelpet	1.000	1.000	1.000	CPC Pasumai	1.000	1.000	1.000
CPC Kanyakumari	0.294	1.000	0.294	CPC Mahima	1.000	1.000	1.000
CPC Pasumai	0.485	1.000	0.485				
Average	0.351	0.766	0.364		0.510	0.810	0.587
Max	1.000	1.000	1.000		1.000	1.000	1.000
Min	0.021	0.050	0.049		0.033	0.262	0.033
St Dev	0.345	0.292	0.287		0.425	0.300	0.397

Note: NA denotes that those CPCs were removed from the DMU set as they were detected as outliers in the model. For detecting outliers, authors used Super efficiency SBM framework.

$$\sum_{J=1}^{N} \delta_{js} = 1$$

This model is known as SBM-VRS model. Once the PTE is estimated for every CPCs, SE can be calculated by dividing OTE by PTE. SBM-VRS model helps us to know the actual reasons behind inefficiency in any CPC i.e., inefficiency arising either due to inefficient production technique or due to the scale size of the CPC. We have also checked the stability of efficiency scores using various methods mentioned by Avkiran (2007)

Results and Discussions

Table 2 shows the estimated efficiency scores of the CPCs for the financial year 2018–19 and 2019–20. SBM-CRS model results show that the relative efficiency values range from 2.12 percent to 100 percent for FY 2018–19. The average OTE of the sampled CPCs in India for the FY 2018–19 is 35.11 percent, with a standard deviation of 34.52. This indicates that on average, CPCs can cut down the inputs or augment the outputs by 64.89 percent of the reported level to become overall efficient. Similarly, the SBM-VRS model results for the FY 2018–19 show efficiency scores that range between 7.6 percent and 100 percent. Under this model, we calculate the pure technical efficiency (PTE) of the CPCs which explains the efficiency of the companies to convert the inputs into outputs. At the same time, their scale size is kept constant. The mean efficiency estimate of the sample CPCs in India under the VRS assumption is 76.59 percent with a standard variation of 29.19. This indicates that, for an average CPC in the model to reach the efficiency frontier where the efficient CPCs lie, it should have adopted 23.41 percent of cost-saving measures (1 - [76.59/100]) during FY 2018–19. We also calculate the scale efficiency (SE) of the CPCs by comparing both model results. SE measures the impact of the scale size of the CPC on OTE. The SE values range from 36.4 percent to 100 percent, with a standard deviation of 28.7. Estimates of both models show that there exists a considerable level of variation in efficiency scores among the CPCs.

Among the sample CPCs, 5 CPCs are found to be efficient under the CRS assumption. Among the efficient CPCs under SBM-CRS, CPC Kalpatheertha and CPC Pudukkottai turn out to be peers and thereby considered role models to those inefficient CPCs to reach the efficiency frontier.

The estimated efficiency scores for the FY 2019–20 can be interpreted similarly. It is visible that the average scores of the sampled CPCs for FY 2019–20 have improved compared with the estimated efficiency scores of FY 2018–19 (See Table 2). Under SBM -CRS, there are 10 efficient CPCs across the sampled CPCs and it is found that majority of the CPCs have improved their efficiency scores compared to the previous financial year. The number of efficient CPCs has further increased to 16 under SBM-VRS assumption. Comparing the estimation results for the study periods, it is visible that efficiency scores have improved from the base period (FY 2018–19). This improvement can be attributed to firm size, proper implementation of business plans, and other factors. But it should be noted that the growth rate in OTE and SE is not found satisfactory. This can be due to the outbreak of covid pandemic as it stroked off the economic activities during the last quarter of FY 2019–20.

Conclusion

In this paper, we estimate the relative efficiencies under CRS and VRS assumptions for sampled CPCs in India using non-oriented SBM model. According to the estimation results of SBM non-oriented models, the average OTE, PTE, and SE scores have improved in FY 2019–20 as compared with FY 2018–19. Similarly, the number of efficient CPCs under CRS and VRS assumption has also improved over the period. Though there is improvement in overall efficiency scores (OTE, PTE, SE), there are still pieces of evidence that the CPCs have scope for reaching their potential efficiency. Our model results suggest input reduction and output reduction for the inefficient CPCs to attain maximum efficiency. The study suggests that governments and other stakeholders must provide additional support both in financial and non-financial terms to the inefficient ones along with establishing new CPCs. The study also conducted necessary stability tests to check the robustness of the efficiency scores on the frontier. The successful results of these tests assure that the respective stakeholders can take managerial decisions based on the analysis.

Limitations in this study include the consideration of CPCs promoted by CDB alone for the performance assessment. An improved version can consider all the stakeholders who promote the FPCs in a specific state or India and make comparisons to get wider understanding of the performances of these institutional arrangements in Indian agricultural sector.

References

Aksoy, E. E., and Yildiz, A. (2017). Applying Data Envelopment Analysis to evaluate Firm Performance. In: Hacioglu U., Dincer H., Alayoglu N. (eds) Global Business Strategies in Crisis. Contributions to Management Science. Springer, Cham. https://doi.org/10.1007/978-3-319-44591-5_22

Avkiran, N. K. (2007). Stability and integrity tests in data envelopment analysis. Socio-Economic Planning Sciences, 41, 224–234.

Boussofiane, A., Dyson, R. G., and Thanassoulis, E. (1991). Applied data envelopment analysis. European Journal of Operational Research, 52, 1–15.

Bowlin, W. F. (1998). Measuring Performance: An Introduction to data envelopment analysis (DEA). Journal of Cost Analysis, 7, 3–27.

Dyson, R. G., Allen, R., Camanho, A. S., Podinovski, V. V., Sarrico, C. S., and Shale, E. A. (2001). Pitfalls and protocols in DEA. European Journal of Operational Research, 132, 245–259.

Golany, B., and Roll, Y. (1989). An application procedure for DEA. Omega, 17, 237–250.

Patino, C. M., and Ferreira, J. C. (2018). Inclusion and exclusion criteria in research studies: Definitions and why the matter. Jornal Brasileiro de Pneumologia, 44(2), 84.

Tone, K. (2001) A slacks-based measure of efficiency in data envelopment analysis. European Journal of Operational Research 130, 498–509.

A Study on the Impact of Artificial Intelligence on Traditional Grocery Stores

Anil Ramesh Bajpai[1,a] and Dipti Anil Bajpai[2,b]

[a]IMED, Bharati Vidyapeeth, Pune, India
[b]Savitribai Phule Pune University, Pune, India
E-mail: [1] anilbaj@gmail.com; [2] diptibajpai.iicmrmba@gmail.com

Abstract

The traditional Indian Grocery Stores although unorganized is gradually modifying the outlook by altering the store layout to make it appear more like a supermarket. The focus has shifted from traditional selling to customer experience, from wastages to LEAN management, manual recording of transactions to software-enabled recording, manual pricing to RFID, and paper money to digital money. The journey emphasizes transformation from initial technology resistance to embracing technology. During post-pandemic world, the ubiquitous Indian grocery stores supported the upsurge in demand and witnessed high sales of essential items. Considering the rising demand adaptation of AI will smoothen the process with efficient inventory management and control and real-time data-driven decision-making. Artificial Intelligence can help such stores to gain operational efficiency that aspires to provide exceptional customer service and bank upon the range of possibilities.

Keywords: Artificial Intelligence, grocery stores, data, operational efficiency

Introduction

The world is gradually ascending towards Industry 4.0. The impact of AI is visible in almost every industry sector, while the quantum of impact varies. Artificial Intelligence or AI can be defined as the capability of machines that enables them to reason, remember information, learn and identify new insights through data discovery. Traditional grocery stores are an indispensable part of our life, this belief got affirmation during the Covid19 pandemic. (Alkhaladi, N, et al., 2021) As per McKinsey's report, during pandemic, 85% of the market share in US was held by physical grocery stores, this shows that despite the rise in online shopping brick-and-mortar stores cannot be eliminated and will still exist as part of the future of grocery shopping. While AI has made its way in supermarkets it still awaits recognition in the traditional Kirana stores. Alkhaladi et al. (2021) Stephen Hawking predicted that artificial intelligence will transform every aspect of our lives including traditional Grocery stores. Future projections of the intrusion of AI in the retail market are expected to grow exponentially to $24 billion by 2027, from a meager $3 billion in 2020. This growth trend shows a growth rate of 29.7% over seven years. AI is emphasizing its indispensable disposition in all aspects of a grocery store, from pricing enhancements to product placements and online order fulfillment.

Prominence of an Indian Grocery Store

According to research firm Red Seer, India is home to around 15 million grocery stores; while a Nielsen report has a different statistic that states the number to be nearly 12 million. Singh et al.

(2020) This number accounts for 90 percent of domestic retail and FMCG sales. Out of which around 8 percent of retail sales come from modern trade, while 2 percent is from e-commerce.

During the Covid 19 Pandemic, the focus from luxury and non-essential buying shifted to essential buying. The dependency on these pervasive grocery stores grew. These stores worked hard to make all the required supplies available for the customers, and the results of this effort were evident in the accelerating revenues earned by these stores. Pune-based Rahul Agarwal, the owner of Rahul General Stores, has been seeing a surge and superb growth since the lockdown. He mentions that the sales of his stores accelerated during the pandemic. Sales have doubled. However, with the rising demand, procurement issues remained a challenge that consequently impacted the fulfillment of orders. The Indian household is a major contributor to the economy and is estimated to account for around 66 percent of total retail revenue by 2020. The majority of the grocery business happens through the unorganized sector.

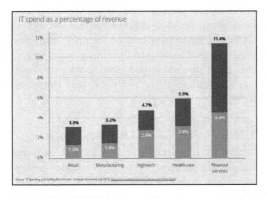

Fig. 1. Deloitte (2019)

Figure 1 graph shows the lack of IT spending and technology adoption in the retail industry.

Artificial intelligence, automation, IoT, Bots, and transformation are the buzz word today. To survive the changing landscape of data, artificial intelligence (AI) based innovation turns out to be part of an indispensable roadmap. The Growth of non-traditional competition from online shopping, the shift from Goods to Services that shifted focus from product to customer experience, and the rise

in social commerce accelerated the need for AI-enabled systems in every aspect of business.

Recent research by Deloitte reveals Global retail spending on IT was at US$196 billion in 2019 and is estimated to grow to over US$225 billion by the end of 2022. (Deloitte, 2019). However, despite this growth, the retail sector still lags in the required IT investment scale of other industries.

Role of AI in Traditional Grocery Stores. In the current scenario, organizations in retail and consumer products are using intelligent automation to perform varied internal processes that depend on existing rich-data sets, such as demand forecasting and customer intelligence. But going forward in the next three years, these organizations aspire to incorporate intelligent automation into more complex processes that would require broader sets of data, external collaboration, and additional system integrations. With the growing awareness about automation and AI it is projected that the penetration of AI could be more than 70 percent across organizational areas including the value chain. Considering the huge market of traditional grocery stores AI can play a revolutionary role in transforming the way traditional grocery stores function.

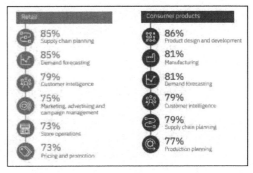

Fig. 2. Deloitte (2019)

Figure 2 justifies the proliferation of intelligent automation in the retail sector and traditional grocery stores cannot be an exception. Starting from customer preference to procurement than to supply chain all the aspects are data intensive So AI predictive analytics with machine learning is the best solution. Identification of key areas for improvements in grocery stores with small investments is important from a feasibility point of view. Considering the above data from the viewpoint of a traditional grocery store the role of AI can be assumed in the following areas.

Inventory Management, forecasting, and Food Waste Reduction for accurate inventory projections thereby minimizing wastages. According to a report published in Financial express in 2019 up to 10 percent of oilseeds, pulses, and cereals grown in India are completely wasted along with 16% of fruits and vegetables being wasted every year (Sharma, 2022). One of the major reasons for this wastage is the lag in the supply chain system in the country. An AI-enabled solution that provides both demand forecasting and inventory replenishment solutions for grocers, can help with proper inventory forecasting and consequently result in reduction of wastage. **Theft Management** One of the major pain areas of the traditional grocery stores is the loss suffered due to thefts. Such thefts can occur at the warehouse or as a result of shoplifting customers. This can be avoided by the use of AI-enabled software that tracks every movement of the inventory and helps maintain a record of it and identify the theft.

Literature Review

According to Statista, the year 2020 witnessed extensive use of AI in the consumer goods and retail sectors for various purposes (Terry Wilson May et al., 2021). The retail business has many stakeholders that include the supply chain ecosystem; the consumer-packaged goods industry and most important stakeholders are the Consumers, equipped with varying purchasing power, consumers who are loyal to brands while some are adventurous while seeking and making choices. These different functions served by the stakeholders emphasize their importance as key decision-makers that will guide the way in which and to what extent artificial intelligence will be incorporated into their retail shopping experience. This means that the future of AI in retail rests on the benefits AI provides to these stakeholders (Srivastava, et al., 2021).

As per a report published by Deloitte, the growing competition, ever-changing customer expectations, and the continuous development of technology is primarily reshaping every aspect of the retail business. Hence, sustaining performance and remaining relevant in the market demands an effective response to the changing landscape and making the best of the available opportunity. The report identified four major trends that enabled the namely, growth of nontraditional competition, a shift from goods to services, hyper-personalization, and a push to modernize technology.

As per a report published by the IBM Institute of Business Value, in the next three years retail and consumer products industries are projected to grow from the existing 40 percent of companies to more than 80 percent. However, it is anticipated that AI will majorly penetrate the supply chain planning aspect of retail businesses.

A recent report about Kirana stores revealed that orders (including grocery and daily essentials) worth $4 billion were sourced directly through online apps, thereby completely avoiding the distributors. Besides, large companies like P&G, ITC, and Marico have created their in-house apps and connected various indigenous grocery stores. This move resulted in accelerating sales by about 1 billion dollars in the last fiscal year (Malviya, 2021).

Research Gaps

During the research, it was observed that the traditional grocery store faces similar issues as the big stores of order fulfillment, customer preferences, and its dependency on manual methods of data gathering and using similar data for procurement, which can lead to inaccuracies.

In 2020, most businesses experienced a shift in customer behavior as consumers were forced to shop online, irrespective of their previous shopping habits, due to global lockdowns. Traditional grocery stores although have embraced technology but the acceptance is at the basic level. Considering the level of competition they face with the big players, to remain relevant in the competition they need to cater to two key stakeholders, firstly the consumer and secondly supply chain management. Adoption of artificial intelligence initially for these two key stakeholders, will not only accelerate their growth but will lead to the efficient functioning of the traditional grocery stores by minimizing wastage and understanding the actual consumer requirement based on real-time data.

Objectives

i. To study the impact of AI within the traditional Indian retail grocery segment.
ii. To study the impact of pandemic on retail grocery store sales and supply management.
iii. To study technology adoption across traditional retail grocery stores.

Research Design

The research focuses on the adoption of Artificial Intelligence in traditional grocery stores. The research methodology used is Descriptive research. To understand the existing technology usage, inventory management, wastage elimination, and theft management, questionnaire was shared with the respondents. The respondents were the owners of the grocery stores in and around Nigdi Pradhikaran, Pune. The questionnaire was prepared using Google forms. The data collected is based on the responses received on the questionnaire and secondary data collected through an extensive search on the websites of articles and research papers written on the said topic.

Sample Design: For understanding the adoption of technology samples were selected based on the convenience of the researcher in the designated area.

Data Analysis

Table 1. Impact of Supermarkets on traditional grocery stores

Survey Questions		Variables and Results		
S. no.		Yes	No	May be
1.	Do you think Big Super Stores like DMart had any impact on your sales	76.9%	15.4%	7.7%
2.	Do you think Covid 19 has revived your sales graph	73.1%	26.9%	
3.	Do you manage stock-in & stock out efficiently	42.3%	57.7%	
4.	Do you use software that gives you proper data about the inventory	30.8%	69.2%	
5.	Do you face inventory wastage annually	69.2%	30.8%	
6.	Are you using Digital money	84.6%	15.4%	
7.	Do you think Digital money is helpful in knowing the actual revenue	69.2%	30.8%	
8.	Are you accurately able to project demands for products in your shop	76.9%	15.4%	7.7%

Table 2. Methods adopted by grocery stores for inventory forecasting.

Question	Software App	Manual Recording & Observation	Intuition	Others
How Kirana stores forecast inventory	18%	68%	11%	3%

Table 2, reveals 18% use Software apps, while 68% resort to intuition or manual records for inventory forecasting.

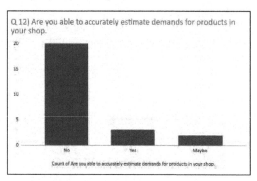

Fig. 3. Emphasizes the accuracy in projecting demands for the products in their stores

Inferences

1. Technology adoption is happening at good pace.
2. Observations reveal that data reliability for inventory or procurement is weak and respondents acknowledge it.
3. So, the trend will be adaptation of new technologies, AI products, theft detection systems, cameras, etc.
4. Most of the early adapters have overcome the problems of forecasting demand, supply chain, and improving consumer digital experience.

Conclusions

The adoption of AI in traditional grocery stores can result in two major benefits for the grocers: Customer Experience and Operational Efficiency. AI can make use of the business's existing data and transform it into information that is easy to understand. This information can provide valuable insights into consumer behavior and purchase decisions. With the shift in focus from sales to customer experience, AI applications can help improve customer experience and customer service.

Implementation of AI applications in these stores is plagued by high costs and willingness to embrace AI and convincing these grocers of the long-term benefits of AI is a challenge. Their acceptance of technology is restricted to RFID or accounting software due to ease of use. Lack of technical know-how could add to the resistance to AI adoption.

References

Alkhaladi, N. (2021). Innovation analyst, Deploying AI in your grocery stores (May).

Singh, A. (2020). In SME Futures (May).

Srivastava, M. M. (2021). In Indianretailer.com, AI the future of retail (March).

In the Instor Blog, Role of AI and IoT in the Retail Industry

Mandel, Pini (2021). In Food Logistics, Putting the AI in Grocery Aisles (January).

Role of International Commercial Arbitration in Resolving WTO Disputes

N. Garg[*,a]

[a]Research Scholar, (Ph.D. Law), School of Law, CHRIST (Deemed to Be University) India
E-mail: [*]neha.shrutika.jain1@gmail.com

Abstract

International organizations flourished with the fragrance of effective interactions between parties and reconciliatory measures promoting amicable dispute settlement. In this scenario International Commercial Arbitration comes to the rescue of international organizations resolving multilateral, bilateral, and investment disputes between the parties, promoting sustainable development. The rationale of the article is to promote the parties towards a system of dispute redressal that is opposite to adversarial form of dispute resolution and thereby promoting an arbitration-friendly economy. Arbitration is not considered to be a preferable dispute resolution and the article provides for importance of arbitration in research gap in various research papers by using doctrinal research, analyzing articles and statutes, and providing an answer to the research question "How to promote sustainable development through International Commercial arbitration" and "How to promote arbitration-friendly environment." This article discusses WTO dispute resolution board's novel approach to promoting arbitration.

Keywords: International Organizations, Dispute resolution, Sustainable development, International Commercial Arbitration, World Trade Organization

Introduction

International arbitration can be traced back to the operation of the United States through Friendship Commerce and Navigation (FCN) treaties, in which arbitration was one of the selected dispute settlement mechanisms. International organizations have proved to be of paramount importance in dispute resolution in the present global scenario. The World Intellectual Property Organization has extended its branches in resolving disputes through the armor of arbitration and conciliation providing for smooth dispute resolution ensuring party autonomy and smooth business relations. The International Centre for Settlement of Investment Disputes is functioning under the aegis of the World Bank, providing a global platform for the resolution of investment disputes through arbitration. It provides various services like legal, paralegal personals, and financial assistance to provide for expenses of the arbitration proceedings and a final financial statement indicating the breakdown of expenses.

Promoting Sustainable Development Through International Commercial Arbitration

The term "Sustainable development" bears a wide impetus. It is not only confined to environment and resources but is playing a vital role in promoting sustainability in International Economic Law. The concept gained global attention in Brundtland Convention in 1987 and became a subject matter of Global agenda emphasizing meeting the present needs without compromising the needs of future generations. It has promoted intergenerational and intra-generational equity resting on pillars of economy, environment, and society. International Investment Agreements, Bilateral investment treaties, and free trade agreements are drafted to promote Foreign Direct Investment but they lack green investment policies to promote sustainability and development simultaneously. International Commercial Arbitration and Investor-State Arbitration may play a pivotal role in combating adverse effects on climate change or the environment by the introduction of such clauses in the arbitrator agreement that protects both the host states and investor states. The 2030 Agenda for sustainable development as adopted by 193 countries in 2015 strives to attain 17 SDG goals and combat the barriers that come in the way of advancing sustainable development.

WTO and Sustainable Development

WTO has a multilateral trade system that promotes sustainable development. (Understanding on Rules and Procedures Governing the Settlement of Disputes, 1994) WTO agreement support transparency and it also concerns the adequate measures that may be adopted to mitigate various environment-related issues. The trade agreements are beneficial for developing countries as it provides flexibility to adopt measures as agreed in the commitments made. The agreements pertaining to the technical requirements, animal and plant developmental health, and food safety measures help in promoting environmental security and it also balances international trade ensuring that these requirements do not become hindrance in hassle-free international trade. The Trade-related Aspects of Intellectual Property Rights promote green technology and also the transfer of technology that is environment friendly. WTO has a green box policy that promotes economic policies protecting environment.

Sustainable Dispute Resolution Mechanism by WTO Through International Commercial Arbitration

Article 25.1 of the WTO provides for the expert resolution of disputes through the procedure of arbitration as an alternative means of dispute settlement which facilitates the solution of certain disputes as per the mutual consent of both parties to the arbitration agreement. (Brack Duncan, 2005) This is an alternative measure that is resorted by the parties; the condition is that the subject matter of the dispute should be arbitrable and relate to the concerned issues that are being addressed by both parties to the arbitration. As per Article 25, the parties can agree to the procedure that may be adopted by them to be followed during the arbitration proceedings i.e., they have the freedom to choose the procedures by which the arbitration proceedings will be carried out and there are no limitations like selection of the arbitrator, hearings, evidence or other relevant matters about the arbitration. As per Article 25.3, an arbitral award given is binding on both the parties to the arbitration and WTO members countries if it is a matter of concern, then is presented and interpretation of the same is being done before the panel otherwise the arbitration award is final and binding on both the parties (Improved Dispute Settlement, 1987). In US section 110(5) of the US Copyright Act, recourse to the arbitration under Article 25 of the DSU WT/DS160/ARB25/1, Nov 9, 2001, was made. There was a complaint that was made by the European communities and the member states against the United States into section 110(5) of the US Copyright Act as amended by the fairness in Music licensing Act enacted on 27th October 1988. It provided for the playing of radio and television music in public places that included shops, bars, and restaurants and it did not require the payment of any royalty fees. The major contention of the European communities was that the statue was inconsistent with the US obligations as per Article 9(1) of the TRIPS agreement which required that the members must comply with Articles the Berne Convention. The dispute concerned the compatibility of the exemptions that were provided under section 110(5) of the US Copyright Act, as it was considered to be inconsistent with Article 13 of the TRIPS agreement which provided for limitations are exceptions to the Exclusive rights of holders and provided limitations only in the special cases. The business examination which was provided under section 110(5) allowed the amplification of the music broadcast and it did not require the payment of royalty fees by the food service concerning the drinking establishment of the retail shops and there was a proviso that their size shall not exceed a particular square footage. New style and examination provided under paragraph a of section 110(5) allowed the small restaurants and retail outlets to amplify the music Rock broadcast without the payment of royalty fees using only the home-style equipment i.e., the equipment used in the private homes in private premises.

Independently, the third-party rights were reserved by Japan, Switzerland, Australia, and Brazil. The major findings of the panel were that the business exception that is provided under subparagraph V of section 110(5) of the US Copyright Act did not comply with the requirements of Article 13 of the TRIPS and was found inconsistent with Article (11) Berne Convention. In 2000 August, United States informed the dispute settlement body that it would implement whatever the recommendations are being given by the dispute settlement body and requested a time frame of 15 months as a reasonable period required for the implementation of those recommendations. But in October 2002, European committee requested that the reasonable period for the implementation of the recommendations may be decided as per the arbitration which is provided under Article 21.3 of DSU. United States contended that it was engaged in productive meetings with the European communities in order to resolve the disputes before the expiry of a reasonable period. But European committee was of the view that it was not possible to conclude any arrangement before the end of the reasonable period that is 15 months' order to implement the recommendations. In 2001, the United States and the European Committee informed dispute settlement body to resort to arbitration as per Article 25 that benefits that may be assigned to the European communities as per section 110(5B) of the US Copyright Act. All members of the panel served as the arbitrators and the Chairman of one of the original panels was no longer available so the Director General, as per the agreed procedures agreed by the parties, replaced them. The proceedings were carried out under

Article 22 of the DSU, which provided that the US bring about the appropriate measures within a reasonable period but since the recommendations were not complied, so the European committee requested to suspend all the concessions that have been given to the US as per Article 22.2 of the DSU. But in January 2002, the US objected to the suspension of the obligations that were imposed upon the European communities and requested the DSB to refer the matters to the arbitration as per Article 22.6 of DSU. The United States also complained that the principles and procedures required under Article 22.3 has not been followed and they were engaged in negotiations. They were hopeful that a mutually amicable and satisfactory solution might be achieved. But in 2003, United States and European communities achieved a mutual temporary agreement in accordance with the US Congress.

The arbitration award is final and binding on both the parties to the arbitration as per Article 25.3 of DSU (Dispute Settlement Understanding)

Article 25.3 also provides that the arbitration award may not be notified to dispute settlement body or any committee arbitration agreement the award in public domain but it also provides that any member of the WTO can ask for the arbitration award to be fully disclosed to the WTO members only. Whenever an arbitration award is made under Article 25 then it is subjected to the same implementation and surveillance as for the other reports of the appellate body. Article 21 and Article 22 of DSU shall apply to the awards that have been passed as per Article 25 which means they must have the same standing as the recommendations of the rulings of the Dispute Settlement Board.

The arbitration award constitutes a final and binding award on the parties to the arbitration but the enforcement of the arbitration award may not be recognized if it is suffering from any substantive irregularities, in that case, the recognition of an award becomes impossible. A respondent declaring a nullity might, in such a situation, refuse to comply with the award and instead initiate new dispute settlement proceedings directed against the complainant's measures to enforce the award (i.e., suspension of concessions), thus subjecting the award to an indirect panel and Appellate Body Review. (Similarly, a complainant claiming that an arbitration award is a nullity would be inclined to initiate subsequent panel proceedings on the basis of its original claim that is a clear case of res judicata.) Although the Award that is rendered under Article 25 is subject to Article 21 and 22 of DSU; it does not provide for a procedure of or a specified procedure for publication and formalization of the award.

WTO has an advanced system of dispute resolution known as "Dispute Settlement Understanding." As codified in the Dispute Settlement Understanding (DSU), any WTO Member may challenge a measure taken by another Member that allegedly violates one or more "covered agreements" (which together comprise the WTO Agreement) before a neutral panel. The parties may give written or oral submissions for this purpose and the timeline for deciding such complaints is 6 months from the date of panel constitution. It provides the right of appeal if a party considers that adverse findings have been made pertaining to the complaint raised by him.

Role of International Commercial Arbitration in Resolving WTO Disputes

The early regime of Uruguay round saw the blooming of arbitration supported by US and European Union. In mid-1987, the US submitted a proposal for the improvement of the dispute settlement system, including binding arbitration as an alternative means of resolution that would co-exist with the panel system. United States suggested that arbitration is the most effective means for amicable settlement of disputes as compared to the traditional panel methods of dispute redressal as it involved laches and immense political influence. The proposal suggested that the process of arbitration may be resorted to as per the mutual consent of the

parties and if any recommendations or award were not implemented by either of the parties then it would involve the right of compensation and retaliation to the aggrieved party. In September of 1987, the EC submitted its proposal, asserting its underlying philosophy that the primary goal of dispute resolution should be a negotiated settlement and that the legal aspects ought not to become the "key element." The proposal set by EU and US provided for the applicability of arbitration that did not involve complex legal issues and was factual in nature. Article 25 provides for expeditiously dispute resolution as an alternate remedy as per the consensus ad idem and mutual agreement between the parties (Pohl Hillebrand Jens, 2018). The agreement for the same must be notified to all the members before the commencement of such proceedings. The award shall be submitted to the Dispute Settlement Body and also to the council. Article 21 pertains to the surveillance regarding the implementation of the rulings of the DSB of WTO. If the recommendations proposed by the DSB are inconsistent with any of the provisions of WTO then the matter shall be referred to the panel which shall furnish its report within the period of 90 days after the matter is referred to it.

Conclusion

International Commercial arbitration is proving to be an effective mechanism in resolution of disputes for inter-state commercial transactions and foreign direct investments. (Carmody,1976) Studies find that more than two-thirds of multinational corporations prefer commercial arbitration over traditional litigation, either alone or in combination with other alternative dispute resolution mechanisms, to resolve cross-border disputes. International organizations are now resorting to arbitration for effective and speedy dispute resolution and WTO has played a major role in this regard in adopting the arbitration approach from its very inception.

References

Understanding on Rules and Procedures Governing the Settlement of Disputes, Annex 2 of Marrakesh Agreement, Uruguay ROUND AGREEMENT Establishing the World Trade Organization, 1869 U.N.T.S.401, April 15, 1994, Article 1, 3, 5.

Brack Duncan (2005). The World Trade Organization and sustainable development: A guide to the debate. Chatham House, Simon Lester.

Improved Dispute Settlement: Elements for Consideration. Discussion Paper Prepared by United States Delegation, GATT Doc No. MTN. GNG/NG13/W/6 (25 June 1987) ("Improved Dispute Settlement").

Pohl Hillebrand Jens (2018). Blueprint for a plurilateral WTO Arbitration agreement under Article 25 of the Dispute Settlement Understanding, SSRN Electronic Journal.

Carmody (1976). Distributive justice calls for relief that is not conceived as compensation for past wrong in a form logically derived from the substantive liability and confined in its impact to the immediate parties; pp. 310–312.

A Study of Supply Chain Management for Manufacturing Industries of Electronic Sector in Pune District

More Hemant Vishwanath,[*,a] *and Kulkarni Milind Audumbar*[b]

[a]Research Scholar, Savitribai Phule Pune University, Pune, Maharashtra, India
[b]Director, Chetan Dattaji Gaikwad Institute of Management Studies, Savitribai Phule Pune University
E-mail: [*]hemantvmore@gmail.com

Abstract

Supply chain management played a critical role in the Electronic industrial chain optimization. This paper comprises an introductory path toward small-scale industries. It content the Background of Study, Small Business in India, Industrial Policy a license to growth, Overall Scenario of SSI and MIDC in Maharashtra, Industrial Scenario of Pune Division. Growth Pattern of Industries in Pune MIDC during 2006–16. Challenges and Opportunities for SSI, financial evaluation, Government Policy, District Industrial Centre (DIC), Small Industries Development Bank of India (SIDBI), State Financial Corporations (SFCs), The Credit Guarantee Fund, Eligible Lending Institutions, Eligible Credit Facility, Guarantee Cover Tenure, Fee, Website of Guarantee, Awareness Programs, Credit Linked Capital Subsidy for Technology Upgradation, Market Development Assistance, Technology, and Quality Upgradation Support, Mini Tools Room and Training Centre, Financial Assistance to SSIs, Short Term Loans, Nonfinancial and Financial Institution for SSI, and finally role of UNIDO towards SSI.

Keyword: Supply Chain, SSI, Electronic Industries, SME

Review of the Literature

Top management always concentrates their views on supply chain management. Initiation of any manufacturing activity begins from the supply chain management and the end of the activity means customer feedback is noted through the supply chain management (Tayur et al., 2012). The hurdles in the path of top management are most of the time solved by the supply chain. Last point of the product cycle is monitored by the supply chain and in reverse cycle, the starting point means planning the product manufacturing also controlled by monitoring of supply chain. Outward quantity of the volumes of the products is maintained in the supply chain cycle and inward raw material and services related also maintained in the supply chain cycles. Investment requirements in the business are easily calculated by monitoring the supply chain. Assets of the organization are fluctuating but good revenue makes the assets stable and confirmed stage. (Thomas and Ross, 1979) Supply acts as an intermediate step for the segregation of production line performance and customer service (Wisner et al., 2014). Supply chain is affected due to many factors but paper considers only two factors when the cost factor considered operational cost, raw material cost, material handling, and distribution cost comes in front. Other factor uncertainty considered then demand forecast uncertainty and lead time uncertainty comes across. (Urban, 2000). Supply chain commitment contracts with the customers always enforce the customers to buy dedicated quantity of the products with certain intervals. (Lundin and Morton 1975) Demand forecast of the products with the batch size or lot model with continuous monitoring systems. (Schwarz 1977) The supply chain also works as per vendor model or the utilization procedure always incorporated in the supply chain. (Chand and Morton 1986, Federgruen and Tzur 1995). Supply chain overcomes in uncertainty with the safety stocks, a certain amount of stock always keeps wide by every manufacturer to maintain the supply flow, during uncertainty flow of products keeps moving towards the customer not as per the requirement but to maintain the critical operations, customer service becomes comfortable due the safety stocks and production planning get some time to overcome against the unpredicting things. Production output level maintain related to the production level of the products. (De Bodt and Wassenhove,1983. Howard 1984). Requirements from the customer is fluctuating when it is on smaller and daily basis. The availability of the products with manufacturer with multiple variations leads to the critical supply chain for the OEM. Then to handle such a complex supply chain new concept arises i.e. Vendor Management Inventory (VMI). These VMI developed smaller supply which arises from the end point of the different vendors and start point of the OEM. Many challenges of supply chain like safety stock of inventory, last-minute delivery to customers, smaller but critical inventory management, and space for every product with the distribution sequence are overcome due to the VMI system. VMI systems have several advantages which are pointed out by many authors (Waller et al.1999, Fox, 1996 and Williams 2000). VMI increases the availability of the material at dedicated time, enrichment of the service to the customers, maintenance of the

optimum level stock of the material, and economic cost of the products which makes benefits the OEM, customer, and vendor. When predecided orders of material are in existence then the supplier can send the material to the customer at predecided location, time with communicated quantity with help of information technology. (Romano, 2014). Implementation of the effective VMI required a good relationship between the producer and chain of distribution. (Taleizadeh et al., 2015). VMI gives benefits on priority firstly to the distribution chain which increases the impact of the manufacturer, distributor, and customer (Southard, 2001). Supply chain gives quick response in delay of the material, an interception in the manufacturing process, and unavailability of the mass products to customers. (Skipworth and Harrison, 2004).

Gap in Literature

Researcher is made some good reviews based on the literature review provided in the previous section. The study being undertaken is considered to meet these needs and is deemed to add to the existing literature on the subject significantly.

Significance of the Study

The study will make a significant impact on the performance of small industries related to supply chain management. Maintenance of the safety stock, space availability with respect to the locations of the customers, prevention of the empty down of the stock, data collection, and analysis for proper interpretation. Government officials survey the implementation of the supply chain modules by small industries and concluded that many organizations get optimum results in profits, customer relationships, and optimum inventory level maintenance. Study will help with the supply chain success and failure analysis of small businesses.

Beneficial for Entrepreneurs

The study analyses the problems of entrepreneurs of the supply chain of small and medium businesses in the various talukas of Pune. Performance of current supply chain and problems for the same and overcoming ideas towards the problem. Financial aspects also get much more help from the study.

Beneficial for Government

Central government and state government always make policies to improve the performance of the micro, small, and medium-scale businesses in the country. Study will help for the implementation of the policies of government which improve the performance of these organizations. Government authorities also get help from the study about the weak and strong points of policies with respect to the respective business units.

Beneficial for Startup Entrepreneurs

The analysis of the research will direct to the business units about the supply chain policies throughout the surrounding areas. Business performance parameters of also be evaluated with respect to the supply chain and its benefits and challenges will be identified.

Beneficial for Academicians and Researchers

Understanding this type of research help academicians and researchers with their future research related to small business units with different topics. Study of a supply chain focuses on the different policies of government with respect to small business units which will help researchers in their further research.

Customer Service Enrichment and Opportunity

Supply Chain Management

Material and information is important for the supply chain systems. Organizations with different profiles and different products can use the same supply chain. The main objective of the supply chain is economic delivery cost of the product to the customers at committed time. Supply chain challenges arise when the multiproduct delivery with multiple locations with smaller margin of the delivery time. Integration model of supply chain works nicely through these critical situations. In this model production facilities, and addresses are defined with information on the distribution

channels and as per the customer requirements, decisions are taken for the initiation of supply chain.

Information and products are the inversely moving parameters within supply chain. Location of customer, quantity of product, and lead time is the various parameters that make supply chain successful.

Transportation Management

Production planning and production control works from raw material procurement to finished products. But during these competitive times deliveries of the products at defined time is also important. For this newly developed challenge the distribution system and transportation management need to be strong. The best quality, optimum quantity, and economic cost are characteristics of the good product and manufacturer but the delivery of the product at defined time is also added same. Transport of the products within time also saves the cost of the manufacturer because lead time defaulters have to pay debits from the customers. Proper transportation system makes better image of manufacturer in market and also saves operational costs for the manufacturers and customers.

Warehouse Management

Warehouse is the place where the finished products or raw materials required for production of the multiple products. Warehouse protects the products from environmental changes, makes instant availability of products, and distribution management initiated from the warehouse and ends in the customers' warehouse but today's technologies do not allow the storage of raw materials or products in the warehouse. Various tasks like packaging, labeling, and branding are carried out in the warehouse. Procurement of products in bulk quantity does not mostly depend on warehouse management.

Vendor Management

A competitive world leads to an increase in the better management of suppliers and vendors. Requirement of the raw material moves up and down as per production requirements. But the supply of the raw material has to manage accordingly, great challenge arises in front of small entrepreneurs who are suppliers of global manufacturers. Sudden change in the quantity and quality of the products. Technological impact and distribution management help the vendor management system in supply chain management.

Returns Management

Difficult and unwanted process for any manufacturer but still this returns management needs to be handled properly. Return management is totally opposite process of distribution management. Planned return management avoids excessive finance transactions, delayed in reworking, excessive utilization of time, and lead time management of the customer. Return management also makes better relationships with the customers.

Customer Support After Sales

Most customer needs support after purchasing the products. Support to the customers leads to the relationship with customers, mouth publicity leads to marketing, and remarks from the customers lead to innovation for new products or better modification in current ongoing products. Supply chain management ends with customer support but it very well helps operational processes.

Partnerships and Collaboration

Supply chain management leads to collaboration with competitors or the producer of the same product. The industries understand the importance of supply chain management. Production definitely important but delivery to the customer in time with economic cost with post-sale service also makes impact on the performance of the industry. Benefits and challenges are handled properly with mutual understanding and mutual benefits. Globally most of the manufacturer and supply chain drivers help each other or collaborate with regional partners which helps to reach remote and isolated areas also.

Mentioned seven parameters of the supply chain to help small and medium entrepreneurs

of the electronic manufacturing sector. Industry handles the situations to achieve the objectives of the industry. Increase in efficiency, quality management, lead time management, customer feedback, and innovation management handle properly.

Objectives

To study impact supply chain for raw materials and final products with respect to production planning and demand of customer.

To study supply chain control on optimum stock of material.

Hypothesis

H1o: - Small and Medium scale enterprises never face Supply Chain management problems significantly.

HA1: - Small and Medium scale enterprises face Supply Chain management problems significantly.

Testing of Hypothesis

H1: - Proposes the differences in small and medium-scale enterprises that face Supply Chain management problems. This hypothesis the study investigates. Whether SME faces Supply Chain management problems.

H2o: - Entrepreneurs of Pune districts are significantly unaware of industrial policies, techniques, and strategies.

H2: - Entrepreneurs of Pune districts are significantly aware of industrial policies, techniques, and strategies.

Testing of Hypothesis

H2: - Proposes the differences in awareness about the policies, strategies, and supportive techniques provided by Industry are low among entrepreneurs in Pune District. Though this hypothesis the study investigates. Whether Supportive techniques provided by Industry is low among entrepreneurs in Pune district.

H3o: - There is no significant difference across the business growth increasing every Financial year.

H3: - There is a significant difference across the business growth is increasing every financial year.

Testing of Hypothesis

H3: - proposes the differences in employee business are increasing every Financial year. Though this hypothesis the study investigates. Whether the business is increasing every financial year.

Scope of the Study

The samples taken for the study are from taluka of Pune and around area. Enterprises selected for the study are registered in the district center for the industries.

Table 1. Revised MSME Classification

Revised MSME Classification, W.E.F. 1 July 2020			
Composite Criteria: Investment and Annual Turnover			
Classification	Micro	Small	Medium
Investment of the Enterprises	Not more Than Rs.1Cr.	Not more Than Rs.10Cr.	Not more Than Rs.50Cr.
Turnover of the Enterprises	Not more Than 5Cr.	Not more Than 50Cr.	Not more Than 250Cr.

As per the Gazette of India, Extraordinary, Part II, Section 3, Sub-section (ii), vide S.O. 1642(E), dated the 30th September 2006 the classification of Micro, Small, and Medium Enterprises (MSME) is changed and made effective on 1 July 2020. The study is limited to Supply Chain management aspect of enterprises selected in study i.e. small, and medium. This research is limited to time horizon 2020–2024.

Research Methodology

Type of Data

Primary as well as secondary data will be used to achieve the objective of the research.

Sources of Data

Primary data will be collected from owners of small and medium enterprises. Supply Chain control techniques used by the company, Supply Chain systems such as perpetual and periodic systems, Stock levels, etc.

Company's website Secondary data will be collected from

Newspapers, Journals and magazines, Books on and medium enterprises, etc., Numerous websites, and Publications from different Government organizations such as:- District Industries Centre (DIC) Micro, Small and Medium Enterprises (MSME), Maharashtra Center for Entrepreneurship Development (MCED). Maharashtra Industrial Development Corporation (MIDC).

Data Collection Method

Survey method will be used to collect primary data. The data will be collected from the owners/managers of the selected enterprises in the Pune district by personal interview.

Sampling Plan

Population/Universe: From the taluka of the Pune district, the enterprises are selected as a population. Registration of the selected enterprises is verified and the enterprises selected as a population which having complete registration formalities till 31 March 2019. Approximately 200 enterprises were selected as a population.

Sampling Unit: One registered enterprise selected as one sample unit.

Sample Size: The size of the sample will be up to 200 SME enterprises.

Sampling Method: Proportionate Stratified Random Sampling Method will be used to collect the data. The population will be divided based on the taluka in which the enterprise is located. Each taluka will be considered a separate stratum. There are 12 talukas in the Pune district.

Table 2. Details of Taluka

Sr. No.	Taluka	Sr. No.	Taluka
1.	Purandar	2.	Baramati
3.	Shirur	4.	Maval
5.	Velhe	6.	Indapur
7.	Bhor	8.	Junner
9.	Ambegaon	10.	Mulshi
11.	Khed	12.	Haveli

1. Research Instrument

Questionnaire, which will include open-ended and close-ended questions

2. Tools for Data Analysis:

Percentage and weighted average methods will be use to analyse the data. Z-test and chi-square test will be used for hypothesis testing. Various charts like Line, Bar and Pie chart will also be used to interpret the data.

Limitation of the Study

The study will be restricted to Pune district only. The analysis is based on figures present in the internal records only. This research is limited to time horizon 2020–2024. The study will be restricted for electrical Industry vendors only.

Table 3. Plan of research

Sr. No.	Topic	Period for Study
1.	Literature Review	1 Months
2.	Research Design	1 Month
3.	Data Collection	15 Days
4.	Data Analysis	8 Days
5.	Hypothesis Testing	7 Days
6.	Report Writing	5 Days
Total Period Required for Study		03 Months 5 Days

Scope of Future Research

Researcher made some good contributions to further scope of study or research work.

Major Electronics Supply Chain Management Challenges

As per the current practices of the market, customer requires the products within less margin of time means it pressurizes the enterprises for a lesser lead time. Many giant industries handed over the raw material supply to suppliers. Minimum lead time is acceptable when the production volume of the products is on higher side. Lead time and volume make an impact on the cost of the products. In 2020 epidemics heats every sector of the globe electric supply chain was also impacted but now upcoming time 2021 rises with new opportunities. The industries related Electrical and electronics faces issues related cheap of the motherboard.

Lack of Diversity

In this the traditional supply chain model is followed by retailer, primary distributor, and main distributor. Many of them are connected to the manufacturer so they can manage the flow of material but the isolated place and distributor from the isolated place are unable to get the products and which leads to the shortage of the material and higher cost of the products in the area. Different models of the supply chain were not adopted by the manufacturer, distributor, and retailers hence the system collapsed during the epidemics. Some of the centers follow the rules of JIT but are unable to maintain in epidemics. Hence the diversity with different models needs to be followed for all levels of the supply chain.

Shorter Product Lifecycles

Demand of the customer changes very fast hence maintaining the inventory with high volume is become risky. Technological upgradation is moving rapidly so the products in the markets may become obsolete within a very short time means lifecycle of the products is unpredictable so the maintenance of the product and necessary utilities may become less useful within a very short time. Inventory management very critical task for every enterprise. Big players in the markets are also played safe against the product life cycle and respective inventories. Central inventory management technique is adopted by many enterprises and is key for successful inventory management with an economical price.

Commoditization and Product Complexity

The product life cycle mostly depends on the technology upgradation. But OEM always thinks about new products with different variations to keep their hold in the market and to attract customers, which makes an impact on the stock of the products in the markets, inventory for the products, and inline finished products. Product complexity challenges the supply from raw materials to finished products. Researchers put the opinion that movement of the supply chain importance increases the importance of Pune with respect to handling of the various products from different suppliers and distribution to different locations of the globe, which may nearby place or different parts of the world. Pune acts as a global manufacturing area of the automobile sector and ESDM sector also growing fast in Pune so products of the industry also need to be a strong supply chain across Asia. Pune becomes a central place for supply chains of various global players.

Expect More Diversified Supply Chains

Worldwide OEM and EMS have got more economic workforce in Pune. But it may lead risk of unacceptable handling of critical products. Value of human lives and environmental conditions also needs to be verified during diversification of the supply chain management.

References

Adhinarayanan, B. (2014). Study on evolution of the problems and prospects of SSI. Chennai: Chennai University.

Government, M. (2017). Retail Trade Policy 2016, Pharmaceutical & Drug Manufacturers, Pharmaceutical, Formulations, Drugs, Medications. Mumbai: Industrial department Maharastra Government.

Kashyap, D. P. (2017). Automobile Industry In India: Innovation and Growth. International Journal of Creative Research Thoughts (IJCRT), 961–964.

Majeed, A. (2009). Cotton and textiles the challenges ahead. Mumbai: Automobile Industry.

Narain, S. J. (2007). Indian society of Gandhian studies. Journal of Gandhian studies, 5, 125.

Sandeep, G. (December–January 2015–16). Small Scale Industries and Indian Industrialisation. Journal of Engineering Technology and Management Science, Vol. I, No. 7.

Statistics, D. O. (2017). Economic Survey of Maharashtra 2015–16. Mumbai: Govt. of Maharashtra.

Vij, Taranjit Singh, and Batra, G. S. (2014). Information Technology and It Enabled Services. International Journal of Computing and Corporate Research.

Vivek, M. (2012–2017). SAGE Publications India Pvt Ltd., Economic Sectors, Volume II,. New Delhi: RECTO Graphics, Delhi and printed at Saurabh Printers.

Assessment of Impact of Pick to Light Systems on Performance of Inventory Management in SME's

Dr. Prashant Kotasthane[a] and Dr. Sachin Vyavhare[b]

[a]Savitribai Phule Pune University, India, [b]Savitribai Phule Pune University, India
E-mail: [a]pmkotas@gmail.com; [b]sachinvyavhare@gmail.com

Abstract

Artificial intelligence is an area of computer science that focuses on the creation of intelligent machines programmed to work and react like humans. Manufacturing companies are fastly adopting AI to streamline business methods and increase efficiency. Pick-by-light systems are widespread in the industry, especially in order picking. Significant improvements have been shown in firms that have already implemented it. The study took place in small and medium-scale enterprises in Pune industrial region in Maharashtra, India with N=76, who recently implemented the systems. Two methods were implemented for the same tasks One of them assisted by pick-by-light and the other by pick-by-paper. Four selected process parameters viz. efficiency in utilization of resources, productivity (in terms of time consumed), material picking accuracy, and total material handling costs were considered for the study. The outcome suggested substantial improvements in all process parameters justifying the installation of the system.

Keywords: Pick-to-Light, Order Picking, Learning, AI

Introduction

Artificial intelligence is an area of computer science that focuses on the creation of intelligent machines programmed to work and react like humans. Manufacturing companies are fast adopting AI to streamline the way they do business and increase efficiency. Few examples of areas where AI is prominently used in industry are:-

- Collaborative Robots (cobots): Unlike autonomous robots, cobots are capable of learning various tasks, detecting and avoiding obstacles creating spatial awareness to work alongside human workers.
- Robotic process automation (RPA): This software is most useful in the back office. RPA software is capable of handling high-volume, repetitive tasks, transferring data across systems, queries, calculations, and record maintenance.
- Predictive Maintenance: Heavy equipment users are increasingly adopting AI-based predictive maintenance (PDM). Optimization of time spent for and between maintenance schedules saves valuable time and money. PDM systems can help companies To predict which replacement parts will be needed and when.
- Machine learning algorithms: AI systems that use machine learning algorithms can detect buying patterns in human behavior and give insight to manufacturers. This ability to predict buying behavior helps ensure that manufacturers are producing high-demand inventory before the stores need it.

- Inventory Management: Some manufacturing companies are relying on AI systems to better manage their inventory needs.
- Detection of errors: Manufacturers can use automated visual inspection tools to search for defects on production lines. Visual inspection equipment — such as machine vision cameras — can detect faults more quickly and accurately than the human eye.

Artificial Intelligence for Inventory Control

Significant improvements have been shown in those companies that have already implemented it. AI can predict scenarios, recommend actions and even act — independently or with human approval (Stockingera et al., 2020).

The numerous ways in which inventory control can benefit from artificial intelligence include advantages such as:

Demand prediction: Demand prediction for inventory control is built around a time series prediction model that can estimate what demand will be like for the coming days across all items of inventory stock. External data sources that impact demand can be incorporated into the demand prediction system. Improving demand prediction can improve inventory management efficiency.

Optimization of inventory stock: AI helps in proposing recommendations for optimum inventory stock levels for thousands of SKUs based on demand predictions. Optimizing inventory control is done by removing the complex data volumes that were traditionally the problem of inventory managers.

Automated ordering: AI automatically orders the correct amount of raw materials to fulfill manufacturing orders enabling distributors to merge datasets for predicting future demand for products. Thus these organizations can take well-informed business decisions, reduce waste and increase profit.

Reinforcement learning systems: Reinforcement learning systems are a more advanced AI methodology that involves a model taking optimal control of the inventory operations, with human checks and balances. It represents a domain in AI where the models don't simply make predictions but act on the predictions.

Pick to Light System

Pick-to-Light system is a popular, cost-effective method of order picking having precision, expediency, and ease of use (https://www.manufacturingtomorrow.com). It is appropriate for a manufacturing organization looking for an efficient order fulfillment system for warehouse that offers higher speed, increased accuracy, and cuts down on manpower and associated cost. Pick to Light system improve productivity by reducing travel time between the picks and boosts order accuracy and throughput using visual, light-directed picking (Randall Patzke, 2008).

Working Mechanism: Light indicator units mounted to the warehouse rack or shelf direct operator to the item location and quantity to be picked (MHI) The operator presses the start button, selects, or scans each item picked, and confirms that the pick is finished by pressing the confirm button (Baechler, 2016). Pick-to-Light systems can be integrated directly with warehouse's inventory system to ensure seamless continuity of movement and fulfillment of orders (River Systems, 2022).

Major Advantages: The major advantages of pick-to-light systems implementation include (https://www.manufacturingtomorrow.com/): - Fast Speeds: A light-directed picking is acknowledged to be the fastest operator-based picking strategy available which acts by lighting the exact locations needed.

High Accuracy: Getting the operator to the right location each time greatly simplifies the picking process. Easy-to-complete tasks that are easily replicated dramatically increase pick accuracy.

Hands-Free & Paperless.: No more pick sheets to handle or tally marks to record. Operators simply scan an order number on the carton or tote — and the system does the rest. Paperless picking reduces costs, reduces errors, and streamlines operations.

Service Providers For Pick To Light Systems In India

The market consists of both Indian and importer service providers. Major service providers for pick-to-light systems in India (www.wonolo.com) are as follows:-

Table 1. List of a few Service Providers Picked to Light Systems in India

International Brands	Indian Brands
Abel Womack	Turck India Automation Pvt. Ltd, Nashik
Acme World	Profusion India Consulting, Chandigarh
ARCO Solutions	Addverb Technologies Private Limited, Noida
Bastian Solutions	Compucare India Private Limited, , Vadodara
BP Controls Conveyors and Drives, Inc.	Compucare India Pvt. Ltd. Alkapuri, Vadodara

Source: www.wonolo.com

Objective of the Study

Small and medium-sized enterprises (SMEs) are non-subsidiary, independent firms that employ fewer employees and have problems. Specifically, if they are connected to large-scale companies in a supply chain. JIT manufacturing is the need of the hour and delivery speed is considered the most important feature of industrial services.

Pick light-to-light systems can reduce the delivery time of the vendor and enhance the efficiency of the supply chain ultimately benefiting the supply chain and ultimately the end consumer. The purpose of the paper was to assess the extent of advantage gained by industries where pick-to-light systems were implemented.

Research Methodology

Research type: The methodology is a combination of primary and secondary research based on the theme. It was descriptive and quantitative research interviews were held. The interviews were to capture insights and understanding about the phenomenon of offshoring of automotive engineering. The interviews helped in developing the idea of various engagement methodologies. The purpose of the paper was to assess the extent of advantage in industries where pick-to-light systems were implemented.

Sampling System: 30 organizations, including manufacturing organizations and organized retail stores in industrial areas of Pune, were selected at the convenience of the researcher who had installed and implemented pick-to-light systems from a period of March 2019 to Oct 2019. Data were collected through observations of various activities in material delivery during the material location process and with a structured questionnaire.

Hypothesis: Research hypothesis formulated was as follows :

Ha: The process performance for material retrieval and issue has changed after the installation and implementation of pick to light system

Analysis

Hypothesis testing: Research hypothesis was tested as follows :

Ha: The process performance for material retrieval and issue has changed after the installation and implementation of pick to light system

H0: The process performance for material retrieval and issue has not changed after the installation and implementation of pick to light system

The purpose of the study was to test the extent of process performance over the period. Respondents had given their opinion about the statements on 5 point Likert's scale.

Parameters were positive indicating a negative difference in selecting choices if there was any difference in the performance parameters.

Results obtained were subjected to paired 2 sample tests. The t-Test Paired Two Sample for Means tool performs a paired two-sample Student's t-Test to ascertain if the null hypothesis (means of two populations are equal) can be accepted or rejected.

This test does not assume that the variances of both populations are equal. Paired t-tests are typically used to test the means of a population before and after some treatment.

Table 2. Paired sample test [SPSS Data Analysis Paired sample test statistics]

S. No.	Parameters	N	Correlation	Sig.
Pair 1	Efficiency	76	0.029	0.801
Pair 2	Productivity	76	–0.012	0.92
Pair 3	Accuracy	76	0.253	0.028
Pair 4	Total costs	76	0.143	0.216

Table 3. Paired sample test [SPSS Data Analysis Paired sample co-relations]

S. No.	Paired Samples Test	Paired Differences $\alpha = 0.05$	
		Lower	Upper
Pair 1	Efficiency	–0.891	–0.503
Pair 2	Productivity	–1.086	–0.677
Pair 3	Accuracy	–1.377	–0.965
Pair 4	Total costs	–1.215	–0.522

Table 4. Paired sample test [SPSS Data Analysis Paired sample tests]

Sr. No.	Paired Samples Test	t	Df	Sig. (2-tailed)
Pair 1	Efficiency	–7.162	75	0
Pair 2	Productivity	–8.597	75	0
Pair 3	Accuracy	–11.341	75	0
Pair 4	Total costs	–4.989	75	0

Result

\# Ha: The process performance for material retrieval and issue has changed after the installation and implementation of pick to light system

\# H0: The process performance for material retrieval and issue has not changed after the installation and implementation of pick to the light system.

a. H0: μd = 0, H1 : μd ≠ 0

b. $t = \dfrac{d - \mu d}{Sa / \sqrt{n}}$ .... (1)

c. α = 0.05

d. t0,05,75

H0 is rejected if the value of t ≤ -1.667 or ≥ 1.667

t cal for all four pairs are as follows :

Table 5. Paired sample test [SPSS Data Analysis Paired sample tests]

Pair1	Pair2	Pair3	Pair4
–7.162	–8.597	–11.341	–4.989

Conclusions

No value is seen to be in the specified range. Hence H0 is rejected and Ha is accepted meaning that the process performance for material retrieval and issue had changed (improved) after the installation and implementation of pick to light system

Recommendations and Suggestions

Medium and small enterprises are most reluctant to install and implement the system though they are quite aware of the advantages basically due to the cost involved with Indian manufacturers venturing into the market the cost has been considerably reduced to be affordable to these manufacturers. Medium and small enterprises are requested to carry out a payback analysis and install these system accordingly.

References

Stockingera, C., Steinebacha, T., Petrata, D., Bruns, R., Zöllera, A. (2020). The Effect of Pick-by-Light-Systems on Situation Awareness in Order Picking Activities, Procedia Manufacturing, 45, 2020, 96–101, 2351–9789.

https://www.manufacturingtomorrow.com/article/ 2021/03/challenges-of-ai/16588/ 09/16/20, 05:31 AM | Automation & Networking, Design & Development | IIOT, logistics, Warehouse Systems Article from Andea Solutions.

Randall, L., Patzke, P. (2008). Key Attributes Used to Compare Pick-to-Light and Put-to-Light Technologies, American Psychological Association, May 2008, 5th Edition ISBN: 978-1-55798-790-7 University of Wisconsin-Stout Menomonie, WI.

MHI - The Industry That Makes Supply Chains Work, /https://www.mhi.org/solutions-community/solutions-guide/pick-to-light

Bächler, A., Bächler, Liane, Autenrieth, S., Kurtz, P., Hörz, Thomas, Heidenreich, T., Krüll, Georg, 2016. Comparative Study of an Assistance System for Manual Order Picking – 49th *Hawaii International Conference on System Sciences (HICSS)*, 5 January 2016, Print ISSN: 1530-1605.

River Systems (2022). https://6river.com/what-is-a-pick-to-light- systems ?Fergal Glynn, updated February 3rd, 2022.

https://www.manufacturingtomorrow.com/ article/2020/ 09/warehouse-automation-pick-to-light-systems/22/7/21/10:00 A.Carlicia.Layosa, Product Marketing Engineer I MISUMI.

www.wonolo.com // Top Pick-to-Light Systems - Wonolohttps://›blog›best-pick-to-light-sys/. Cristine Hanks, Warehouse Operations, Aug 7 2018.

Impact of Covid-19 on Large Scale Industries of Information Technology Sector in Pune Zone

More Hemant Vishwanath,[a] and Kulkarni Milind Audumbar[b]

[a]Research Scholar, Savitribai Phule Pune University, Pune, Maharashtra, India
[b]Director, Chetan Dattaji Gaikwad Institute of Management Studies, Savitribai Phule Pune University
E-mail: [a]hemantvmore@gmail.com

Abstract

The first Covid-19 patient was found in November 2019 globally, on 30th January 2020 in India, and on 9th march 2020 in Pune. Lockdown procedure makes an impact on all sectors of the economy and lives. In this study, IT industries were selected location-wise from Pune and studied with their work domain. The study crosschecked the working tenure of CEO with the industry, age of the industry was also verified. Revenue and profit of the organization is vital parameter of the study, which is crosschecked with the three previous year's performance. The growth total revenue performance of selected industries tallies with the previous year's revenue performance, which indicates the performance of 16.6% of the selected industry reflects negative revenue in 2019 which increased 33.3% in 2020. Profit margin criteria i.e. 10% of selected industries reduced in 2020 as compared to 2019, number of industries reduced from thirteen (54.1%) to eight. (33.3%) industries.

Keywords: Covid-19, Information Technology, Large Scale Industries, Pune, Age, Revenue and profit of industries

Introduction

Covid-19 is a big disaster happening these days. Human lives became uncertain due to the pandemic. All the nations around the world are heavily impacted due to Covid-19, in the medical background, economical conditions, and mental health conditions. Developed nations like Japan, Singapore, United States of America, European nations are also affected heavily. The medical infrastructure and economic conditions of these nations are stronger throughout the globe but Covid-19 made an impact on the same. Initiation of social distancing, traveling restrictions, and closing of industrial activity directly impacted tourist movements, and the hospitality industry, fuels, and engineering functions were also impacted. Near around 2.7% of revenue decreased in industries that are related to information technology functions only due to the delay in business decisions. (ASSOCHAM, April 2020). Information technology sector has a working vital role during this critical situation of the pandemic. Google meets, Zooms, and Microsoft Teams are supporting working professionals for routine meetings and maintaining the work-life balance. Children are attending school lectures and online examinations through various software applications. Social distancing is most important during this pandemic which is supported by Information Technology applications. This technology is helpful into the procurement of grocery items, E-passes required for emergencies, and essential services which are authorized by Government officials. Aarogyasetu app is helpful for Indian citizens for medical status and vaccination requirements, an exact situation about the patients around the users. Payment modes are shifted to digital payment modes like G-pay, Phone-pay, etc. by common users. Approximately 191 billion $ in revenue calculated for the Information Technology industry in 2020 which leads to 350 billion $ up to 2025 are the expectations.

Literature review

Pandemic is increasing these days SARS (Serve Acute Respiratory Syndrome), Influenza A (H1N1), Middle East respiratory syndrome (MERS), and Avian Influenza A are examples of epidemics. Epidemic management needs to handle critically because human lives and economic situations mostly depend on it. Currently, Covid-19 virus infected a lot of people globally. In a very short period large populations of the glob infected, initiation of the infection is from China as a starting point. Then many areas of the world face infection of Covid-19 (He, Zuopeng, and Wenzhuo, 2021). Social distancing is identified as prevention against Covid-19, which is possible with usage of the different technological aspects. Understanding of the need of human lives and technology usage coming in front of the world. All technological experts are designing, developing, and implementing technological modules for the safety of human lives. Information technology and information systems usage avoid overcrowding and implement isolation with the availability of various requirements (Ye, Jin, and Hong, June 2020). Various applications, modules, gadgets, instruments, and systems are involved in the fight against the epidemics of Covid-19. This all is useful for communication, knowledge exchange, and analysis of data for conclusions and implementation. Robotics in manufacturing, Augmented reality, and Simulation models are vastly used in current situations (He, Zuopeng, and Wenzhuo, 2021). Epidemic moves very fast around the world, and prevention systems are looking so weak. Data collection instruments, systems, and applications are integrated but unable to conclude the final decisions. In short, the medical systems integration is required properly

which is helpful for the hospital management, medicine suppliers, pharmaceutical, biological and vaccine industry. Inference calculating by decision-making authority to do with the help of this system. (He, Zuopeng, and Wenzhuo, 2021) Information technology and information systems are assured tools and techniques for the functioning of the other modules of business. Usage of IT increases tremendously after COVID-19 infections time. Information Technology supports business and other functions (Modi Bimal, Pandey, and Dusija, April 2020). India has a vast sector of software, automation, services, and hardware. Revenue of the information technology sector supports the Indian economy majorly, but differentiation within a sector are following services 50.7%, engineering services along with products of software at 21.1%, Business process management at a share of 19.8%, and Hardware at 8.4% share.

The collective review says that the information sector has to work for technology to fight against COVID-19 and they have a bright future. Some statements from the annual report show the impact of COVID-19. Restrictions, lockdowns, and laws by local authorities in most places of the world lead to the growth of the information technology sector. The work passion, working style, and function of working lead the growth of the IT industries. Big Data, Cyber security, and integration of systems work in favor of the IT industries. Impact that the COVID-19 pandemic had on the world's capital markets infrastructure (CMI). Pandemic makes an impact on industries, import, and export of many things, orders from customers are also affected and supply to customers is also impacted. Demand and supply ratios are mismatched due to the Covid-19 situation, overall business planning moves towards noncompliance, regulatory norms, and guidelines changes to protect against covid-19 but impacted most of the business plans. But during this period customer relationships, health safety of the workforce were maintained with the revenue of industry. The industry's total revenue declined; revenue can be attributed to the pandemic as well as the divestiture of its transportation systems business.

Objective

- To study the geographical location and profile of large-scale industries.
- To verify the tenure of CEO with industry and age of the large scale industries.
- To estimate revenue and profit in Covid-19 on of the large-scale industries.

Scope of the study

The study is restricted to the large-scale industries of information technology located in the geographical location of Pune. The study concerned to profile, age revenue, and profit/operating margin/operating income of selected large-scale industries in the information technology sector.

Hypothesis

H1:- Revenue and profit impacted significantly in Covid-19 of large-scale industries.

H0:- Revenue and profit not impacted significantly in Covid-19 of the large scale industries.

H2:- There is a significant correlation between the age of large-scale industries and the impact of Covid-19.

H0:- There is no significant correlation between the age of large-scale industries and the impact of Covid-19.

Research Methodology

The data referred to in this review was collected from the annual reports of the selected Industries, websites, and economical applications like Money control (Prabhakar, 2021), Google Finance, Yahoo Finance. Reviewer also visited the individual website of particular Industries to get the exact information about the location, profile, CEO, his tenure with the industry, number of employees, age of industry, and most vital factor revenue. The reviewer identified the large-scale industries located in Pune from Information technology and software industries which are listed in Indian and global stock markets. Pune has established, in recent years as IT hub so selected as a sample location place. The impact of Covid19 was decided based on the revenue generated and operating margin of the industry so actual economic data was collected from annual reports of the industry with respect to the particular year. Selected industries are subsidiaries of the parent industry so the revenue generated and operating margin calculated are from all group industries. Selected Industry is mostly subsidiaries of the parent industry or working as a supportive industry for the parent industry so the number of employees considered with the whole group of the industry. Covid19 creates critical conditions for every industry so experience matters so the age of Industry is taken for calculations. Pune works as a global hub of IT industries in recent years so Pune and related geographical areas were selected as sample locations places.

Data Analysis

Pune was selected as a geographical location for the studies and from various locations of Pune, large-scale industries from the information technology sector.

Location of Industry

In recent years, many IT industries started their subsidiaries or supportive software concern industry in Pune. Geographical location, Infrastructure, connectivity to other IT hubs, and power supply are the main reasons why Pune has become IT hub of the information technology and software Industry. Selected industries for study are located as follows, In Hadapsar, Magarpatta city is the main location where most of the IT Industry are situated which are 42%, Kharadi 21%, Hinjewadi 17%, Talawade 8% and the other 12% includes Shivaji Nagar, Karve Road, Senapati Bapat road, etc.

Table 1. Location of industries

S. No.	Location of Industry	No. of Industry out of 24	Percentage of Industry
1.	Hadapsar	10	41.7
2.	Hinjewadi	4	16.7
3.	Kharadi	5	20.8
4.	Talawade	2	8.3
5.	Others	3	12.5

Profile of Industry

In this study, IT industries are segregated on their work domain or profile i.e. the core area on which they work mostly. The study sample indicates Technology, Services, and solution industries are around 42% which is followed by the Information technology industries with 29%, media, communication, and networking with 17% of the total population, and Operation Management services with the least count mentioning 12%. Most of the industries are also working as service industries which is supportive to the other manufacturing, automobile, Pharma, electrical and mechanical industries also which is not mentioned below chart.

Tenure of CEO with Industry

The future and present of every industry depends on the top management's decision, implementation, and projection towards the upcoming time for the industry. Chairman, Managing Director, CEO, CFO, CS are working as decision-makers for the industry. Expansion, merging, splitting and critical situations are handled by these positions. Covid-19 is also a critical situation and unknown to all of us. Covid-19 impacts on Industry and role of CEO for Industries are vital, but 5 Industries' CEO has a tenure period of fewer than 2 years. But the revenue and profit percentage of the mentioned 5 industries for the year 2020 which exact Covid-19 period dose do not show a remarkable negative performance of the Industry.

Table 2. Tenure of CEO with Industry

S. No.	Tenure of CEO in Years	No. of Industry out of 24	Percentage of Industry
1.	Less than 2 years	5	20.8
2.	2 - 5 year	10	41.7
3.	6 - 10 year	4	16.7
4.	11 - 15 year	2	8.3
5.	More Than 15 Years	3	12.5

Age of Industry

Covid-19 is an unexpected critical situation that is handled by industry. Selected Industries have good experience in the IT sector. Though the Covid-19 situation is unexpected to all from selected industries, only 4% of the Industry have less than 15 years of age and the remaining 96% have more than 15 years of age. In short, all selected industries are capable to handle the situation.

Table 3. Age of Industries

S. No	Location of Industry	No. of Industry out of 24	Percentage of Industry
1.	Less than 15	1	4.2
2.	15 - 20	2	8.3
3.	21 - 30	7	29.2
4.	31 -45	10	41.7
5.	More than 45 years	4	16.7

Revenue of the Industry

Covid-19 impacted every sector of the economy. But IT sector has less impact than any other sector because for maintaining social distancing, IT software application plays a fantastic role for the same. The Graph reflected the decreasing trends of revenue. In the year 2018, one industry have negative revenue performance followed by four industries in 2019. Eight industries shows negative revenue in 2020, which means four industries increased in negative revenue performance. Though industries already struggling due to the economical conditions and Covid19 also hit them hard with the rising of number of industries 16.6 % to 33.3% from year 2019 to 2020. In year 2018, twenty-three industries in positive revenue performance, in the year 2019 twenty industries have positive revenue performance. But in the year 2020 sixteen industries have positive revenue, which means four industries decreased from positive to negative revenue performance, which means positive revenue performance shifted to 83.3% to 66.6%, which shows the impact of Covid-19.

Table 4. Revenue of Industries

S. No.	Revenue Generated	No. of industry out of 24		
		Year 2018	Year 2019	Year 2020
1.	More Than −5.0%	0	3	3
2.	−4.0 to 0.0	1	1	5
3.	0.1 to 5.0	7	5	4
4.	5.1 to 10.0	8	7	3
5.	More Than 10.0	8	8	9

Profit of the Industry

Profit makes the industry more comfortable to grow and live. Organizational behavior changes timely as per requirements of clients, orders, internal changes, economical changes, norms, and regulatory guidelines of governmental authorities. All the changes of the industries to gain profit. Epidemic affects the performance of the most every industry which leads to affection of profit. When negative profit band is considered four, nine and seven industries reflect with respect to the period i.e. years 2018, 2019, and 2020. Increase in always keep industry in comfort zone. When the profit band considered against the percentage with respect to year, eight industries in band of profit of more than 10% for year 2020. When 2019 was considered thenthirteen industries found in band of more than 10.0% profit. Which remark the covid-19 impacted profit of the industries. Digital calculations show approximately 25% of industries are affected. 54.1% of industries come in band of profit having more than 10% in 2019 and 33.3% of industries in band of profit had more than 10% in 2020.

Table 5. Profit of Industries

S. No.	Profit Earned	No. of industries out of 24		
		Year 2018	Year 2019	Year 2020
1.	More Than −5.0%	3	8	6
2.	−4.0 to 0.0	1	1	1
3.	0.1 to 5.0	7	1	3
4.	5.1 to 10.0	1	1	6
5.	More Than 10.0	12	13	8

Results & findings

Industries are located in five different locations of Pune zone in the Industrial area, Knowledge park, and special economic zone. Many giant players of software industries have subsidiaries in Pune. Industries profiles are related to the service, technology, business management module, and software products. Five industries' CEO positions were handed over within two years. Covid-19 impacted the revenue of the industries, reflected by decreasing trend of revenue with respect to the year. Trending figure of the industries indicates that the negative revenue industry percentage 16.6% rises to 33.3% from 2019 to 2020. Indutries come down from a profit band of more than10%, near around 20% reduction shows in number of industries. In 2019 around 54.4%industries is in band of more than 10% profit but in 2020 it is down to 33.3% profit margin increases in 2020 as compared to 2019 but less than 10% never shows a good comfort zone for the industries.

Discussions & suggestions

Paper speaks about the industrial locations in Pune area, working profile mentioned with category and working functions of the industries. Three years of revenue and profit of industries were considered for the calculations and analysis. Industries mostly declared the revenue and profits/operating margin/operating income in the Millions or Billions of USD, Millions or Billions of Euros, Millions or Billions or Corer of Rupees. Analysis of revenue and profit are done with consideration of base value of USD is considered as 75 Rs and value of Euro considered as 90 Rs. Trend of three-year revenue and profit estimated to calculate the impact of covid-19. One preventive measure for covid-19 is social distancing, which makes impact the saving of transport cost, power utilization cost, and hospitality cost.

Conclusion

Estimation of data reflects that Information technology from large-scale classification also impacted. Covid-19 spread throughout all places so no support from other location of industries. Decreasing number of industries with respect to revenue in the year 2019 to year 2020 concludes revenue of 25% of industries was reduced. Profit margin having more 10% number of industries reduced by around 20%.

Limitations & future study

Selected industries are from Pune, as Pune is IT hub but Bengaluru, Noida, Mumbai have good numbers of IT industries. Selected industries are subsidiaries of the parent industry or supportive of the parent industry so no separate revenue or profit margin is declared. Separate and individual study with respect to the industry required. Selected industries have stocks in markets; the private industry also needs to be selected for the study. Revenue and profit/operating margin/operating income are the only two variables are considered for the verification of economic performance which needs to be increased i.e. per capita income, sales, orders, and infrastructure also gives concrete support to the study. Only four profile industry selected for study, different profile industry also needs to be included. A quarterly review of revenue and profit for the year 2020 needs to be verified which gives the exact impact about steps in Covid-19, i.e. spread of infection, lockdown period, decreasing trends of infection, and launching of various vaccines.

Acknowledgment

I would like to thank Mr. Sumit Nisal, a working professional in the Biotechnology sector, and Mr. Jaidev Gujar, working as an IT professional with rich experience of 20 years.

References

Ajay, T. (2020). An analysis of the Impacts of Covid-19 on Large Scale Firms in Nepal. Nepal: PEDL Themes.

ASSOCHAM. (April-2020). COVID-19 Impact on Indian Industry. Primus Partners, 1-4.

He, W., Zuopeng, Z., and Wenzhuo, L. (2021). Information technology solutions, challenges and suggestions for tackling the COVID-19 pandemic. International Journal of Information management, 57.

Shen, Huayu, Mengyao, F., Hongyu, P., Zhongfu, Y., and Yongquan, C. (2020). The Impact of the COVID-19 Pandemic on Firm Performance. Emerging Markets Finance and Trade, 2213–2230.

Kumar, A., Luthra, S., Mangla, S. K., and Yigit, K. (2020). COVID-19 impact on sustainable production and operations management. Sustainable Operations and Computers 1, 1–7.

Modi Bimal, Pandey, S., and Dusija, D. (April 2020). COVID-19 impact IT due diligence considerations and technology enablement for the future. Deloitte Touche Tohmatsu India LLP., 1–18.

Murthy, C. (2016). Management Information System . Pune: Himalaya Publishing House.

Prabhakar, B. (2021, October 20). Moneycontrol. Retrieved from www.Moneycontrol.com.

Sanjoy, P., and Chowdhury, p. (June 2020). A production recovery plan in manufacturing supply chains for a high demand item during COVID-19. International Journal of Physical Distribution & Logistics Management, 51(2), 104–125.

Ye, Q., Jin , Z., and Hong, W., (June 2020). Using Information Technology to Manage the Covid-19 Pandemic; Developmental of a technical framework based on Practical experience in China. JMIR Medical Informatics, 8.

Optimizing Operational Cost and Delivery of Online Food Delivery Apps Using High-Tech Vending Machines

Deyon C. Vincent,[a] Dr. Justin Joy,[a] Pretty Deena Mathew,[a]
Paul Thomas,[a] P. R. Rahul,[a] and Hanna Maria George[a]

[a]Christ Deemed to be University, Bangalore, India
E-mail: *deyon.vincent@mba.christuniversity.in

Abstract

Consider the present scenario of placing online food orders while traveling, waiting for them, and struggling to collect them on time. This issue can be addressed by creating and deploying a fully functional high-tech vending machine. With the evolution of technology and the necessity for constant improvement in service quality, customers are thriving for a better customer experience. This article aims to design and methodologically assess the importance of installing vending machines around the most crowded public transportation hubs by dispensing purchased food or beverages online. It focuses on providing a convenient delivery mode for online-ordered food at travel boarding points and public gatherings. Vending machines at these locations gather and distribute food to consumers based on orders from online food delivery apps such as Swiggy, and Zomato, thus optimizing and improving the delivery experience. It focuses on optimizing the operational cost associated with online food delivery platforms and reducing the carbon emissions contributed by multiple deliveries that happen towards the common drop-off points.

Keywords: Vending Machine, Public Transport Hubs, Online Ordered Food & Beverages, Carbon emission, Optimization, Sustainability

Introduction

This paper aims to design an efficient food delivery system allowing online food delivery platforms to execute daily operational tasks more quickly while enhancing customer experience. Using the conventional method of ordering food from an online food delivery platform (Swiggy, Zomato) and receiving the service from one delivery partner to one customer arise a lot of energy and time for both company and customers. While each online delivery platform employee deals with a large number of customers, it creates a lot of human error and may significantly impact profitability. The analysis shows that rising demand for delivery services will result in a 36 percent increase in the number of delivery vehicles in inner cities by 2030, resulting in increased CO_2 emissions and traffic congestion.

Traffic congestion is predicted to increase by nearly 21% due to increased carbon emissions. The proposed system will significantly reduce carbon emissions by optimizing the delivery service in online food delivery apps. Over the past years, people have moved to a faster world. People have busy schedules, so they prefer online services in all manner which make their life easier. This includes online food delivery and purchasing of online goods, which has moved us all to a digital world. Even though an online food delivery service is available, consumers still face delays in receiving the products due to various travel constraints. Apart from that, the online delivery platform is also not interested in wasting time waiting for a delayed customer. And the most important focus should be shifted to striving for operational excellence in the company. Operational excellence and optimization are essential strategies to sustain growth while staying ahead of competitors. Achieving operational optimization in an organization is like a never-ending quest for hidden treasures.

Thus, this is to propose a suitable food delivery system for the food and beverage industry to solve the problem mentioned above. The system will be a valuable tool for restaurants and online food delivery platforms to improve their customer experience and satisfaction by applying the latest technology and the conventional delivery method to coordinate the receiving action of food delivery services. The prime objective of our paper is to lower the overall operational cost of the online food delivery platforms (Swiggy, Zomato) and to achieve optimization in the overall process through the fully functional high-tech vending machine. It could apply this innovation to improve the customer buying experience simultaneously.

Methodology

The proposed vending machine is connected to a mobile or web-based system (Swiggy, Zomato) which has every feature and information of the current approach in managing, ordering, and delivery activities in online food delivery apps. The Proposed system focuses more on the Order with Delivery or a Fully Integrated Business Model. In the conventional method, the customer logins into the application through mobile or laptop. The user will direct to the home page, and the customer will get a list of several cuisines, meals, unique dishes, and combos. The customer can use the filter for specified food or the restaurant. The user adds the meal to the cart. Then purchaser gets a detailed description, price, etc., of the ordered meal. After quantifying the ordered meal, the customer should give the system a preferred

address. Finally, the purchaser should choose the payment method. The delivery employees will deliver the food.

The paper presents a different approach to delivering online ordered food through the vending machine. After quantifying the meal, the system generates two other delivery methods. The user should choose one delivery method, *Deliver through the Vending Machine or Deliver through the Delivery Employee*. When the user decides to *Deliver through the Vending Machine*, the user receives a list of vending machine locations and must choose one from the list. After selecting the payment methods, excluding the pay-on delivery, a unique QR code is generated for each customer. It also generates a navigation map directing to the location of the preferred vending machine. The vending machine scans the QR code, and the user can receive the online ordered food item from the vending machine without any delay.

Software Prototype

QR Code: The online delivery platform generates a unique QR code for each customer once the payment is made. The QR code contains detailed information and data on the online ordered food. The user should ensure that the vending machine scans the QR code to dispense the ordered food.

Navigation System: The online delivery platform allows customers to locate the preferred vending machines to receive their ordered product. For customers who order the food item through the "Delivery through Vending Machine" option, a navigation system will direct them to the vending machine, which makes the customers reach it in the least time.

Hardware Prototype

Temperature Regulators: The ambient temperature in a hot holding vending machine must not be less than 57°C for more than 120 minutes after the machine is filled. Whenever the customer order is placed and the system receives payment, it initiates the manual delivery of the ordered food items to the vending machines by the delivery employee. The vending machine is incorporated with advanced technology to maintain the temperature inside remotely. The temperature regulating system in the vending machine will keep the food warm and fresh. It enables the customer to receive the order without compromising on the quality.

QR Code Scanners: The vending machine contains a QR code scanner. The generated QR must be scanned using the QR code scanner in the vending machine.

Display Screen: The vending machine contains a display screen. The screen shows all the information to the customers. When the QR code scanner scans the QR code, all the details of ordered food items are shown on the display screen. And in a few seconds, the user receives the ordered food through the dispenser. There is a lighting system to have clear visibility at night time.

2D Barcode Scanners: The 2D barcode scanners are used inside the vending machine to scan the products placed inside the vending machine. The system records the order when the user places the order using delivery through the Vending Machine option in the online delivery platform. It delivers the product to the vending machine via the delivery partners. The packed food item contains a unique QR code similar to the QR code generated in the user's device. The 2D Barcode scanner scans the products in the vending machine, and the user receives the product when both the QR code and bar code matches.

Cancellation Process

The user can cancel the order at any point in time. But the refund policy is constrained by a time window. The policy defines that the customers need to collect the order within a time window of 2 hours. If they fail to collect within 2 hours, they must cancel the order beforehand to get the refund after reducing the cancellation fee. If the customer fails to cancel or pick up the order within 2 hours, they would be charged the whole amount.

Data Collection and Analysis

Today, people live in a fast world where different technologies make movement smart and quicker.

Air, water, and land transportation, which includes rails or railways, as well as road and off-road transportation, are the different means of transportation used by man to move from one destination to another. Roads, airlines, and trains well connect India's metropolitan cities.

The Transport Network in Metro Cities of India

Roads, airlines, and trains well connect India's metropolitan cities. People are having a rat race between Nine-to-Nine without being able to maintain a work-life balance. The city has a vast transportation network for the convenience of the fast-moving city.

Food Delivery Apps in India

In India, the Food Tech companies' reach has risen six-fold in the last years, riding the wave of fast digitization and continuously increasing consumption. It will continue to develop at a faster pace. Overall internet expenditure in India is rapidly increasing, with over $130 billion estimated to be spent in the next five years. India's online food business will develop at a compound annual growth rate (CAGR) of 25–30% to become an $8 billion market by 2022. Variety of cuisines (35%) was one of the top reasons for regular usage of online food ordering applications, followed by attractive discounts and convenience, according to a Google and Boston Consulting Group (BCG) survey. Food technology is already present in over 500 cities across India, and as consumer confidence grows, new opportunities for firms to "win with the customer" in an evolving market emerge.

Market Opportunity

There is a market opportunity for food delivery platforms in the metropolitan cities of India. Table 1 Transport usage in a year (millions) passengers using the Transport network in the metropolitan cities of India. Most of the passengers in these transport networks are commuters or travelers

Table 1. Transport usage in a year (millions)

Metro Cities in India	Metrorail Ridership	Suburban Rail Ridership	Airways Ridership
Mumbai	164.25	2226.4	45.24
Delhi	365	4780	63.75
Kolkata	255.5	1200	20.00
Chennai	73	912.57	20.82
Bangalore	167.9	–	30.46
Hyderabad	109.5	800	19.57

The common among these passengers is they either skip their meals or order the food in their comfort places. Still, some of them depend on conventional methods for meals. The time constraints restrict them to not to having alternate or favorite meals. Most of the traveling points will be crowded during peak hours, crowding nearby restaurants and hotels. This causes a delay in the takeaway process. It leads customers to skip the food or order them in their homes. The vending machines of online food delivery apps can utilize it to explore hidden market opportunities.

The machine is installed around the most crowded public transportation hubs to make the customers receive the food. These passengers order the meals at their stations and can receive the food without delay.

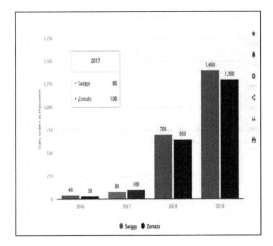

Fig. 1. Number of online food delivery orders in a day (in thousands)

Operational Optimization

Swiggy currently has over 1,30,000 delivery partners, and Zomato has around 1,62,000 delivery employees to maintain the delivery service. The wage of a delivery employee is based on their time and distance. They get about 8–20 orders per day. On average, one gets 14 orders per day. The policy defines delivery employees who are paid an estimated Rs. 40 for 4 km. Extra will be paid on behalf of several factors such as waiting time, customer pay, delivery during rain, etc. They will be paid Rs. 30–120 for one delivery service.

Estimated Calculations

On an estimation, an employee gets an average of Rs. 60 per order and 14 no. of orders per day. Rs. 60*14 gives Rs. 840 per day. For a month, it will be Rs. 25200. Considering Swiggy, pay about Rs. 3,276,000,000 for 1,30,000 delivery employees. Zomato should also pay similar to these values. The vending machine can streamline operational delivery costs by replacing an average of 5% of the total delivery employees. A single delivery employee can fill the vending machine. 5% of 1,30,000 employees is 6500, which would save Rs. *163,800,000 in a month*. In a year, Rs. *1,965,600,000* will be saved.

On an estimation, Swiggy has 14,00,000 orders per day. The vending machine can streamline operational delivery costs by replacing around 10% of the total order through the vending machine. The 10% of 1400000 is 140000, Thus,140,000*Rs 40 will give Rs.5,600,000 per day. For a *month* it will be *Rs.168,000,000* and Rs. *2,016,000,000* for a year.

Lowering Carbon Emission and Operational Sustainability

This approach also put forward the sustainable approach of online food delivery. The average emission of motorcycles in India is about 27gCO$_2$/KM. The analysis reveals that the average distance traveled by the delivery agent in India per order is 5 to 6 Km. Then there will be around 162 g of CO_2 emission per order (27*6) since we expect around 1,40,000 orders to divert through the vending machines, which helps us to minimize the carbon footprint of 22.68 MT of CO_2 emissions per day.

Conclusion

There is considerable scope available for the food delivery players to tap into the new possibility with the commuting people, where the immediate food requirement can be met by using our high-tech vending machine. Placing the vending machine at the crucial and busiest hotspot will help meet the requirement without the need for a delivery boy for food delivery platforms, which saves vast sums for the firm and reduces carbon emissions. To be successful, the food delivery firm should develop rules and technologies and build an ecosystem for using the system.

References

Sibanda, V., Munetsi, L., Mpofu, K., Murena, E., and Trimble, J. (2020). Design of a high-tech vending machine. Procedia CIRP, 91, 678–683. https://doi.org/10.1016/j.procir.2020.04.133

Alam, W., Sarma, D., Joyti Chakma, R., Alam, M., and Hossain, S. (2021). Internet of Things Based Smart Vending Machine using Digital Payment System. Indonesian Journal Of Electrical Engineering And Informatics (IJEEI), 9(3). https://doi.org/10.52549/.v9i3.3133

Tourist's Perception of Handicraft Tourism Development: A Study on Channapattana Toy Town in Karnataka

Mr. B. G. Mukunda,[*,a] *Dr. Shwetasaibal Samanta Sahoo,*[b]
Dr. Giridhari Mohanta,[b] *and Dr. Ravish Mathew*[b]

[a]Karnataka College of Management and Science, Bangalore, India, [b]Sri Sri University, Cuttack, India
E-mail: [*]bgmukunda@gmail.com

Abstract

Tourists' perceptions are very often regarded as an essential tool in planning and developing tourism in any tourist destination. The study predicts tourists' perception toward handicrafts in Channapatna toy town. Further, the study explores the important variables indicating good craftsmanship and value for money leading to tourists' perception of handicraft tourism. Data were obtained from 372 respondents employing the convenience sampling technique using the five-point Likert scale. The principal Component with the Varimax rotation method was performed while employing exploratory factor analysis to identify the predictors. Further to validate the predictors, confirmatory factor analysis was also performed in AMOS 21.0. The study reveals that tourist perception is determined by two factors, Good Craftsmanship, and Value for money. Further, it reveals that developed infrastructure is also the main contributing factor that affects tourists' perceptions leading to the fulfillment of the SDGs.

Keywords: Good Craftsmanship, Value for money, Socioeconomic status, Culture, Handicraft tourism, Tourist perception

Introduction

Handcrafts serve a significant role in reflecting the culture and traditions of a nation. The preservation of traditional skills and abilities that are linked to people's lives and histories, as well as our traditions, heritage, and culture, is facilitated by crafts. This labor-intensive, decentralized, and artisanal industry is currently present across the nation, mainly in rural and urban areas. Small towns and rural locations house the majority of craft manufacturing facilities, which have considerable market potential both within India and beyond. Over six million people are employed in the artisanal industry, many of whom are women and belong to the most disadvantaged social groups. As per the report furnished by the Ministry of Textiles, Government of India, this sector is a major source of income for rural areas. The Geographical Indication Tag (GI) is a special honor for the community to preserve its culture and customs. The Geographical Indications tag, if effectively implemented, can be a significant tool for increasing the value of MSMEs, especially in the handicraft sector with financial benefits in the form of increased sales, price level changes, quality enrichment, and profitability. The handicraft industry has experienced sustained trade growth over the past 20 years as a component of the wider global economy, and it has demonstrated the ability to establish even greater ties with the tourism industry. Handicrafts are increasingly seen as being a crucial component of the tourism industry because they serve as significant keepsakes, expanding the visitor's experience both physically and temporally. It is also widely seen as a chance to broaden the offerings in the tourism business. It is an essential component of the tourist's experience, showcasing regional customs, the skills, talents, and employment prospects of the local community. Handicrafts have improved the national economy, and identity, and encouraged cultural heritage, distinctiveness, and authenticity as a unique visitor experience (Mehta and Bisui, 2021; Upadhya, 2020). Tourists that buy handicrafts as souvenirs are buying something to keep in mind about their trip and the experience they had while there. Tourists frequently buy souvenirs either for themselves or as gifts for family and friends back home. There are many studies previously conducted relating to tourists as souvenir consumers. However, studies emphasizing tourist perception of handicrafts as a souvenir is limited. The more we understand the tourist, the better services could be provided to tourists for the development of tourist destinations. Thus, understanding tourists' perceptions are frequently viewed as a crucial instrument in planning and developing tourism in any tourist destination. (Mehta and Bisui, 2021).

Literature Review

Good Craftsmanship and Tourist Perception

The country's economic development depends heavily on the handicraft industry, which also preserves the nation's cultural legacy while employing a large number of artisans in rural and semi-urban areas and generating significant foreign exchange. Handicrafts have a great deal of potential to support the country's millions of established artists as well as the increasing number of newcomers to the crafts industry. Today, handicrafts contribute significantly to job creation and exports (Dash and Mishra, 2021). Handicrafts depict the local culture and tradition that attracts tourists to the destination thus developing the tourism industry. Rather than relying exclusively on government initiatives, it promotes entrepreneurial skills and abilities among individuals to upturn their long-term socio-economic benefits (Mohanta et al., 2020; Haliza et al., 2017). As a result, many social entrepreneurs emerge to preserve their culture and tradition to support the lives of craftspeople. The number of traditional craftspeople in Channapatna has been steadily expanding since the early 2000s. There are roughly 3,000 toy-making families and small businesses participating in selling in a tourist destination and a few of them have adopted e-commerce platforms also (Manohar and Manshi, 2018). While making any decision about a tourist destination tourists try to imagine and evaluate the tourism images about the services and environmental conditions in which they act Further, handicrafts are one of the pull factors of tourists the key challenges that the souvenir suppliers and merchandisers in the destination encounter i.e., authenticity, place of origin, quality certification, and geographical indications should be well-taken care of to improve the tourist's image of the handicrafts (Xun and Xu, 2021). To truly understand the skill and craft of the artisans and the distinctive tradition of handicrafts, it is important to connect tourists with craftspeople. This will help tourists better understand and appreciate the art. Handicraft is more than just the commercialization of cultural arts or the display and sale of products. (Upadhyay, 2020). Tourists purchase local art and craft as souvenirs because they keep their distinctiveness and demonstrate an intuitive link to the tradition and culture of the area. This has increased the intrinsic value of local art and craft to international tourists. Being a part of Indian culture, handicrafts are considered cultural tourism attractions. Many tourist destinations are promoting handicrafts through their tourism websites. A better image of the handicrafts of a destination influences tourist buying decisions (Fangnan et al., 2016). Hence, it can be said that good craftsmanship influences tourists' perceptions.

Value for Money and Tourist Perception

Various independent consumer perception factors, such as value-for-money service, quality, and performance, have an impact on customer satisfaction. The tourism industry shares the association between value for money and customer satisfaction, and value for money is a key factor in determining consumer satisfaction. (Kansal et al., 2015). Value for money (VFM) is a word that, in reality, has been poorly understood. The use of VFM in decision-making is not sufficiently explicit, especially when differentiating value from money. (Barton et al., 2019). An object's value shouldn't be determined by its cost but rather by the service it provides. The importance of tangible products or intangible services to people, the advantages they receive, and the prices they incur become crucial and delicate aspects when choosing tourist destinations. When people believe that the quality of the services provided to them is higher than what they paid for it, tourists' perceptions of a destination have an impact on their satisfaction. (Operinde and Praise, 2020). Hence it can be stated that value for money is one of the key factors that affect tourist perception. Based on the above theoretical argument, it is seen that value for money and good craftsmanship play a vital role in affecting tourist perception towards handicrafts toys, and did not find that any study has considered these predictors that ground tourist perception. Hence, the researcher got an opportunity to throw more light on this issue and for this, the following model has been proposed and which is presented in Fig. 1.

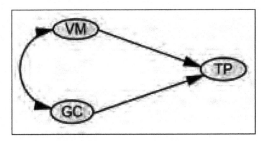

Fig. 1. Tourist perception model
Source: Author's own

Objectives of the Study

- To find out the predictors affecting tourists' perception of handicraft tourism in Channapatana.
- To understand the tourists' perception towards their buying decision for handicrafts in Channapatana.
- To explore the underlying variables indicating Good Craftsmanship and Value for Money leading to tourists' perception of handicraft tourism in Channapatana.

Research Methods

Research Design

A structure was proposed to examine the predictors of tourist perception towards handicrafts in Channapatna toy town in Ramanagara District of Karnataka handicrafts which are presented in Fig. 1. Further, an examination of the connection has been made between exogenous (value for money and good craftsmanship) and endogenous (tourist perception) variables. The reason for the inclusion of such a district is that the craftsmanship and tourist visit rate is high as compared to other districts of Karnataka and that was the main motivating factor of inclusion such a district in the study.

Sample Frame and Data

The convenience sampling method was used to collect the data since it has the highest response rate and uses less time and resources. Data were gathered over six months from June 2021 to November 2021 through a structured survey instrument consisting of 103 items on a five-point Likert scale from 372 samples. The reliability and validity of the survey instrument were checked through a pilot survey by taking 30 respondents and found adequate and satisfactory. Internal consistency of the scale was measured using Cronbach's alpha and found the value to be superior to 0.7 which proves the reliability of the survey instrument (Nunnally et al., 1967). Further, the scale was validated through confirmatory factor analysis (CFA) for its reliability and validity and was found acceptable and satisfactory.

Result Analysis

The collected data were analyzed with the help of SPSS and AMOS.

Reliability Analysis of Scale

The examination of the reliability of the scale was made using Cronbach's alpha and values for all the constructs were observed to be fulfilling the cut-off value of 0.7. The Cronbach's alpha values for the three constructs value for money (VM), Good Craftsmanship (GC), and tourist perception (TP) were 0.848, 0.876, and 0.822, respectively. To check the theorized model a two-stage SEM was performed (Hair et al., 2013). In the initial stage, confirmatory factor analysis was executed to check the theorized model (MM) and the common method bias (CMB) was validated through Harmon's single-factor method in factor analysis by applying the principal component extraction method without applying any rotation. The extracted single factor has explained 48.900 percent of the total variance and that is within the suitable boundary i.e., less than 50 percent. Therefore, common method bias was omitted. Through EFA three factors have been extracted and the extracted three factors are explaining 70.048 percent of the total variance. Additionally, the adequacy of the sample for factor analysis was measured through Kaiser–Meyer–Oklin (KMO) and Bartlett's Sphericity tests and observed the value to be 0.828 which is more than 0.50 (Hair et al., 2013).

Measurement Model

Goodness-of-fit indices are calculated for the whole three latent constructs in confirmatory factor analysis in a single model to measure its goodness of fit (Schreiber et al., 2006) which is shown in Fig. 2. The goodness-of-fit for three constructs in the model was checked by comparing the values of model fit indices. It is found that the values of p = 0.05; CMIN/DF = 1.738; GIF = 0.950; NFI = 0.956; CFI = 0.981; TLI = 0.971; AGFI = 0.909 and RMSEA = 0.060, are within the suitable ranges representing the well fit of data for the model (Hair et al., 2010).

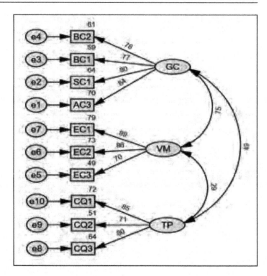

Fig. 2. Measurement model

Source: Authors' own

Convergent Validity Test of Measurement Model

The Construct Soundness of the measurement model (MM) is being measured through convergent validity and discriminant validity. The convergent validity is measured from the value of average variance extracted (AVE) and the composite reliability from the result of CFA. The MM model was observed to be convergently validated as AVE > 0.5 and CR > 0.7 (Fornell and Larcker, 1981). The AVE for TP is 0.610, GC is 0.600 and VM is 0.669, and CR for TP is 0.823, GC is 0.856 and VM is 0.857 respectively.

Table 1. Discriminant Validity

Construct	TP	GC	VM
TP	0.823*		
GC	0.510	0.856*	
VM	0.296	0.753	0.857*

Source: Authors' own

Note: * Square root of AVE

Table 1 shows the discriminant validity of the constructs and seen that the diagonal discriminant values are larger as compared to correlation values by the side of the relevant row and column values hence the discriminant validity of the constructs is validated.

Structural Model

The structural model (Fig. 3) shows that the exogenous variables (GC, VM & TP) have explained 62 percent of the variance on the endogenous variable (ID) as its R^2 value is 0.62. All the values are floating around 0.9 indicating a well fit for the proposed model. The RMSEA is 0.075 (<0.08), X^2/df is 2.145 (<5.0), and GFI, NFI, CFI, TLI, and AGFI are 0.942, 0.946, 0.970, 0.955 and 0.893 respectively indicating the well fit of data for the model.

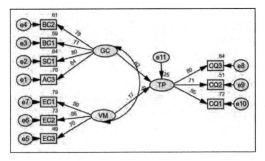

Fig. 3. Structural model
Source: Authors' own

Table 2. Path Analysis

	Estimate	S.E.	P-value	f-squared	Effect size **
TP ◄— VM	0.254	0.130	0.050*	0.1867	Medium
TP ◄— GC	0.565	0.112	0.001*	0.0400	Small

Source: Authors' own

Table 2 depicts the path details of the latent constructs and, observed significant influence on endogenous variables i.e., tourist perception. R^2 included is 0.25 and R^2 excluded for the first path (TP &VM) is 0.11 and for the second path (TP & GC) is 0.22 in the model. Value for money (VM) has a medium effect on tourist perception (TP) whereas good craftsmanship (GC) has a small effect size on tourist perception. Therefore, the influence of predictors i.e., value for money (VM) and good craftsmanship (GC) on tourist perception cannot be refuted at all.

Discussion

The proposed model has been examined by the outcomes of the study. And observed that the model fit indices of the MM and SM models are quite good to validate the model as they are in line with the prescribed standard indices. Further, it is seen that the CFA also authenticates EFA in terms of research output. It is also observed that these two predictors i.e., value for money (VM) and good craftsmanship (GC) have a significant positive impact on the tourist perception (TP) while making economic decisions. However, value for money plays an important predictor in decision-making when it is compared with good craftsmanship. R^2 in the model is 0.25 which implies that 25% of the variance is explained by the two predictors' value for money and good craftsmanship. F- squared values of VM and GC about tourist perception are 0.1867 and 0.0400, implying a medium and small effect size respectively.

Conclusion

The study was conducted to examine the influence of value for money and good craftsmanship on tourist perception. And, it is observed that these predictors have a significant impact on the tourist's perception about making an economic decision in the tourist destination. In this connection, data were collected through a well-designed survey instrument and validated through MM and SM models and observed well fit of data as per fit indices values. As it is a new concept and emerging discipline a thorough study is needed by taking more predictors and risk management mechanisms in a tourist destination that is supported by local and government bodies. The study will be more beneficial to young entrepreneurs to explore more opportunities to establish their businesses and will attract the attention government to focus on the development of this industry. The major hurdles are the accessibility of different national and international tourists and the inclusion of fewer predictors in the study.

References

Barton, R., Ajbinu, A. A., and Romero, J. O. (2019). The Value for Money Concept in Investment Evaluation: Deconstructing its Meaning for Better Decision Making. Project Management Journal, 50(2), 210–225.

Dash, M., and Mishra, B. B. (2021). Problems of Handicraft Artisans: An Overview, International Journal of Managerial Studies and Research (IJMSR), 9(5), 29–38. https://doi.org/10.20431/2349-0349.0905004

Fangnan, C. Yaolong L., Yuanyuan, C., Jin D., and Jizu L. (2016). An overview of tourism risk perception: Natural Hazards: International Society for the Prevention and Mitigation of Natural Hazards, 82(1), 643–658.

Kansal, P., Walia, S., and Goel, S. (2015). Factors Affecting Perception of Value for Money and Customer Satisfaction for Foreign Tourists in Goa. International Journal of Hospitality & Tourism Administration, 8, 93–102. 10.21863/ijhts/2015.8.2.017.

Manohar, R., Anitha M., and Munshi, Reshmi (2018). Consumers' Views on Handicraft Preferences: A Case Study On Channapatna Turns Wood Lac Ware Handicrafts. International Journal of Sales & Marketing Management Research and Development (IJSMMRD), 8(4), 1–6.

Mehta, L., and Bisui, R. (2021), Measuring Tourist's Perception in Tourism: Tools and Techniques, International Journal of Transformation in Tourism & Hospitality Management and Cultural Heritage, 5(2), 62–67.

Mohanta, G., Sahoo, S. S., Mukunda, B. G., and Subbarao, P. S. (2020). Sustainable Rural Tourism Development as a tool for reducing rural migration: A case study on selected districts of Odisha. Gazi University Journal of Science, 33(11), 10–21.

Operinde, O. H., and Praise, E. E. (2020). Assessment of Tourists' Perception and Satisfaction in Agodi Park and Gardens Ibadan as A Nature-Based Tourism Attra0ction. International Journal of Research - GRANTHAALAYAH, 8(7), 144–159. https://doi.org/10.29121/granthaalayah.v8.i7.2020.653

Upadhyay, P. (2020). Promoting Employment and Preserving Cultural Heritage: A Study of Handicraft Products Tourism in Pokhara, Nepal. Journal of Tourism & Adventure, 3(1), 1–19.

Xun, Lin, and Xu, Y. (2021). Tourists' Perception Evaluation of Red Tourism Attractions Based on Grounded Theory. Academic Journal of Humanities & Social Sciences. 4(12), 109–112. DOI: 10.25236/AJHSS.2021.041220

Green Purchase Behaviour: Effectiveness of Reference Group Influence for Explaining Green Purchases in Emerging Market

Balween Kaur[a] and Veer P. Gangwar[b]

[a,b]Mittal School of Business, Lovely Professional University, Phagwara (Punjab) India
E-mail: [a]balweenkaur4@gmail.com (Corresponding author); [b]veer.23954@lpu.co.in

Abstract

This present study has endeavored to investigate the precursors of customers' buying behavior with regard to green personal care products in an emerging economy. An empirical test of the model has been conducted where the influence of reference groups (social and peer influence) on green buying behavior of Indian millennials has been analyzed. Primary data collected from 385 respondents belonging to the age group of 24 to 40 years was analyzed via Partial Least Square- Structural Equation Modelling (PLS-SEM) approach to examine the posited research hypotheses. The PLS-SEM revealed that both the constructs of reference group influence are positively related to Indian millennials' buying behavior in respect of eco-friendly personal care products. The scant literature in respect of sustainable buying behavior of consumers of an emerging economy has been enlarged as well as marketing theory being underpinned with inputs delivered by findings of the study.

Keywords: Consumer behaviour, Green personal care products, Reference group, Social influence, Peer influence

Introduction

The unending degradation of the environment has thrown the biggest challenge in front of mankind. People all over the world have become aware of the threats of continuous environmental deterioration. As a consequence, they have already started following the green path of sustainability to avoid further destruction of the planet and contribute to its healing process. Commercial organizations noticed this behavioral change among consumers and modified their operations to support pro-environmental practices (Sharma et al., 2020). Therefore, manufacturers have launched a wide range of green products in different categories such as green electronics, organic food, green beauty, and makeup products, green IT products, green vehicles, green convenience and personal care products, and green cosmetics. Recently, consumers' interest in green products has enhanced due to environmental and health benefits (Lin et al., 2020). In emerging economies like India, consumers have shifted their buying preference from conventional products to eco-friendly products which do not cause any harm to environment (Kautish et al., 2019; Sadiq et al., 2020). Over the period, awareness of Indian consumers concerning sustainability has enhanced, as a result, consumption of green products in India has risen significantly (Sharma and Sharma, 2013; Wang et al., 2014). Researchers all over the world have done numerous research studies to detail the factors which stimulate consumers to adopt green products for meeting their routine needs and demands (Sadiq et al., 2021; Al-Swidi and Saleh, 2021; Wang, 2020). The global researchers have elucidated the impact of various psychographic (Wei et al., 2018; Walia et al., 2019; Verma, 2017; Sun et al., 2018), socio-demographic (Ifegbesan and Rampedi, 2018; Nath and Agarwal., 2015; Jaiswal et al., 2020), social (Persaud and Schillo, 2017; Suki and Suki, 2019; Varshneya et al., 2017) and green marketing factors (Kaur et al., 2022; Mahmoud et al., 2018; Rahahleh et al., 2020; Mehraj and Qureshi, 2020) on purchasing preferences of consumers belonging to diverse cultures. As far as pollution is concerned, India is behind the two most polluting countries in the world; therefore it has become urgent to save the country from a huge ecological crisis. But the research on the pro-environmental behavior of public in Indian context is still in infancy stage (Bukhari, 2011; Khan et al., 2016; Narula and Desore, 2016). To the best knowledge of researchers, there is hardly any research study in Indian context elucidating the impact of reference group influence on green buying behavior of young population of India concerning eco-friendly products used for personal care. To fill these lacunae and to make significant contributions to the extant literature, this research study responds to the call for a stronger focus on the adoption behavior of the developing nation's millennial population towards sustainable personal care products. Taking this as motivation, the authors have formulated a conceptual framework to corroborate the impact of reference group influence on consumers' behavior to buy green personal care products. Therefore, the authors envisage determining whether the reference group influence (social influence and peer influence) impacts the buying behavior of Indian millennials for eco-friendly products used for personal care. Therefore, the present research study aims to 1. To reveal the impact of social influence on the pro-environmental behavior of consumers. 2. To determine the impact of peer influence on the pro-environmental behavior of consumers. Section 1 gives a detailed account of the background, rationale, and novelty of the study. Section 2 narrates the review of literature, theoretical framework, formulation of hypotheses, and

conceptual model. Research methodology describing the measures adopted, sample, and procedure have been detailed in Section 3. Section 4 elucidates the data analysis and findings of the study. Discussion and implications have been presented in Section 5 and at last, Section 6 presents the conclusion, limitations of the study, and directions for future researchers.

Literature Review and Hypothesis Development

For understanding and determining the behavior of human beings, the most widely used theories in past research studies are Theory of Reasoned Action (TRA) and Theory of Planned Behaviour (TPB). Both these theories were propounded by Icek Azjen and Martin Fishbein in 1975 and 1980 respectively. These theories underline that attitude, subjective norms, and perceived behavioral control of an individual determine buying intentions, and buying intentions are then converted into buying behavior. Global researchers have conducted an unending number of empirical research studies elucidating the effect of different psychographic, social, personal, and cultural factors and consumers' socio-demographics on sustainable adoption behavior of consumers by adopting such constructs as extensions to TRA and TPB model (Wang, 2020; Sharma et al., 2020; Liu and Liu, 2020; Zhang et al., 2019, Yarimoglu and Gunay, 2019; Sharma and Foropon, 2019; Emekci, 2019; Sun et al., 2019). Human beings are social animals and always prefer to associate with their social groups (Clark et al., 2019). The influence exerted by persons important to us or by society at large upon an individual to behave in a specific manner is termed social influence. As a consequence, the purchasing intention and behavior of an individual get modified when society expects him to make his consumption decision in conformity with them (Sadiq et al., 2021). The green consumption behavior of Chinese consumers (Geng et al., 2017; He et al., 2016) and Turkish consumers (Midilli and Tolunay, 2021) has been significantly impacted by social influence. As suggested by previous literature, social influence has been found as a significant precursor of buying behavior of consumers belonging to diverse nations and cultures (Hosta and Zabkar, 2020; Joshi and Rahman, 2015; Ferguson et al., 2017; Johnstone and Hooper, 2016; Persaud and Schillo, 2017; Varshneya et al., 2017; Mosavichechaklou and Bozbat, 2020). As a consequence, the following hypothesis has been proposed: *H1: Social Influence significantly impacts green behavior of consumers.* The motivation and encouragement from peer groups to perform certain actions and behave in a specific manner is referred to as peer influence. The consumers undergo modifications in present behavior and confirm their behavior of adopting certain products for consumption with their peer group members and embrace green principles when green agenda is endorsed by the peer group. The green behavior of consumers is spurred by peer interaction as the members of peer groups are generally consulted by them before adopting new kinds of products and making any changes in product preference. Lee (2009) and (2010) has revealed that peer group influences green behavior of consumers directly or indirectly (Sheoran and Kumar, 2020). Past studies in global context have highlighted statistically significant connection between the influence of peer groups and green behavior of consumers regarding various categories of green products (Bertrandias and Gambier, 2014; Connolly, 2019; Suki and Suki, 2019; Khare, 2020; Prasher, 2020). Thus, the following hypothesis has been formulated: *H2: Peer Influence significantly impacts green behavior of consumers.*

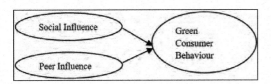

Fig. 1. Conceptual Model

Research Methodology

We formulated a well-structured questionnaire with 10 variables. Despite developing a suitable scale for data collection, we have adopted the statements from various scales used by researchers in earlier studies published in journals of good repute (Khare, 2020; Clark et al., 2019; Kautish

et al., 2019; Sharma et al., 2020; Lee, 2010; Tang, 2014; Suki and Suki, 2019; Wang, 2014; Persaud and Schillo, 2017). The questionnaire was divided into 2 sections, where, Section 1 collected information regarding respondents' demographic characteristics, and in Section 2, the respondents were asked questions regarding social influence, peer influence, and green buying behavior. The responses have been asked on a 7-point Likert Scale with two extremes of "strongly disagree" and "strongly agree." This part has seven statements 'relating 'to two variables of reference group influence, and three statements relating to green behavior of consumers.

Measures adopted: The first construct was a social influence (SI) having 3 variables: conformity with group mates, social persuasion, and guidance from referents (SI1, SI2, and SI3). Another construct was peer influence (PI), with 4 variables: green agenda endorsed by peers, choice as per peer opinion, obtaining environmental knowledge from peers, and parental motivation (PI1, PI2, PI3, and PI4). Further, green consumer behavior (GCB) was measured using three variables: ensure to be eco-friendly, eco-friendly packaging, and products with no harm to environment (GCB1, GCB2, and GCB3).

Sample and procedure: The authors prepared a cover letter explaining the purpose of research and protection of privacy. The questionnaire along with a covering letter was distributed to respondents using both online and offline methods. The primary data was collected from Indian millennials from January 2022 to April 2022. According to Pew Research Centre, people who have taken birth between 1981 and 1996 are defined as millennials. Therefore, for collection of data for the present study, Indians belonging to the age group 24 to 40 years who were either using green personal care products or were not using but aware of such products, were contacted. As the millennial generation is the primary bread earner in most Indian families, therefore, the major decisions for buying different kinds of products for meeting family's routine needs and luxurious demands are generally taken by them. Moreover, the persons of millennial generation are more qualified and aware of ecological crisis than older generations. This generation has understood the responsibility of using scarce resources judiciously to save such resources for future generations and promote sustainability. Therefore, the concept of sustainability is quite popular among Indian millennials and adolescents. The total study population (27,51,49,600) belonging to the age group 24–40 years has been retrieved from the Sample Registration System Report of Government of India published on the structure of the population based on 2011 census. Sample size has been calculated as 385 using the Raosoft sample size calculator available on Raosoft website with 5% margin of error and 95% confidence level. For data collection, the questionnaire was distributed to 450 respondents. The data was collected but 65 incomplete and unusable responses were eliminated and finally descriptive statistics was formulated using 385 complete responses.

Table 1. Socio-demographic characteristics of respondents

Descriptive Statistics		N	%
Gender	Male	130	33.76
	Female	255	66.23
Age	24–30 years	181	47.01
	31–35 years	135	35.6
	36–40 years	69	17.92
Educational Qualifications	Undergraduate	153	39.74
	Graduate	124	32.20
	Post Graduate	85	22.07
	Doctorate	23	5.97
Occupation	Student	136	35.32
	Homemaker	43	11.16
	In Service	87	22.59
	Professional	69	17.92
	Own Business	50	12.98
Monthly income	< Rs. 20000	157	40.77
	Between Rs. 20000 and 50000	132	34.28
	Between Rs. 50000 and Rs. 100000	73	18.96
	> Rs. 100000	23	5.97
Marital status	Unmarried	238	61.81
	Married	147	38.18

Source: Authors' calculation

Data Analysis

Internal reliability and convergent validity checks have been employed to analyze the outer model specifications. To check the reliability of data, Cronbach's Alpha (CA), RhoA, and Composite Reliability (CR) are used where the threshold limit of more than 0.70 shows that the data is reliable (Hair et al., 2017 and 2020). Average variance extracted (AVE) has been used to check convergent validity which should be greater than 0.50 for the main constructs of the proposed model (Fornell and Larcker, 1981; Hair et al., 2019). The outer loadings (L) along with internal reliability and convergent validity have been discussed in Table 2. Rule of thumb for factor reliability for outer loading values is 0.708 or greater (Hair et al., 2019).

Table 2. Reliability & Reflective Model Assessment

Construct	Items	L	CA	RhoA	CR	AVE
SI	SI1	0.930	0.886	0.894	0.930	0.815
	SI2	0.897				
	SI3	0.880				
PI	PI1	0.911	0.935	0.935	0.953	0.837
	PI2	0.916				
	PI3	0.911				
	PI4	0.921				
GCB	GCB1	0.918	0.905	0.905	0.940	0.840
	GCB2	0.914				
	GCB3	0.918				

Source: Authors' Calculations

According to Hair Jr et al., 2016, the uniqueness of one construct of the conceptual model from all other constructs is ensured by employing discriminant validity assessments. It is established when the shared variance of one construct is higher than that of all other constructs (Hair et al., 2019).

Table 3. Discriminant Validity of Constructs (Fornell-Larcker Criterion)

	GCB	PI	SI
GCB	0.917		
PI	0.909	0.915	
SI	0.410	0.395	0.903

Source: Authors' Calculations

Heterotrait Monotrait ratio of correlations (HTMT) criterion has been employed to achieve efficiency in measurement model. HTMT values should be less than 1(Henseler et al., 2015), but as stated by Voorhees et al., 2016, HTMT should be less than 0.85. But Gold et al., 2001, has set the permissible limit at 0.9. Table 4 highlights that HTMT values are less than threshold limit.

Table 4. HTMT Ratio & Discriminant Validity Assessments

HTMT	GCB	PI	SI
GCB			
PI	0.846		
SI	0.457	0.433	

Source: Authors' Calculations

The association among various constructs and their predictive relevance is determined by structural model assessments (Hair et al., 2017). The multi-collinearity issue is being checked by tolerance and Variance Inflation Factor (VIF) for which the most conservative threshold limit is VIF value less than 3.33 (Diamantopoulous et al., 2008). The inner VIF values were found within the threshold limit as PI (1.185) and SI (1.185); proving no collinearity issues were involved in the study (Hair et al., 2017). After satisfying the collinearity issue, the empirical analysis has been executed with 5000 bootstraps without change to arrive at p-values for testing of hypotheses (Hair et al., 2020). The coefficient of determination (R^2) of GCB has been found high at 97.9%. According to Rasoolimanesh et al. (2017), a value of R^2 0.20 and above is considered to be good. SRMR value less than 0.08 is considered to be a good fit for testing the fitness of the model (Hair et al., 2020). The SRMR value for present study is 0.058 revealing a good fit and suggests for a higher explanatory power of the model (Henseler et al., 2016; Hu and Bentler, 1999). It has been unearthed that both hypothesized independent constructs affect the dependent variable but PI (b=0.980) is playing a more significant role in impacting green buying behavior of consumers as compared to SI (b=0.023). Therefore, both H1 and H2 are supported.

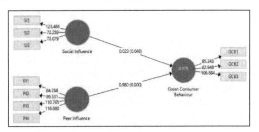

Fig. 2. Structural Model

Table 5. Structural Model Assessments

Hypotheses	H1	H2
Path relationship	SI->GCB	PI->GCB
Std. Beta	0.023	0.980
Sample mean (m)	0.024	0.980
T-Value	1.984	181.210
CI 2.5%	0.001	0.969
CI 97.5%	0.045	0.990
P-value	0.048	0.000
Decision	Supported	Supported

Source: Authors' Calculations

Discussion and Implications

The impact of reference group influence on buying behavior of Indian millennials about green personal care products has been elucidated in the empirical analysis of the present research study. The authors have made a significant contribution to the extant literature on green consumer behavior by providing empirical pieces of evidence of the significant impact of consumers' social and peer groups in shaping and molding buying behavior. The descriptive statistics of sampled respondents highlight that unmarried, female consumers of the age group 24–35 years, working as professionals, students, or in service having monthly income less than Rs. 50000 are found to be using green personal care products more frequently than male, married consumers having more income and are either homemakers or doing own business. Both hypotheses are accepted as social influence and peer influence are found to significantly impact green purchasing behavior of Indian millennials. These findings are congruous with the empirical evidence provided by similar research studies recently conducted by researchers in India (Kaur and Gangwar, 2022) and in western developed countries (Sadiq et al., 2021; Clark et al., 2019; Persaud and Schillo, 2017; Johnstone and Hooper, 2016; Khare, 2020; Prasher, 2020). The researchers have highlighted that the sapience of Indian millennial customers for eco-friendly products becomes strong if the decision gets assent of near and dear ones (Khare, 2020). Therefore, the study has made significant contribution to literature by corroborating that opinion of relevant others and near ones improved customers' appraisal of attributes of eco-friendly personal care products.

Theoretical and Practical Implications: An important contribution has been made by the present study in literature by elucidating the favorable impact of influence of referees of the consumers on their buying behavior in relation to sustainable personal care products. The descriptive statistics also provide insight into the socio-demographic features of Indian millennials who are more aware of and result in buying green products for meeting their routine needs. It guides marketers about the basis of selecting a target group for their future marketing endeavors as female, unmarried Indians of less than 35 years of age who are either students or doing job and having monthly income upto Rs 50,000 are more resulting in buying green personal care products than male, married persons doing business and having more income. It is noteworthy that no previous research study explored the effect of reference group influence on adoption behavior of Indian millennials concerning eco-friendly personal care products and the present study has attempted it for the first time, to the best of researchers' knowledge. Therefore, this research study makes an addition to the extant literature as empirical pieces of evidence plays a vital role in theory formulation around the sustainable behavior of consumers.

Conclusion, Limitations, and Future Research

Social influence and peer influence can be utilized by marketers in enhancing the sales and market share of their products. The present conceptual model has been framed to provide valuable guidance to marketers as Indians are generally interested in consulting with known others while making different kinds of decisions. It has emerged as the key study which provides valuable insights to marketers that reference group influence can work as a multiplier in capturing a larger market by the sellers. When green personal care products are preferred by customers, they become the brand ambassadors of the seller company and endorse the benefits of such products to their social and peer groups, thus resulting in higher sales and increased revenue at no additional cost. As far as limitations are concerned, the practical value of findings of study is limited for marketers as it has used cross-sectional research design. Moreover, the present study has been conducted in an emerging economy; therefore, the findings cannot be generalized to the developed countries of western world. Also, the present research study has ignored the total working population of the country and tested the hypotheses only on millennial population of India. As the present study was confined to a developing nation only, this opens the gate of opportunity for future academicians to test the conceptual framework in different geographical locations and cultural contexts. Future researchers are suggested to study the moderating effect of experience of using other categories of green products on the relation between reference group influence and green buying behavior of consumers.

References

Clark, R. A., Haytko, D. L., Hermans, C. M., and Simmers, C. S. (2019). Social influence on green consumerism: country and gender comparisons between China and the United States. Journal of International Consumer Marketing, 31(3), 177–190.

Ghose, A., and Chandra, B. (2020). Models for Predicting Sustainable Durable Products Consumption Behaviour: A Review Article. Vision, 24(1), 81–89.

Jaiswal, D., Kaushal, V., Singh, P. K. and Biswas, A. (2021). Green market segmentation and consumer profiling: a cluster approach to an emerging consumer market. Benchmarking: An International Journal. 28(3), 792–812.

Kaur, B., and Gangwar, V. P. (2022). Identifying the Predictor Antecedents of Green Purchase Intentions of Millennials in India. ECS Transactions, 107(1): 8121.

Kaur, B., Gangwar, V. P., and Dash, G. (2022). Green Marketing Strategies, Environmental Attitude, and Green Buying Intention: A Multi-Group Analysis in an Emerging Economy Context. Sustainability, 14(10), 6107.

Khare, A. (2020). Antecedents to Indian consumers' perception of green apparel benefits. Research Journal of Textile and Apparel. 24(1), 1–19.

Sheoran, M., and Kumar, D. (2020). Modelling the enablers of sustainable consumer behaviour towards electronic products. Journal of Modelling in Management. 15(4), 1543–1565.

Yarimoglu, E., and Gunay, T. (2020). The extended theory of planned behavior in Turkish customers' intentions to visit green hotels. Business Strategy and the Environment. 29(3), 1097–1108.

Study of Impact of GST on Real Estate Industry in Pune City

Shilpa R. Kulkarni[a] and Sudhindra Apsingekar[b]

[a]Matrix school of Management studies, Pune, India
[b]SIOM Research Centre Pune, India
E-mail: Sudhindra.apsingekar@gmail.com

Abstract

The real estate industry is considered the most important pillar of the Indian economy, contributing 6–8% to India's Gross Domestic Product (GDP). Under the old value-added tax (VAT) regime, many taxes such as services-related tax and VAT were related to the real estate sector. The new GST system would completely modify the indirect tax structure. This white paper attempts to provide insight into the impact of the GST on the real estate industry and highlight key issues. The study also found that the GST has simplified the tax treatment of the property sector, solving long-standing problems such as valuation and nature of supply. However, there are still certain gray areas that have yet to be evaluated. GST has increased transparency and fluency in the real estate area for all stakeholders and helped a wide range of ancillary industries by increasing openness and accountability.

Keywords: GST, Real estate Industry, Indian Economy

Introduction

According to the new standards under Goods and services tax (GST) for the Real Estate area, GST will accrue when selling a property under construction before receipt of the Certificate of Occupancy (OC). Under GST, the valuation rate is set at 18% (12% for certain affordable housing projects), allowing a standard reduction of 33% in property value. The successful GST rate available for the sale of properties under development is therefore 8% to 12 % of the total contract value, as opposed to around 5.5% (including limited credits) under the previous indirect tax regime. Additionally, the GST requirement far exceeds the stamp duty (approximately 5%) payable on the contract value. This results in a typical tax rate of around 13% to 17%. A high tax on each transaction will result in an acceleration of the last buyer's cost. In any case, GST is allowing more ITCs and henceforth in a perfect world arrangement value of property should decrease. The various GST rates (5%, 12%, 18%, and 28%) on purchases of inputs and input services are another view that complicates the tax collection framework and leads to unfair classification disputes. The government continues to challenge the justification of tariffs, signaling its willingness to move towards a simplified tariff structure. This was partially improved by the rate cut from 28% to 18% on 27 July 2018.

Need for Study

Introduction of GST is one of the most important initiatives in India since 2017. The GST is a national concern and has received considerable attention from tax professionals, the business community, consumer groups, politicians, etc. The purpose of this study is to identify the effect of the introduction of GST on the real estate industry. The details below highlight the need for systematic research in this area.

1. The successful implementation of GST in the real estate industry has received mixed reviews from experts and tax professionals. Significant analysis is required to assess the short-term trade-offs and long-term payback of GST implementation.

2. GST is a nationwide tax. However, all state-specific taxes are aggregated, so a detailed area-specific analysis is required to examine the effectiveness of GST implementation in a particular area.

3. GST faced some implementation challenges after its introduction in India. It is therefore important to weigh the benefits of GST in the real estate industry against near-term implementation challenges.

Objectives

1. Understand the effect of GST on the real estate industry.
2. Understand the difference between the traditional indirect tax system and GST for real estate area.
3. A presentation of the benefits and difficulties of GST in the real estate industry.

Literature Review

New GST rates for residential real estate were announced at the 3rd GST Council Meeting, which was held on February 24, 2019, and will take effect on April 1. The following new GST rates have been put forth for residential real estate

Residential properties that are not included in the affordable housing sector will be subject to a 5% GST charge without an input tax credit (ITC).

Residential properties covered in the affordable housing segment will be subject to 1% GST without ITC.

Following the implementation of the GST, state government taxes such as real estate registration fees and stamp duties have not changed. These fees can differ from one state to the next and even within the same state from one circle to another. While GST will only apply to properties that are being sold while they are still under construction, stamp duty and registration fees will still be levied on both completed and unfinished properties in India.

A Study "An Enquiry into the Effect of GST on Real Estate Sector of India," points out that the transition from the current indirect tax regime to GST will help the Indian economy prosper in the long term (Dubey et al., 2017)

A research paper on "Impact of GST on India's Real Estate Secto," stated that the introduction of GST would not only help India's economic development but also improve the country's GDP by more than 2% (Shalini, 2016).

Ehtisham Ahmed and Satya Poddar review Goods and Services Tax Reform and Government-to-Governmental Considerations in India and found that although the benefits of GST are tightly linked to the concept of GST law, however GST would result in a simpler and more evolved framework for evaluation (Ahamad et al., 2009).

Nitin Kumar examined goods and services taxes and showed that the GST would be the most useful framework to eliminate financial harm and ensure a consistent and simple customs framework across the country (Kumar, 2014).

A study "Research paper on the impact of the Goods and Services Tax (GST) on the Indian economy," argues that with introduction of the GST, government is targeting India's huge population to the potential inflation. Research further points out that efforts should be made to mitigate potential inflation (Dani, 2016).

Sampling Design

Sampling Technique: Researcher has used Purposive sampling technique for this research.

Universe and sample size: All people aged 25 years and living in and around the Pune area were considered part of the population for this study. A sample size of 200 is chosen for this study.

Sampling unit: Population units belonging to the age group of 25 years and older were treated as the sampling unit.

Sampling Frame: Selected areas of Pune City such as Dhayari, Sinhgad Road, and Sadashiv Peth are considered sampling frames.

Data Analysis

After surveying with a sample size of 200, a data analysis of the responses received from customers and industry representatives was conducted. For this study, researcher prepared a 25-item questionnaire and employed an interview technique. Data analysis and interpretation of some of the key questions are provided below.

Table 1. Do you think Implementation of GST in Real estate industry results in increased cost of residential properties

S. No	Response	Numbers
1	Yes	90
2	No	74
3	No Idea	36

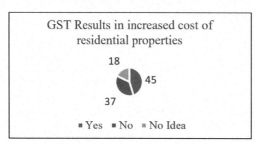

Fig. 1. Perception of GST in real estate Industry

Figure 1 indicates perception of real estate customers post-GST implementation. Approximately half of the population confirms that GST will result in increased cost of residential properties.

Table 2. Do you think GST will lead to easier administration & simplified procedures for business

S. No	Response	Response in Numbers
1	Yes	162
2	No	25
3	No Idea	13

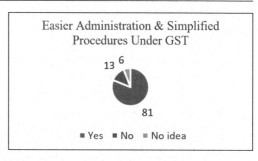

Fig. 2. GST Procedures & Administration

Figure 2 shows 81% of the sample population believes that GST will reduce cascading of taxes, which will simplify taxation procedure and administration.

Findings

1. GST has simply changed the view of individuals in Real Estate industry. Increased rate of GST for under development Residential properties is compensated by Input Tax Credit (ITC) facility given under GST to Real estate sector.

   This Study additionally observed that average citizens are not completely mindful of itemized changes that GST made in this area. Particularly the distinction in tax rate dependent on category or housing segment.

2. No stamp duty deduction – perhaps the biggest test real estate area will seek under GST is 6% stamp duty to the state government, for which no deduction can be guaranteed. The double taxation rate of GST and stamp duty increases the overall cost for the buyer.

3. Positive effects on ancillary industries – The real estate business benefits from various rates charged for raw materials required for the development of structures. The tax rate may be the same or higher than the previously imposed tax rate, but logistical burden of moving concrete, iron bars, and blocks has decreased since GST removed some cascading taxes.

4. Tax cuts for low-cost housing projects – To facilitate the prime minister's housing

plan, the public agency recently reduced the real GST rate for home purchases under Credit-Linked Subsidy Scheme (CLSS) of Pradhan's Mantri's Awas Yojna (PMAY) to 8%. This concession has also been extended to apartments up to 60 square meters of carpet area.

5. Improved perception – Despite the RERA (Real Estate Regulatory Authority) law, the execution of the GST will help improve the image of the sector as builders and developers work under greater responsibilities and stricter solvency requirements.

6. The impact of GST on the property was negligible in terms of tax rates. In the long term, businesses and customers will benefit from increased transparency and accountability that sector functions have so far avoided.

7. This research Study notices some viable difficulties in Implementation of GST in real estate industry. Some of them are as beneath.

8. Each input cost as far as materials, work, and so forth must be independently and altogether broken down to give appraisals of absolute GST due.

   • It is challenging to provide accurate evaluations of upfront costs and records for input tax reduction since the cost of materials is susceptible to change over the lifecycle of the project.

   • Advantage of input tax credit applies to GST paid, however, currently, no system is set up to analyze an increase in other non-GST costs as a result of transition.

9. Summary of advantages and difficulties of GST in Real estate industry is as follows

Advantages of GST in Real Estate Industry

• GST removes cascading impact of indirect taxes in real estate.

• Simple methodology for tax computation & fewer compliances.

• Defined treatment for E-business administrators.

• Improved transportation & logistics proficiency.

Difficulties of GST in Real Estate Industry

• Increased costs because of software product buying.

• Not being GST consistent can draw in penalties.

• Increase in operative expenses.

• Difficulty in adapting to a total online tax collection framework.

• SMEs will have a higher taxation rate.

10. **Difference Note: Earlier Indirect Tax Regime vs. GST**

   In the previous expense system, when property under development was bought, the buyer was exposed to the payment of VAT, Service charge, stamp duty, and registration charges. Property bought after completion was subject to stamp obligation and registration charges only.

Table 3. Pre-GST tax on Real Estate Transactions

Tax Details	Tax rate
VAT	1 to 4 %
Service tax	4.5%
Registration charges	0.5 to 1%
Stamp Duty	5 to 7 %

Table 4. Post-GST tax on Real Estate Transactions

Property Details	Tax Rate
Ready to move properties with completion certificate issued	No GST
Under development properties (Credit-Linked Subsidy Scheme)	8%
Under development properties (Not linked to Subsidy scheme)	12%
On Resale of properties	No GST
Works Contract	18%

Conclusion

Policy changes such as GST, the introduction of Real Estate Regulatory Authority (RERA), and the relaxation of standards for foreign direct investment are beginning to have a significant impact on the investment situation, bringing much-needed transparency, consistency, and corporate Governance to the framework. The real estate sector is expected to develop steadily in the coming years. GST implementation is truly an amazing achievement for the government and the country as a whole.

GST is off to a good start and the benefits for all stakeholders are clear. The GST has improved the valuation process for the real estate sector, solving long-standing problems of valuations and the nature of supply, etc. However, there are still certain unclear circumstances that have not yet been evaluated by developers to withstand litigation. Some of these are, the taxability of land value when developer has contracted separate arrangements for supply of land and construction property.

References

Dubey, N., Kumar, D., and Pandey, S. (2017). An Enquiry Into the Effect of GST on Real Estate Sector of India." International Journal of Trend in Scientific Research and Development, 1(6), 1001–1005.

Ms. Shalini, R. (2016). Impact of GST on Indian Real Estate Sector. International Journal of Business and Administration Research Review (IJBARR), XII (III), 1251–1257.

Ahamad, Ehtisham and Poddar, Satya (2009). Goods and Service Tax Reforms and Intergovernmental Consideration in India. Asia Research Center, LSE, 2009.

Kumar, Nitin (2014). Goods and Service Tax in India-A Way Forward. Global Journal of Multidisciplinary Studies, 3(6), May 2014.

Dani, Shefali (2016). A Research Paper on an Impact of Goods and Service Tax (GST) on Indian Economy. Business and Economics Journal. vol. 7(4).

Comparative Study of the Adoption of Marketing Practices in Selected SMEs

Sonal Muluk[*,a] *and Rajesh N. Pahurkar*[b]

[a]Dr. Vishwanath Karad MIT World Peace University, Pune, India
[b]Savitribai Phule Pune University, Pune, India
E-mail: [*]sonalmuluk@gmail.com; pahurkarrajesh@gmail.com

Abstract

Marketing efforts may help businesses to maximize profits. Medium and small-scale industries have started focusing on marketing activities to increase their revenue. The study includes a comparative analysis of the adoption of marketing activities by different sub-sectors in medium and small-scale industries to understand the linkage of marketing practices with the performance of SMEs. This study focuses on a case-based research approach. In-depth interviews were taken with SME owners or managers. In this research, both qualitative and quantitative methods are used for data collection. Research findings of the survey, small and medium-sized businesses (SMEs) are increasing their profitability through creative marketing strategies. Achieving success has been favorably impacted by marketing practices implemented by SMEs. This research can contribute to SME owners for framing marketing strategies and helping society with employment creation, a well-distributed economy across the population which contributes to comprehensive growth, hence enhancing the standard of living.

Keywords: Marketing practices, Marketing mix, Marketing orientation, Production orientation, etc.

Introduction

The Indian economy's Micro, Small, and Medium Enterprises (MSME) sector has grown significantly during the last several decades. It encourages entrepreneurship and produces a huge number of job possibilities at a cheaper capital cost than agriculture, both of which contribute considerably to the country's economic and social development. The MSMEs serve as ancillary units for the larger businesses, and this sector makes a substantial contribution to the nation's inclusive industrial growth.

Planned and concentrated marketing operations contribute to success for any firm, and this has been shown in several situations, particularly in large-scale businesses. Numerous studies have shown a correlation between a company's competitive strategy and marketing strategy and its degree of success. Market orientation refers to the implementation of a marketing idea (Kohli, 1990); hence, a business that adopts the marketing philosophy is said to be "market oriente." This study is divided into three parts to understand the level of marketing activities adopted by SMEs. First part consists of understanding whether an organization has defined marketing vision and mission statements or not. Second part is to check the adoption of short-term and long-term marketing planning for their products and services. This is because most of the SMEs are not having marketing department as a separate operational unit but many of them have assigned responsibility to one of the person or managers. Third part is focused on how many

SMEs are using Strategic Marketing activities like Marketing mix strategies and STP strategies.

Need for the Research and Research Objectives

Small-scale industries were reluctant to use or adopt marketing practices earlier may be one of the reasons, most of them are focusing on production orientation than marketing orientation. Most of the Managers of SMEs are deciding their business strategies based on their perception and they do not engage in strategic marketing activities. Competition by new entrants' decision-makers in SMEs has identified the need for marketing orientation.

This brings up the question of whether the financial resources of SME management impact their decision to begin marketing activities to gain a competitive edge. And what is the influence of the nature of industry on adopting marketing activities by SMEs? In this research, SMEs are selected from different sectors, which focus on business-to-business (B-to-B) and business-to-customer (B-to-C).

The research objectives are

1. To understand the acceptance of marketing activities in Small-scale businesses
2. To compare Marketing strategies adopted by different business sectors in SMEs

Scope

This study will examine the strategic marketing methods used by small and medium-sized enterprises (SMEs) and their influence on their performance. This study is to compare adoption of marketing activities by different SMEs, which belong to the business-to-business market and business-to-customer market. Survey locations include Maharashtra's main industrial belt of Pune.

Literature Review

Marketing Goals and Objectives

It has been proved over research that, setting marketing objectives and goals have a positive impact on the performance of the organization.

Brooksbank (1991) discovered that the more successful medium-sized enterprises have longer-term profit goals. Medium and large-scale organizations are adopting long-term profit goals, but this vision may lack in small-scale industries. According to (Numazaki, 1997), Taiwanese owner-managers participate in a variety of short-term profit-seeking actions.

Marketing Strategies

According to Taiwo (2010), strategic marketing techniques have a substantial effect on performance characteristics and interact with the many components to enhance performance. (Brooksbank, 1991) agreed with the assertion that better performance Small and medium-sized enterprises (SMEs) are more likely to adopt a strategic focus that is based on increasing the volume of production.

David et al. (2013) find that there is a high association between the marketing strategies of small and medium-sized enterprises (SMEs) and their success in terms of revenue growth, job creation, increased efficiency, and a broader client base. In addition, it helps SMEs to compete fairly with bigger firms. "Marketing mix" is a collection of pertinent elements and strategies that help users satisfy wants and fulfill firm objectives (Pruskus, 2015). Traditional marketing mix consists of four factors, generally known as the four P's of marketing (Golden and Zimmerman, 1980). According to Gronroos (2006), "marketing is a function of the entire business, not simply the marketing department." STP (Segmentation, targeting, and positioning) strategies are the focus of this research. According to Webster (1992), the positioning strategy adopted includes choices on how to segment the market, which market segment(s) to target, and how to accomplish that target. To apply the positioning strategy, the marketing strategy must be operationally implemented.

Research Framework

Research Design

This research is aiming to understand the level of implementation of strategic marketing activities in small-scale companies, hence both exploratory and descriptive research designs are used, depending on the requirements of the investigation. The results of exploratory research may bring new insights and a deeper understanding of a problem or situation. Because an issue has not been precisely outlined, a certain kind of study known as exploratory research is carried out. Exploratory research is useful for determining the most effective study design, data-gathering technique, and topic selection. In contrast, descriptive research, which may also be referred to as statistical research, consists of describing the facts and features of the population or phenomena that are being researched.

Data Collection

Structured questionnaire was framed and utilized during interviews to ensure easy and quick information to get answers to research questions. Case study research was used to undertake an in-depth investigation of the marketing practices of small-scale industries.

Based on a concept of small and medium-sized business marketing strategies and compatible with the theoretical constructs, this study will employ the qualitative technique, which is applied for exploration. According to Harker and Harden (2000), articulating the research topic in the form of a case study facilitates higher comprehension.

The study includes a detailed study of Manufacturing, Engineering workshops, water processing industries, chemical manufacturing units small-scale units. Table 1 represents primary data collected by SMEs and profiles of selected SMEs for the study.

Table 1. Primary Data-Profile of SMEs selected for the study

Group of SMEs Selected for the Research Study

Group	Sector	Frequency
Group-I	Manufacturing or Engineering (B-to-B)	16
Group-II	Chemical industries or pharma (B-to-B)	12
Group-III	Water Processing Industries (B-to-C)	15
	Total	43

Research Hypothesis

The study is divided into three major parts to understand marketing objectives, marketing planning, and implementation of strategic marketing activities which are based on marketing mix and STP strategies (segmentation, targeting, and positioning). Four research hypotheses have been framed which are represented in table 2.

Table 2. Research Hypothesis

Marketing goals and objectives	
H01	The organization has clearly defined goals for marketing activities
Marketing Planning	
H02	Most SME managers have a well-defined marketing strategy with allotted resources.
Strategic Marketing Planning	
H03	SME managers are increasingly engaged in strategic marketing planning

Data Analysis and Findings

Marketing Goals and Objectives

According to research results (Hooley and Lynch, 1985; Wong et al. 1994), effective businesses have long-term profit targets. Cox et al. (1994) discovered that small businesses with superior performance take a more aggressive approach to market and substantially more often embrace marketing dominance objectives.

Hypothesis H01: The organization has clearly defined goals for marketing activities

Table 3 indicates the results for hypothesis 1. Statements are indicating the degree to which marketing objectives and goals are set in an organization.

Table 3. Hypothesis H01: Results

Marketing Activities	Group-I		Group-II		Group-III	
	N	N %	N	N %	N	N %
1	7	44	3	25	2	13
2	4	25	5	42	6	40
3	5	31	4	33	7	47

Marketing activities are based on following statements

1. There are no objectives or goals defined for marketing
2. Marketing Short-term goals are defined
3. Marketing long-term goals are defined

Group I indicates the manufacturing or engineering small-scale companies. It has been found that more than 44 % of industries have not defined their marketing goals.

Group II consists of chemical industries or pharma companies where it has been found that those companies are having short-term goals clearly defined (42%). Comparatively very few have not defined their goals.

Group II belongs to the business-to-customer sector, and it has been visible from the result that water processing companies and solar panel companies have clearly defined long-term goals more than the other two sectors. Almost 47 % of SMEs are defining long-term goals.

Marketing Planning

Hypothesis H02: Most SME managers have a well-defined marketing strategy with allotted resources.

Table 4. Hypothesis H02: Results

Marketing Activities	Group-I		Group-II		Group-III	
	N	N%	N	N%	N	N%
1	3	19	3	25	1	7
2	7	44	4	33	5	33
3	6	37	5	42	9	60

Marketing activities are based on following statements

1. There is no Marketing department
2. There is one manager has allotted for handling Marketing activities
3. Team of salespeople is available for marketing activities

The study shows that SMEs have started taking efforts for marketing activities. Very few SMEs from group I, (19 %) are still focusing on production and they are not having a marketing department. 44 % of companies have one person assigned to marketing. Most of cases this person is either manager or owner of the small business unit. Those companies needed installation for their products and service providers are having entire sales team. 42 % of the SMEs from Group II have well-set marketing teams along with salespeople. Still, 33 % of them are working with managers' or decision makers instructions.

Group III indicates that having a sales team is essential for B-to-C sector. Re than 60 % of SMEs are having salespeople in their marketing department.

Strategic Marketing Planning

Strategic marketing includes marketing mix and STP strategies. **Hypothesis H03:** SME managers are increasingly engaged in strategic marketing planning

Marketing activities are based on the following statements

- SME Managers do market research
- Managers have segmented the market
- Before coming up with the product /service, SME Managers had a clear idea about the target market.
- While pricing, and profit objectives of the company are considered
- SME managers are influenced by competitors for setting prices for their products
- Developing quality products is the key
- SME managers allocate budget for promotional activities
- Managers use personal contacts and networks to expand our business
- Firm prefers to direct the distribution of products and services to customer

The study shows that the adoption of the marketing mix and STP is uneven based on the type of organization. Group-wise results varied, and it has been visible that group-I is more influenced by production-focused activities. Producing quality products is the key for them. Group II is following market segmentation based on customers and type of products. Group-III dividing the market based on geography. Cost plus pricing is common for most of SMEs.

Promotional efforts are not as like large-scale organizations. Positive word of mouth is the most common method in SMEs.

Conclusion

Manufacturing or engineering small-scale industries have reluctant to define their marketing goals and follow marketing practices. One of the reasons is these units are auxiliary units and produce goods and services as per the requirement of their client company. Production orientation is more than marketing orientation. Repeat sales and referrals are an important source of business for the salesperson

Chemical or pharma industries need to target regular customers along with a few testing labs and research and development centers. It has been found that they are clearly defining their marketing objectives and short-term goals, also they are following marketing strategies.

Water processing industries assembly companies are directly providing services to individual customers and hence they required more marketing efforts than the other two sectors.

According to the findings, marketing efforts in small and medium-sized enterprises (SMEs) tend to be informal and responsive to market opportunities, and in most cases, managers have a significant effect on the marketing decision-making process. Small and medium-sized businesses (SMEs) have a distinct style of marketing that differs from that of large corporations.

References

Muluk, S., & Pahurkar, R. (2022). The Implementation Of Marketing Practices Based On Marketing Philosophies In Small And Medium-Sized Enterprises. Journal of Positive School Psychology, 6(8), 2015-2022.

A Study of Circular Economy and Sustainability in Fashion Industry

Rohini Sawalkar,[a] Sonal Muluk,[b] Vimal Deep Saxena,[c] Vishal Gaikwad,[d] Girish Mude,[e] and Vinita Kale[f]

[a,b,e,f]Dr. Vishwanath Karad MIT World Peace University, India
[c,d]Sinhgad Business School, India
E-mail: [a]rohinisawalkar143@gmail.com

Abstract

This paper aims to study the principles of circular economy sustainable practices in the fashion industry. Study also aims to understand the how fashion sector has been shifting from linear economy towards circular economy. This research includes a comprehensive literature review to understand the various aspects of circular economy and sustainability in the fashion industry. The study briefly emphasize how brands and fast fashion industry is turning towards circular systems to contribute to sustainability and environmental responsibility. Primary objective of study is to understand the holistic overview of the circular economy in the fashion industry. The study also gives an overview of various challenges and drivers of circular economy. This research study is a fresh attempt to re-investigate understanding of the circular economy and sustainability in the fashion industry through a review of the literature.

Keywords: Circular economy, sustainability

Introduction

In terms of global pollution, the fashion sector is currently the world's second-largest polluter, trailing only aviation, which accounts for around 10% of all worldwide pollution. When compared to two decades ago, the amount of apparel consumed worldwide has surged by 400 percent (Jia, 2020; Shirvani Moghaddam 2020). To keep up with demand, the amount of energy required to run production must increase as well as the number of materials in circulation and the techniques used to handle the materials throughout their life cycle. All these factors, which appear to be harmful to the environment over the long term, are mandated by the increase in consumption (Chae and Hinestroza, 2020; Chen 2021; Sadeghi, 2021).

A large amount of clothes is wasted and burnt in landfills every year, including microplastics, which are ubiquitous in the ocean and contribute to environmental degradation in a variety of ways, with just one percent of them being recycled. (Fixing fashion: clothing consumption and sustainability 2022) (Sixteenth Report of Session 2017–19, Fixing fashion: clothing consumption and sustainability 2022). By introducing at least one new "collection" every week, fashion meets the insatiable hunger of client demand (Stanton, 2018).

Textile manufacturing is the second most polluting industry on the planet, behind only the oil industry in terms of greenhouse gas emissions, with roughly 1.2 billion tonnes (Change, 2018). According to some projections, the fashion industry would consume up to a quarter of the world's carbon budget by 2050. (Pandey, 2018). By utilizing circular economy principles, we can alleviate the detrimental consequences of the fashion industry on the environment's long-term survival (Saha, 2021).

The fashion and textile industry has a long and complex supply chain that is related to water, chemical, and energy consumption on a large basis which leads to waste generation and pollution. (Niinimäki et al., 2000). This leads to a significant negative influence on the environment and society as well. Despite the known environmental impacts, the fashion industry continues to grow at a fast pace with recent trends of low-cost cheap clothes and products, and has worsened in the last few years. Recently, the destructive manufacturing and production process undertaken in the fashion sector has gained popularity and revealed how badly it impacts the environment. Also, Consumers are getting increasingly concerned and knowledgeable about environmentally friendly items. As a result, the fashion industry must undertake substantial adjustments in the fashion business model, such as slowing down the manufacturing process and incorporating sustainable practices across the business model, to respond to the new client behavior. The fashion industry needs to move from a linear to circular economy. A sustainable and ethical fashion arena should be adopted to maximize the benefit to the fashion industry along with environment and society at a large.

Sustainable fashion refers to the movement that promotes and encourages environmental friendliness as well as social responsibility. It means manufacturing, production, and usage of fashion products in the most sustainable manner possible to minimize the impact on the environment and consider society at large.

A circular economy model should be adopted by the fashion sector to promote sustainability and reduce the waste and pollution produced during production. Core principles of circular economy are Reduce, Reuse and Recycle. Circular economy is a feasible substitute to replace the take-make-waste model with recover-recycle-reuse model.

Objectives

1. To understand the circular economy and waste management in the fashion segment.
2. To study the influence of circular economy on the fashion industry.
3. To have a better understanding of the fashion and textile industry's sustainable practices.

Scope

This study is mainly done to understand the concept of circular economy and how it is important in fashion industry is the major sector in contributing towards pollution. Study is done to gain a holistic view of the circular economy and its impact of it on fashion and textile industry. Along with this, it also gives an overview of various challenges in implementing circular production models. The research has been done based on secondary data including recent studies, journals, papers, articles, and information available on internet.

Literature Review and Background

Circular Economy

In a traditional corporate arrangement, things are generated, used, and discarded, resulting in a linear economy with a clear beginning and end. Almost 60% of extant fibers come from nonrenewable resources like fossil fuels and are used to make apparel (using polluting processes) that is only worn for a short time before being tossed in landfills or burned after being discarded. Every year, clothes worth more than $183 million are disposed of away in landfills (Wrap, 2020). Reduce, reuse, and recycle are three important actions that are simply summarized as the driving principles of the circular economy. All these actions and methods are common in conventional trash management (Manickam and Duraisamy, 2019).

Circular Economy and Fashion Industry

A circular economy for trend results in better products and services for clients, as well as a more robust and vibrant fashion sector, as well as the regeneration of the surrounding environment.

The fashion industry uses significant number of sources which creates a negative impact on the environment and generates large amounts of waste which ultimately end up in landfills and oceans. Because of shifting fashion and client preferences, as well as the mass production process, the fashion industry has seen significant growth and expansion, increasing environmental challenges. The fashion apparel and textile sector are ranked second in terms of land usage, fourth in terms of water consumption, and fifth in terms of greenhouse gas emissions, all of which have a significant influence on the environment. Without a doubt, the fashion and textile industries have a global economic and environmental influence. Fashion product manufacturing and consumption have expanded dramatically in recent years, resulting in a greater environmental effect. This is mostly due to the rise in the middle-class population and the consequent increase in demand. Even though there is still a lack of understanding regarding sustainable fashion and circularity, sustainability has remained a focus of popular interest. Slowly but steadily, shifting consumer behavior and demands for low environmental and social impact are gaining traction.

As a result, the circular economy concept has received a lot of attention to tackling the unsustainability that comes with linear production systems, such as waste creation, material scarcity, natural resource depletion, climate change, and so on. The fundamental of the circular economy lies in a closed loop where the resource flows are improved, and the materials can be used repeatedly reducing the resource use and depleting the waste generation at the same time.

Sustainability in Fashion Segment

The term "sustainability" derives from the French verb soutenir, which implies "to hold up or support" (Brown, 1987) Sustainability is a vast area that cannot be met single-handedly and requires the efforts of every individual to contribute and practice sustainability.

Sustainability is described as a condition in which human activity is carried out in such a way that the earth's ecosystems' functions are preserved (ISO, 15392, 2008). McMichael (2003) proposed a transformation of human lifestyle that maximizes the likelihood of living conditions that will continuously support security, well-being, and health, particularly by maintaining the supply of non-replaceable goods and services, or an indefinite perpetuation of all life on the planet (McMichael (2003) proposed an indefinite perpetuation of all life on the planet (McMichael (2003) proposed an indefinite perpetuation of all life on the planet.

Sustainability is very important and the need of the hour in the fashion industry. This means not only the businesses need to adopt the sustainable manufacturing process but as a consumer of the same, it is important for us to practice sustainability to encourage environment-friendly goals and safeguard the future. The fashion industry needs to be revamped keeping in mind the planet. We need to understand a consumer how to encourage sustainability.

Research Design

Secondary research has been done to conduct the study of this paper. Secondary research involves the collection and analysis of existing published research to get a thorough knowledge of the topic based on in-depth analysis.

Data Collection

The data is collected based on secondary data from various journals, research papers, articles, books, and information available on the internet. The thorough reading was done based on information available on various sites, literature reviews, reports, journals, articles, and sources available on Research Gate, Science Direct, Google Scholar was done to collect the data.

Linear Economy and Waste Production in Fashion and Textile Industry

The fashion sector's destructive supply chain is now coming under increased inspection around the world. The linear economy model has certain limits and does not fulfill the sustainable development goals that are now taken into consideration despite the harmful impacts on the environment, the fashion industry continues to grow at a fast pace and hence it is very much required to slow down the production process and implements sustainable practices across the business model.

Linear Economy Model

Modern-day fast fashion is based on a linear business model which follows the "take-make-use-dispose" system which ultimately allows the companies to produce in mass amounts, mass market and allows consumers to purchase trendy clothes heavily for cheap.

This pattern of manufacturing depends on easily available and accessible resources and energy in large quantities which becomes more unfit when used in further processes. The fashion industry is based on a straight line from extraction, and manufacturing to finished products and sales.

The first step in this starts with the extraction and gathering of raw materials which is the "Take" which can be various natural fibers such as cotton, wool, silk, etc., and synthetics generated from crude oil. This leads to issues such as the use of pesticides, labor abuse, farming, and other environmental issues making it unsustainable. Then the next part is the "Make" in which the actual production of goods takes place, after which use the product ("Use") and then it turns into waste that is the "Dispose." In this linear system, the clothing is not made to be recycled, it is just a simple process to mass produce the clothes, use them, and dispose off causing pollution. This encourages consumers to buy cheaper clothes heavily and more quickly which ultimately leads to shorter product life cycles and increased waste with less consideration for the environment and society.

Waste Generation

Polyester dominates the production in the fashion industry followed by cotton. The stages in the manufacturing clothes starts from the collection of fibers and raw materials such as cotton, wood, and chemical synthesis to yarn manufacturing followed by fiber production. Yarn when turned into textiles consumes a lot of water and energy along with excessive waste produced during the process and is labor intensive. Finally, the finished products are transported to the retail shops to retailers, and finally to the end consumers. The main environmental issue connected with the fashion industry is energy consumption, water usage, chemical use, and waste production. Most of the water usage relates to cotton cultivation and the wet processes in textile manufacturing such as bleaching, dyeing, printing, and finishing. Also, the fashion industry produces almost 8% of greenhouse gases and 20% of wastewater globally. The most amount of carbon footprint comes from the high energy usage during the manufacturing and distribution process. The production stages involve chemical, water, and energy consumption to CO_2 emissions which are harmful to the environment and are referred to as pre-consumer textile waste. Almost 15% of fabric used to get wasted during the manufacturing process. In line with this, post-consumer textile waste refers to the clothing that has been thrown out by the consumers, of which only 1% is recycled which counts as negligible.

Energy consumption is mostly through fiber production, yarn manufacturing, finishing, washing, and drying processes of the clothes.

Water and chemical usage in fiber manufacturing, pretreatment and wet processes, dyeing, washing, and finishing.

Solid waste mainly comes through the discarding of products at the last stage of clothes manufacturing.

Direct CO_2 emissions are produced during the transportation process and global supply chain in the fashion industry and through the coal-based plants in the manufacturing process which again increases the carbon footprint.

There are mainly three stages in the waste generation

1. Post manufacturing waste which is the result of manufacturing and production process.
2. Pre consumer waste comes from the inferior quality garments in manufacturing and distribution along with the unsold stock in the retail center.

3. Post-consumer waste arises from the consumers themselves through worn out, unwanted, and damaged clothes.

Hence, it's required to implement sustainable practices and a circular economy keeping in mind the environmental impact of the same.

Circular Economy Model

The circular model works on the "recover-recycle-reuse" pattern. This model has emerged as an option for linear model and an innovative strategy to achieve sustainability. The circular economy model's objective is to address the environmental, waste, and social impacts as a whole along with the change in consumer behavior and habits for a better sustainable future. The concept of circular economy revolves around 3R's that is, Reduce, Reuse, and Recycle.

Circularity in fashion segment refers to making a circle of the whole business process in a closed loop which doesn't require making new garments massively and recycling and reusing the same over time. This system helps to limit the extraction of raw materials and minimize the production process and reduces the generation of waste materials and disposals by recovering or reusing the products and clothes as much as possible.

This model in fashion industry is a regenerative system in which clothes and other fashion products are used and circulated for as long as possible to skim out the maximum benefit and then be exposed to the environment safely. Fashion clothes and other products are to be designed and produced with resource efficiency, biodegradable, or recyclable. With the eco-friendly nature of the circular system, the products are retained in the ecosystem for as long as possible with minimal waste. The products should be made in such a way which can be recycled and reused again and again to minimize the harmful effect on the environment. The goal of the circular system is that the products and garments should bring no harm to the environment as well as society.

Principles to Implement Circular System in Fashion Sector are

1. Producing and designing the products with sustainable purposes in mind such as longevity, recyclability, biodegradability, and resource efficiency by using high-quality fibers which makes it easy to repair and design timeless designable products and recyclability can be met through using sustainable materials which are easy to disassemble.
2. Production and sourcing of materials without toxicity, efficiency, renewables, and ethical practices.
3. Also collaborating and supporting on long life services of the products.
4. Reusing, recycling, and composting all the remains of the products wisely.
5. Regenerate natural system.

Businesses should take into consideration the following while implementing a circular business model:

1. Are the clothes and other fashionable products durable to last long for years to come?
2. The style of the clothes should be timeliness.
3. Picturing how long will consumers willing to wear these clothes.
4. Whether the raw materials are sustainable and can be reused or are recyclable or disassembled etc.
5. Ethical fashion and fair wages are taken into consideration.
6. Are the clothes or garments can be repaired or resigned or donated or reused instead of disposing off the clothes?

As of now, companies can slowly adopt sustainable ways to encourage circularity such as offering rental services may be on a subscription basis or discounts, creating timeless and durable products with sustainable raw materials, offering tips and services for care and repair, facilitating returns and recycling, leverage technology and find out ways to filter and sort the garments or enable retailer curated collections, can build collection points and enable home pick up facilities to improve accessibility, eradicate single-use packaging, try to reduce manufacturing waste and improve on supply chain impact by training and incentivizing suppliers to reduce and reuse fiber and chemicals, etc. while consumers are constantly supporting these initiatives at the same time.

Drivers of Circular Economy

It has become very evident that the fashion industry which contributes so much to pollution needs to change and adapt to a circular and environmentally friendly industry. The drivers which are key to promoting the importance of a circular economy are:

Environmental change is the need of the hour as sustainability is coming to an important area of consideration nowadays. Almost 92 million tonnes of waste are being generated currently by the fashion and textile industry which is estimated to increase by up to 60% by 2030. This is a huge area of concern as all these wastes end up in landfills and oceans polluting the environment. Circularity is needed in the fashion industry as this helps to close the loop by minimizing waste with the use of sustainable, renewable, and regenerative materials and encourages recyclability.

The next important aspect is the increased level of global policies and regulations like the Paris agreement and the UK government's vision of net zero emissions by 2050. Also, the UK government is introducing a Plastic Packaging tax for more than 10 tonnes of plastic used within a year. The increased level of rules and regulations and the rise in various environmental and social issues are why the fashion industry needs to adapt to a circular system to keep in line with government regulations. To achieve this, everyone starting from companies, manufacturers, suppliers, consumers, and government needs to commit to a common objective.

Another key driver is the rise in the need of reducing the social impact of fashion in this sector. The workers and laborers in the fashion manufacturing industry are often exploited and suffer a lot starting from low wages, child labor, health issues, etc has led to an increased importance of why the fashion industry needs to change. The circular system focuses on sustainability and ethical behavior throughout the operations and supply chain reducing the impact on society.

One of the most important drivers of circular economy today is the increased level of awareness and the change in the behavior and attitude of consumers. Consumers are getting more aware by-passing days and demands, and needs are shifting towards more environment-friendly products and services. The power of social media plays a great importance to spread awareness and hence forcing brands to shift to a sustainable system. With the growth in consumer awareness and demand the fashion industry ultimately needs to change and take a step forward toward circular fashion models.

Findings

1. Circular economy is a regenerative system that minimizes resource input, emission, waste, and energy use, closes the loop, and lowers the impact on the environment.
2. Circular economy mainly revolves around three basic principles that are the 3R's – Reduce, Reuse and Recycle.
3. Circular economy mainly relates to the minimum resource use and other ways to minimize waste production in operations and supply chain while sustainability is wide concept that involves the planet, people, and profit and circular economy is a part of it.
4. Circular economy can be regarded as a mere business trend as of now, as much in-depth analysis and research are yet to be done in this area.
5. Transition and implementation of circular economy require much effort and has a lot of challenges on the way, yet it holds a lot of opportunities for businesses in future and has a lot of benefits too.
6. Fashion industry holds a lot of potential to create various ranges of circular business models, however, the implementation of the same is yet at a very early stage and needs more future research.
7. The fundamental of circular model revolves around redesigning the whole supply chain along with a collaborative approach from all the stakeholders involved in the process.

Conclusion

Textile and clothing are quite important for human beings as it's an external expressions for an individual but the speed at which they end up in landfills poses a severe threat to the environment. It's not a secret that the fashion industry needs to be revamped with the environment and society in mind. As the impact of fashion and textile industry is entering into a serious matter of concern, the linear system is no longer working. With the global population to reach 9 billion by 2030, nature will struggle to fulfill the demands like never before. It's time to reimagine the business model of fashion through circularity and disrupting the status quo to shape a sustainable industry. Circularity and sustainability are quite evident to become one of the key business trends of the next era. However, it's not a mere thing that can be led by a few people rather this requires a collective effort of everyone starting from business, consumers, suppliers, government, and all other stakeholders involved in the supply chain. The value created in circular model is completely different in that the single garment can generate value repeatedly with resale, rental, repair, restore, recycling or refurbishing of the product minimizing the harmful impact on the environment. Although, implementing a circular system faces various sets of challenges that need to be solved for taking a step toward sustainability. As consumers become more aware and engaged about sustainability and its issues, circular system will become one of the keys to a more sustainable future.

References

Chae, Y., and Hinestroza, J. (2020) Building circular economy for smart textiles, smart clothing, and future wearables. Mater Circ Econ, 2(2).

Change, N. C. (2018) The Price of fast fashion. Nat Clim Chang.

Chen, X., Memon, H. A., Wang, Y., Marriam, I., and Tebyetekerwa, M. (2021). Circular Economy and Sustainability of the Clothing and Textile Industry. Mater Circ Econ, 12.

Dissanayake, Kanchana, and Weerasinghe, Dakshitha (2021). R E V I E W P A P E R Towards Circular Economy in Fashion: Review of Strategies, Barriers, and Enablers. Circular Economy and Sustainability.

Geissdoerfer, Martin, Savaget, Paulo, Bocken, Nancy, and Hultink, Erik Jan. (2017). The Circular Economy – A new sustainability paradigm? Journal of Cleaner Production, 143(6), 757–768.

Jacometti, Valentina. (2019). Circular Economy and Waste in the Fashion Industry. Laws.

Jia, F., Yin, S., Chen, L., and Chen, X. (2020) The circular economy in the textile and apparel industry: a systematic literature review. J Clean Prod

Manickam, P., and Duraisamy, G. (2019). 3Rs and circular economy. In: Muthu, S. S. (ed), Circular Economy in Textiles and Apparel. Woodhead Publishing, Cambridge, pp. 77–93.

Sadeghi, B., Marfavi, Y., AliAkbari, R. et al. Recent Studies on Recycled PET Fibers: Production and Applications: a Review. Mater Circ Econ 3, 4 (2021).

Saxena, Dr. Vimal Deep, and Sawalkar, Dr. Rohini Suresh (2020). A Study to Understand the Enterprise Challenges in Small Scale Agro-Food Processing Firms (April 4, 2020). International Journal of Management (IJM), 11 (3), pp. 186–192.

Pandey, K. (2018) Fashion industry may use quarter of world's carbon budget by 2050.

Predicting Young Consumers' Purchase Intention Towards Green Products

Dipandi Mishra,[*,a] *Amarendra Pratap Singh,*[a] *and Alok Tewari*[b]

[a]Indira Gandhi National Tribal University, Amarkantak, India
[b]Dr. A. P. J. Abdul Kalam Technical University, Lucknow
E-mail: [*]dipandimishra2013@gmail.com

Abstract

University students are potential consumers who have all the necessary knowledge, which could be the reason to switch toward ecological behavior. Therefore, it is critical to understand factors that could influence their green product purchase behavior in the future. This study, conducted with 279 university students, analyzed the variables that impact the purchase intention of students for green products. The results indicated a positive impact of environmental knowledge, social media, attitude, perceived severity, and perceived consumer effectiveness on students' intention to purchase green products.

Keywords: Environmental knowledge, Green behavior, Perceived severity, Perceived consumer effectiveness, *Green* purchase intention

Introduction

The conservation of the environment could be done by educating people about the severity of the need for protecting the environment. Regarding education, individuals with university degrees are more concerned about the environment than individuals without university degrees (Gifford and Nilsson, 2014). Previous studies have proved that younger individuals are more worried about their impact on the environment (Van Liere and Dunlap, 1980). An innovative education system is required for the upcoming generation to know more about environmental development (Nasibulina, 2015). The research and education on sustainable development are in their primary stage in universities (Farinha et al., 2018). Still, deep education would direct a positive attitude toward green products to encourage environmental conservation.

Green Products emerged as an opportunity to make conscious choices that can help with the ecological damage caused to the planet (Gherhes and Fǎrcasiu, 2021). University students symbolize a tremendous and significant segment for many marketers (Mokhlis, 2009). The current study concentrates on the factors that affect the purchase intention of university students toward green products as they cover a large segment of consumers.

Literature Review

Environmental Knowledge

The knowledge of the reality and crucial associations which affects the environment, which results in a sense of responsibility towards the environment, ultimately guiding toward sustainable development, is considered environmental knowledge (Taufique et al.,

2016). Many previous studies have observed that environmental knowledge is also dominant among several significant factors impacting consumers' green product purchase intention positively (Mostafa, 2009). Therefore, we hypothesize:

H1: Environmental knowledge impacts green products' purchase intention positively.

Social Media

Social Media has improved the communication between consumers and firms helping consumers in collaborative and interactive purchase experiences (Singh and Sonnenburg, 2012), especially young consumers who are more familiar with digital media usage. The purchase decision of consumers is in a straight line, impacted by social media (Wang et al., 2012). User-generated content on social media has been found to significantly affect consumer purchase intention (e.g., Mathur et al., 2021). Therefore, to understand whether similar behavior exists among young consumers in the context of green products we propose that:

H2: social media positively impacts green purchase intention.

Throughout the purchase decision stages, social media has an extensive impact on the attitude, purchase decision, and perception of consumers toward that product (Mangold and Faulds, 2009). And therefore, our study posits:

H3: social media affect the green product purchase attitude among university students.

Attitude

Attitude towards a behavior is an individual's positive or negative evaluation of the conduct in question (Azjen, 1991). Attitude is known to impact consumers' green product purchase intention. Recent studies (e.g., Tewari et al., 2022a, b) have reported a significant impact of attitude on young consumers' green purchase intention. It was found in previous research that students who have environmental studies

in their curriculum have a positive attitude in the direction of protecting the environment (Chuvieco et al., 2018). Thus, attitude posits the second hypothesis of the study:

H4: Attitude toward green environment impacts the purchase intention positively.

Perceived Consumer Effectiveness

Consumers' evaluation of the extent to which their consumption can make a difference in the overall problem" is said to be perceived as consumer effectiveness (Webster, 1975). There is a positive correlation between the purchase intention of green products and perceived consumer effectiveness (Gupta and Ogden, 2009). It has been observed that the higher the consumers' PCE level, the higher they display a positive attitude and intention to purchase sustainable products (Webb et al., 2008). Therefore, it is hypothesized that:

H5: Perceived Consumer Effectiveness (PCE) positively influences the purchase intention among university students regarding green products.

H6: Consumers' attitude toward green products is positively affected by PCE.

Perceived Severity

University students are in the age interval to understand how the environment is in danger (Shafiei and Maleksaeidi, 2020). "Perceived severity of the threat means the degree of seriousness of the possible harms that is perceived by an individual" (Janmaimool, 2017). Rainear and Christensen (2017) researched a sample of college students and found that perceived severity is a predictor in assessing pro-environmental behavior of students. Perceived severity was an essential factor in understanding the students' intention to purchase green products (Kim et al., 2013). Thus, our study posits that:

H7: Perceived severity positively impacts the university students' green purchase intention.

Table 1. Correlation Matrix

	CR	AVE	ATT	SM	PCE	EK	PS	PI
ATT	0.828	0.617	0.785					
SM	0.802	0.576	0.291***	0.759				
PCE	0.865	0.617	0.310***	0.286***	0.786			
EK	0.825	0.544	0.162*	0.210**	0.200**	0.738		
PS	0.804	0.579	0.095	0.210**	−0.011	0.097	0.761	
PI	0.839	0.635	0.535***	0.581***	0.575***	0.376***	0.283***	0.797

Methodology

Questionnaire Design

The development of the questionnaire for the current study was through sourcing items from applicable literature. Attitude was measured using three items from Wang et al. (2013). Purchase intention was assessed using three items from Yadav and Pathak (2017). Environmental knowledge was measured utilizing four items from Zsoka et al. (2013). Perceived Severity was assessed using three items from Shafiei and Maleksaeidi (2020). Perceived Consumer Effectiveness was assessed using four items from Joshi and Rahman (2019). Social media was measured using three items from Pucci et al. (2019).

Data Collection

To approach the required population of the study purposive sampling technique was employed. Students of the university who were familiar with green products were selected to respond to the questionnaire. All the items were measured on a 7-point Likert scale in which "1" was assigned with strong disagreement and "7" was assigned strong agreement with the measured item. The web-based questionnaire was sent through e-mail to 325 students, out of which 298 responses were received. After eliminating incomplete responses, a final sample of 279 was used for further analysis. Our study had 20 items therefore a minimum sample of 200 was required according to the criterion of 10 responses per item (Kline, 2011). Thus, the size of sample 279 was sufficient for this study.

Data Analysis

AMOS 23 and SPSS 23 software were employed for data analysis. The normality of data was tested using Skewness and Kurtosis indices which were inside the range of ±3 and ±10, respectively, implying that the data set was typically distributed (Kline, 2011). To check the internal consistency of the measurement items, Cronbach's alpha values were calculated. The alpha values of all the constructs were above 0.7 thresholds which is recommended (Nunnally and Bernstein, 1994).

Measurement Model

The measurement model displayed an excellent fit through confirmatory factor analysis with values CMIN/df = 1.529; CFI = 0.966; SRMR = 0.049; RMSEA = 0.044. The estimates for the subscales of reliability and validity exceeded the cutoff criteria. For all the constructs, factor loadings were discovered above the 0.6 mark. The values of composite reliability were above 0.7. Good convergent validity was also found with the help of AVE values above 0.5. A comparison was conducted between the square root of AVE and factor correlations to find discriminant validity (see table 1). The obtained value of former exceeded the latter, reflecting adequate discriminant validity (Fornell and Larcker, 1981).

Structural Model

The model fit was found to be good with the values CMIN/df = 1.519; CFI = 0.966; SRMR = 0.051; RMSEA = 0.043. Furhter, all the hypothesized paths were observed to be statistically significant as shown in Table 2.

Table 2. The path coefficients of the structural equation model

Paths	Path Coefficients	t-value	p-value	Hypotheses
EK ⟶ PI	0.175	3.197	0.001	H1—Supported
SM ⟶ PI	0.322	4.999	***	H2—Supported
SM ⟶ Att	0.225	2.983	0.003	H3—Supported
Att ⟶ PI	0.288	4.839	***	H4—Supported
PCE ⟶ PI	0.363	5.91	***	H5—Supported
PCE ⟶ Att	0.247	3.378	***	H6—Supported
PS ⟶ PI	0.177	3.229	0.001	H7—Supported

Note: *** $p < 0.001$; EK = Environmental Knowledge; PI = Purchase Intention; SM = Social Media; Att = Attitude; PCE = Perceived Consumer Effectiveness; PS = Perceived Severity

Findings and Discussion

The result from the hypothesis testing shows that when the environmental attitude of the students improves, the pro-environmental behavior among them also increases (Shafiei and Maleksaeidi, 2020). The pro-environmental behavior of students, which includes green purchasing too, is positively associated with attitude as confirmed by several previous pieces of research by Cottrell (2003). It is significant to increase positive attitudes toward green products at an individual level (Saqib and Majeed, 2020).

The first item considered for measurement in the study was environmental knowledge. The hypothesis of the study was accepted and proved that students with environmental knowledge are more involved in sustainable practices, which encourage them to purchase green products creating positive purchase intention toward green products among them.

Hypothesis 3 of our study was supported and proved that social media has numerous effects in forming purchase intention for green products. The first consideration of the study was that social media played an essential role in developing a positive attitude among young consumers toward green products. Responses from the questionnaire indicated that consumers consider information on social media reliable and credible, which impacts their green purchase intention. The next factor that impacted the green product purchase intention is the attitude towards the environment. Our study's empirical findings indicated a positive relationship between attitude and green product purchase intention. These young consumers are more considerate toward green products while purchasing than older consumers (D'Souza et al., 2007).

The following hypothesis that showed a positive relationship between perceived consumer effectiveness and attitude towards the environment was also supported and proved that they have a significant relationship. The hypothesis' conclusion shows a significant relationship that shows a positive impact of PCE on attitude. Therefore, their positive attitude due to PCE transforms into eco-friendly behaviors (Altinigne and Bilgin, 2015), which include purchasing green products for a sustainable earth. Therefore, the study shows that if students are taught that their behavior might bring significant change in the earth's environment, they would have a positive attitude towards the environment. The current study and Rothbaum et al. (1982) also concluded that PCE and environmental attitudes have positive relationships.

Likewise, attitude, PCE impacts the purchase intention of young consumers directly also towards green products, which our study's hypothesis has supported. In simpler terms, young consumers having PCE are more inclined to purchase green products. One of the reasons for such inclination could be that PCE is responsible for impacting intention and behavior, especially if individuals feel that their intention would bring expected results. These young consumers believe that intention to purchase green products is going to prevent ecological hazards.

This research exhibits the impact of perceived severity on green product purchase intention. The last hypothesis of the current study proved that perceived severity impacts the purchase intention of young consumers toward green products. Previous studies have also found that perceived severity impacts purchase intention positively (Wang et al., 2021).

Implications

The study's hypotheses were supported, and the results could help marketers form policies. Marketers could use social media to improve the environmental knowledge of young consumers through social media, which could increase their intention to purchase green products. Social media could be employed to communicate with consumers regarding their product experience and other information related to green products through user-generated content, emotional marketing, storytelling, and powerful advertisements. This will also help create a positive attitude among consumers toward green products. Consumers with a positive attitude toward green products are more likely to purchase these products. Marketers could take help from the government in creating a positive image of these products, which will provide more authenticity to these products, resulting in consumers' positive attitudes.

Conclusion

It was evident from the results that factors mentioned in the current study impact the purchase intention of young consumers toward green products. This is because, through these factors, students could get informed about the severity of the ecological condition and how green purchases will help them practice sustainability. Despite their positive attitude towards these factors and green products, this attitude is not transforming into practice. This study highlights the factors that could practically impact consumers' intention to purchase green products. First, the literature review signifies that focus on the young generation has been limited in previous studies.

Second, this study has highlighted the novel and more acceptable factors that could impact young consumers' green purchase intention. It has also depicted how social media could influence attitudes and intentions to purchase green products. Then, PCE, perceived severity, being novel factors, could target young consumers for green products purchase. Consumers will practice green purchases when they perceive the severity of saving the ecology. Finally, the study has focused on only a single group of young consumers to build a comprehensive understanding of factors impacting their purchase intention.

References

Ajzen, I. (1991). The theory of planned behavior. Organizational behavior and human decision processes, 50(2), 179–211.

Altinigne, N., and Bilgin, F. Z. (2015). The Relationship between Perceived Consumer Effectiveness and Environmental Attitudes of University Students. In International Marketing Trend Congress.

Chuvieco, E., Burgui-Burgui, M., Da Silva, E. V., Hussein, K., and Alkaabi, K. (2018). Factors affecting environmental sustainability habits of university students: Intercomparison analysis in three countries (Spain, Brazil, and UAE). Journal of cleaner production, 198, 1372–1380.

Cottrell, S. P. (2003). Influence of sociodemographics and environmental attitudes on general responsible environmental behavior among recreational boaters. Environment and behavior, 35(3), 347–375.

Farinha, C. S., Azeiteiro, U., and Caeiro, S. S. (2018). Education for sustainable development in Portuguese universities, International Journal of Sustainability in Higher Education, Vol. 19.

Gherheş, V., and Fărcaşiu, M. A. (2021). Sustainable behavior among Romanian students: a perspective on electricity consumption in households. Sustainability, 13(16), 9357.

Gifford, R., and Nilsson, A. (2014). Personal and social factors that influence pro-environmental concern and behaviour: A review. International journal of psychology, 49(3), 141–157. No. 5, pp. 912–941.

Gupta, S., and Ogden, D. T. (2009). To buy or not to buy? A social dilemma perspective on green buying. Journal of consumer marketing.

Hu, L. T., and Bentler, P. M. (1999). Cutoff criteria for fit indexes in covariance structure analysis: Conventional criteria versus new alternatives. Structural equation modeling: a multidisciplinary journal, 6(1), 1–55.

Janmaimool, P. (2017). Application of protection motivation theory to investigate sustainable waste management behaviors. Sustainability, 9(7), 1079.

Joshi, Y., and Rahman, Z. (2019). Consumers' sustainable purchase behaviour: Modeling the impact of psychological factors. Ecological economics, 159, 235–243.

Kim, S., Jeong, S. H., and Hwang, Y. (2013). Predictors of pro-environmental behaviors of American and Korean students: the application the theory of reasoned action and protection motivation theory. Sci. Commun. 35(2), 168e188.

Kline, R. B. (2011). Convergence of structural equation modeling and multilevel modeling.

Liere, K. D. V., and Dunlap, R. E. (1980). The social bases of environmental concern: A review of hypotheses, explanations and empirical evidence. Public opinion quarterly, 44(2), 181–197.

Mangold, W. G., and Faulds, D. J. (2009). Social media: The new hybrid element of the promotion mix. Business horizons, 52(4), 357–365.

Mathur, S., Tewari, A., and Singh, A. (2021). Modeling the Factors affecting Online Purchase Intention: The Mediating Effect of Consumer's Attitude towards User- Generated Content, Journal of Marketing Communications.

Mokhlis, S. (2009). An investigation of consumer decision-making styles of young-adults in Malaysia. International journal of Business and Management, 4(4), 140–148.

Mostafa, M. M. (2009). Shades of green: A psychographic segmentation of the green consumer in Kuwait using self-organizing maps. Expert systems with Applications, 36(8), 11030–11038.

Nunnally, J. C., and Bernstein, I. (1994). Elements of statistical description and estimation. Psychometric theory, 3(127).

Pucci, T., Casprini, E., Nosi, C., and Zanni, L. (2019). Does social media usage affect online purchasing intention for wine? The moderating role of subjective and objective knowledge. British Food Journal. http://www.sciencedirect.com/science/article/pii/S0272494401902270.

Rainear, A. M., and Christensen, J. L. (2017). Protection motivation theory as an explanatory framework for proenvironmental behavioral intentions. Commun. Res. Rep. 34 (3), 239e248.

Rothbaum, F., Weisz, J. R., and Snyder, S. S. (1982). Changing the world and changing the self: A two-process model of perceived control. Journal of personality and social psychology, 42(1), 5.

Shafiei, A., and Maleksaeidi, H. (2020). Pro-environmental behavior of university students: Application of protection motivation theory. Global Ecology and Conservation, 22, e00908.

Singh, S., and Sonnenburg, S. (2012). Brand performances in social media. Journal of Interactive Marketing, 26(4), 189–197.

Taufique, K. M. R., Siwar, C., Chamhuri, N., and Sarah, F. H. (2016). Integrating general environmental knowledge and eco-label knowledge in understanding ecologically conscious consumer behavior. Procedia Economics and Finance, 37, 39–45.

Tewari, A., Mathur, S., Srivastava, S., and Gangwar, D. (2022a). Examining the role of receptivity to green communication, altruism and openness to change on young consumers' intention to purchase green apparel: A multi analytical approach. Journal of Retailing and Consumer Services, 66, 102938. https://doi.org/10.1016/j.jretconser.2022.102938

Tewari, A., Srivastava, S., Gangwar, D., and Verma, V. C. (2022b). Young consumers' purchase intention toward organic food: exploring the role of mindfulness. British Food Journal, 124(1), 78–98. https://doi.org/10.1108/BFJ-12-2020-1162.

Wang, X., Yu, C., and Wei, Y. (2012). Social Media Peer Communication and Impacts on Purchase Intentions: A Consumer Socialization Framework. J. Interact. Mark., 26, 198–208.

Wang, H., Ma, B., Cudjoe, D., Bai, R., and Farrukh, M. (2021). How does perceived severity of COVID-19 influence purchase intention of organic food? British Food Journal.

Webster Jr., F. E. (1975). Determining the characteristics of the socially conscious consumer. Journal of consumer research, 2(3), 188–196.

Yadav, R., and Pathak, G. S. (2017). Determinants of consumers' green purchase behavior in a developing nation: Applying and extending the theory of planned behavior. Ecological economics, 134, 114–122.

Supply Chain for a Transition Toward a Gas-Based Economy in India: Customer Acceptability of Natural Gas

Asim Prasad,[*,a] *Anita Kumar,*[b] *and Niti Nandini Chatnani*[c]

[a]Amity University, Noida, India
[b]Amity University, Noida, India
[c]Indian Institute of Foreign Trade, New Delhi, India
E-mail: [*]asimprasadiitk@gmail.com

Abstract

Enhancing clean energy usage is the universally accepted pathway toward carbon neutrality, to which global leaders are jointly committed. India is no exception and has set stiff targets to contribute to stabilizing the global climate with local solutions, promising to achieve net-zero carbon emissions by 2070. The first step in this effort is a gas-based economy. India aims to increase the share of natural gas (NG) in its energy mix from the present 6.3% to 15% by 2030. However, access to a local distribution network for its affordable, reliable doorstep delivery is a prerequisite. Moreover, a positive response to accepting NG as a clean energy option in the domestic and transport segment is critical to enhancing its consumption. With extensive primary data analysis, the exploratory research examines the various factors for NG consumer acceptability in India's domestic and transport segments to accelerate the transition to a gas-based economy promoting SDGs.

Keywords: Natural Gas, Gas Based Economy, Net Zero, Carbon Neutrality, SDG

Introduction

Climate change and international energy policy are on the discussion agenda among global leaders to save humanity from the uneven adverse impact of global warming. Various countries have committed specific targets to achieving carbon neutrality and advancing clean energy usage. India propagates a sustainable and healthy lifestyle with specific energy sector targets under its INDCs to reduce its GDP emission intensity by 33 to 35 % and create additional carbon sinks of 2.5 to 3 billion tonnes of CO_2 by 2030. The medium-term focus is enhancing natural gas (NG) consumption while achieving net-zero carbon emissions with renewables by 2070.

India focuses on NG attributes which provide clean cooking solutions (Patnaik, Tripathi and Jain,

2018) and sustainable mobility (Demirbas, 2012). These relate to the economic and environmental benefits of NG. While India is struggling to enhance the share of NG consumption from the present 6.3 percent to 15 percent by 2030 to achieve a gas-based economy, the NG share in the global primary energy consumption mix is 24.2 percent. In the last seven years, the CAGR in NG consumption in India has been 3.8 % against 1.4 % globally. NG finds usage in power, fertilizer, refineries, petrochemicals, and other industrial sectors. The relatively new City Gas Distribution (CGD) sector, with a 20 % share of NG consumption, has registered the fastest CAGR of 7.5% in the last five years in India, manifesting the quest for growing clean energy demand (Kumar, Shastri and Hoadley, 2020). The government has implemented policy reforms to enhance domestic production and provide NG access to retail customers in domestic household and transport segments by creating local distribution networks across the country. The sector-wise NG consumption trend in million metric standard cubic meters per day (MMSCMD) in India (PPAC, 2022) is in Figure 1.

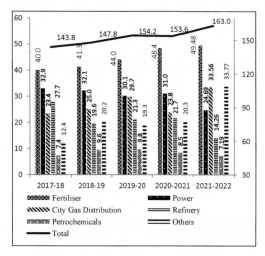

Fig. 1. NG Consumption trend in MMSCMD

Some important research questions that emerge in the context of the domestic household and transport segments are

1. what are the different factors for NG consumer acceptability in these segments
2. which statistically significant factors will lead to its consumer acceptability as a clean energy source
3. does NG consumer acceptability makes a positive difference in enhancing its consumption to accelerate the transition to a gas-based economy in India

Literature Review

In the "World Commission on Environment and Development: Our Common Future" report, Brundtland argues that the environment and development are inseparable from humanity. The industrialization path dominating dirty fossil fuels like coal and oil is unsustainable as it puts tremendous pressure on the planet to self-decarbonize. Therefore, it is necessary to identify common goals to jointly implement sustainable actions to manage environmental and development concerns concurrently. The commission defines sustainable development as the "development that meets the needs of the present without compromising the ability of future generations to meet their own needs." An adequate energy supply at affordable prices holds center stage towards achieving socioeconomic development. An argumentative review was conducted to identify variables for NG consumer acceptability for promoting it over polluting fossil fuels as the benign option to manage the environmental crisis by reducing greenhouse gas emissions while development needs are not compromised.

Most environmentally acceptable and with a high calorific value, NG is the first choice for global transport, and industrial and domestic applications due to the significant difference in carbon emissions compared to coal or oil (Demirbas, 2012). On an energy equivalent basis, it produces 40 % less CO_2 emission, sulfur-free, with low nitrogen oxides than coal. Due to its high burning efficiency and low emission, it substitutes gasoline as an automotive fuel. Therefore developing counties acknowledge NG as the safest option to meet their developmental needs while managing environmental concerns concurrently (Safari et al., 2019). Considering lifecycle costs, visible environmental issues, and technical characteristics like the highest H/C ratio, low contaminants, and 26% less CO_2 emission than petroleum derivatives, Natural Gas Vehicles (NGV) have attracted global interest over gasoline vehicles (Di Pascoli, Femia and Luzzati, 2001). However, NGVs running on Compressed Natural Gas (CNG) are refueled periodically since the energy stored in gaseous CNG is low than liquid fuel for the same unit volume but emits 70% lower CO than equivalent gasoline vehicles. In the residential sector, NG proves economical compared to commercially available alternatives like heating oil, kerosene, electricity, and propane (Demirbas, 2012). Residential and commercial sectors attract higher demand for NG in developing countries like India and China due to the fast development of CGD projects (IEA, 2020), affordability, and awareness about the benefits that NG provides to improve well-being and living standards over fossil fuel derivatives that it replaces. The International Gas Union lists the globally recognized advantages of using NG considering its availability for over 250 years, acceptability due to clean burning, accessibility due to ease of movement in gaseous and liquid form, adaptability to new technologies, affordability due to competitive energy options manifested by efficient end-use (Gladkova, 2013). CGD companies in India identified the benefits of using PNG and CNG listing attributes like uninterrupted supply, safety, convenience, versatility, environment friendly, and economical for different categories of customers. The variables summarised from the literature identified as factors for NG acceptability are in Table 1, while the variables grouped into four constructs are in Table 2.

Table 1. Variables for NG acceptability

Sl.	Variables	Reference
1.	Reduces harmful emissions, no residue, high burning efficiency, safety	Demirbas (2012)
2.	Technical characteristics; reduced emissions	Di Pascoli, Femia and Luzzati (2001);
3.	Economic, affordability	Demirbas (2012)
4.	Awareness about clean energy systems, raising living standards, well-being, sustainability, SDG	IEA (2020)
5.	Availability, acceptability, accessibility, adaptability, affordability,	Gladkova (2013)
6.	Uninterrupted supply, safety, convenience, versatility, environment friendly	Ishwaran et al. (2017)

Accordingly, the research objectives for answering the research questions are:

1. to examine the awareness levels about accepting NG as a clean energy source, the global choice to promote sustainability
2. Significant factors for consumer acceptability to promote NG in the domestic

household and transport segments that will enhance its consumption to accelerate the transition to a gas-based economy in India

Table 2. Constructs and corresponding variables

Sl.	Construct	Variable
1.	NG Awareness (NGAW)	• Clean Energy Systems based on NG(NGA1) • Raising living standards with PNG, CNG(NGA2) • Sustainability, SDGs(NGA3)
2.	Benefits of PNG(BPNG)	• Convenience(BPNG1) • Safety(BPNG2) • Versatile(BPNG3) • Environment Friendly(BPNG4) • Affordability(BPNG5)
3.	Benefits of CNG(BCNG)	• Reduces harmful emissions(BCNG1) • No residue(BCNG2) • Economical(BCNG3) • Safety(BCNG4)
4.	Acceptability for enhancing gas consumption (ANG)	• Acceptability PNG(ANG1) • Acceptability CNG(ANG2) • Consumption (ANG3)

The research hypothesis are:

1. H1: The awareness levels of promoting NG are usual among the populace
2. H2: PNG benefits make a difference in accepting it as a domestic household fuel
3. H3: CNG benefits make a difference in accepting it as a transport fuel
4. H4: The NG acceptability makes a difference in enhancing NG consumption in India

Research Methodology

A survey method was used with an exploratory research design to find answers to the research questions. A structured questionnaire based on research objectives with dependent and independent variables were created to collect primary data for testing the hypothesis. On a Likert scale of five points, the response was obtained. The respondents with minimum high school qualifications were randomly chosen to represent the population. The strong agreement received a high score of 5, whereas the strong disagreement received a low score of 1. IBM SPSS 26 was used for coded primary data analysis. The Cronbach Alpha was found to be 0.910, above threshold limits, manifesting internal consistency.

Results and Analysis

Descriptive statistics present the sample characteristics for the responses. Respondent's regional distribution in Table 3 manifests diversity in representation from a learned sample in Table 4. The average age of the respondents is 43.8 years, with 24.8 percent of respondents between 51 to 60 years, while 46.7 percent are over 40 years of age. The respondents strongly agree with their awareness of NG and acknowledge the benefits of PNG and CNG while accepting it as a clean energy source for enhancing its consumption to accelerate the transition to a gas-based economy in India.

Table 3. Regional Distribution of Responses

Location	Frequency	Valid Percent	Cumulative Percent
East	23	21.9	21.9
West	7	6.7	28.6
North	59	56.2	84.8
South	16	15.2	100.0
Total	105	100.0	

Source: Author's analysis

Table 4. Educational qualification of respondents

Qualification	Frequency	Valid Percent	Cumulative Percent
High School	1	1.0	1.0
Graduate #	49	46.6	47.6
Masters #	33	31.4	79.0
Doctorate #	22	21.0	100.0
Total	105	100.0	

Source: Author's analysis

(# completed/pursuing)

Inferential Statistics: Hypothesis Testing

The results for hypothesis testing are in the following tables.

1. H1: The awareness levels of promoting NG are usual among the populace

Table 5. ANOVA- Awareness

		Sum of Squares	df	Mean Square	F	Sig
Between People		124.171	104	1.194		
Within People	Between Items	1.505	2	0.752	1.852	0.159
	Residual	84.495	208	0.406		
	Total	86.000	210	0.410		
Total		210.171	314	0.669		

Grand Mean = 4.41

Source: Author's analysis

The p-value (p>0.05) suggests that the awareness levels of promoting NG are usual. The respondents strongly agree on all attributes about promoting NG usage to raise the living standard, which is the objective of SDG 3, and to enhance sustainability for meeting the SDG targets by 2030.

2. H2: PNG benefits make a difference in accepting it as a domestic household fuel

Table 6. ANOVA- Benefits of PNG

		Sum of Squares	df	Mean Square	F	Sig
Between People		196.526	104	1.890		
Within People	Between Items	11.478	4	2.870	9.090	0.000
	Residual	131.322	416	0.316		
	Total	142.800	420	0.340		
Total		339.326	524	0.648		

Grand Mean = 4.38

Source: Author's analysis

The p-value (p<0.05) suggests that the PNG benefits make a significant difference in accepting it as a domestic household fuel. The respondents strongly agree on all five attributes that benefit a household customer using PNG as cooking fuel.

3. H3: CNG benefits make a difference in accepting it as a transport fuel

The p-value (p<0.05) suggests that the CNG benefits make a significant difference in accepting it as a transport fuel while reducing environmental pollution by replacing gasoline or diesel. The respondents strongly agree on all four attributes that benefit a customer to use CNG as a clean and economical transport fuel.

Table 7. ANOVA- Benefits of CNG

		Sum of Squares	df	Mean Square	F	Sig
Between People		131.390	104	1.263		
Within People	Between Items	19.819	3	6.606	18.879	0.000
	Residual	109.181	312	0.350		
	Total	129.000	315	0.410		
Total		260.390	419	0.621		

Grand Mean = 4.46

Source: Author's analysis

4. H4: The NG acceptability makes a positive difference in enhancing NG consumption in India

The p-value (p<0.05) suggests that NG acceptability positively enhances NG consumption in India. The respondents strongly agree that they accept NG as a household cooking and transport fuel. Increasing its consumption in these segments will accelerate India's transition to a gas-based economy.

Table 8. ANOVA- Acceptability of NG

		Sum of Squares	df	Mean Square	F	Sig
Between People		126.921	104	1.220		
Within People	Between Items	3.149	2	1.575	4.667	0.000
	Residual	70.184	208	0.337		
	Total	73.333	210	0.349		
Total		200.254	314	0.638		

Grand Mean = 4.46

Source: Author's analysis

Promoting cleaner energy systems for carbon neutrality is critical for achieving India's SDGs while PNG and CNG are vital energy inputs for socio-economic well-being, providing sustainable benefits to society. The PNG attributes make it the perfect cooking fuel for urban Indians, where it replaces costly LPG, leading to household savings. In rural India, PNG replaces polluting firewood, coal, and kerosene, providing solutions to health challenges and improving well-being. CNG burns completely, enhancing efficiency by reducing harmful emissions retarding global warming. It does not ignite accidentally, making it a safe automotive fuel. As a clean energy source, the PNG benefits (p=0.000) and CNG benefits (p=0.000) are the significant factors for NG consumer acceptability. Analysis of results reveals that NG consumer acceptability (p=0.010) in the domestic household and transport segments enhances NG consumption, accelerating the transition to a gas-based economy in India.

Conclusion

Energy Access, Availability, and Affordability are critical to India's National Energy Policy. Creating a local distribution network, the supply chain, for NG retailing in India to provide 42.3 million domestic PNG connections and set up 8181 CNG stations under 9th and 10th CGD bidding round authorizations ensures ease of NG access supporting SDG 7. Preferential allocation of cheap domestic gas ensures its affordable availability to PNG domestic and CNG transport segments. Concurrently, increasing the NG consumption by replacing polluting fuels in the domestic and transport segments promotes a healthy lifestyle that enhances productivity supporting SDGs 3 and 8, leading to sustainable, inclusive growth in India.

Developing economies with diverse national priorities can adopt the Indian model for promoting PNG and CNG among the populace to address the interrelated twin challenges of socio-economic development and climate change to promote carbon neutrality collectively.

References

Demirbas, A. (2012). Green Energy and Technology, In Green Energy and Technology. 2010th edn. London: Springer, p. 186. doi: 10.2174/9781608052851112010l.

Gladkova, K. (2013). Natural Gas Supporting Social and Economic Development, In 18th Turkmenistan International Oil and Gas Conference.

IEA (2020) Gas 2020.

Ishwaran, M., et al. (2017). Environmental and Social Value of Natural Gas, In Advances in Oil and Gas Exploration and Production, pp. 101–111. doi: 10.1007/978-3-319-59734-8_4.

Kumar, V. V., Shastri, Y., and Hoadley, A. (2020). A consequence analysis study of natural gas consumption in a developing country: Case of India, Energy Policy. Elsevier Ltd, 145 (July 2019), p. 111675. doi: 10.1016/j.enpol.2020.111675.

Di Pascoli, S., Femia, A., and Luzzati, T. (2001) Natural gas, cars and the environment. A (relatively) "clean" and cheap fuel looking for users, Ecological Economics, 38(2), pp. 179–189. doi: 10.1016/S0921-8009(01)00174-4.

Patnaik, S., Tripathi, S., and Jain, A. (2018). A Roadmap for Access to Clean Cooking Energy in India, Asian Journal of Public Affairs, 11(1). doi: 10.18003/ajpa.20189.

PPAC (2022). PPAC Archives, PPAC. Available at: https://ppac.gov.in/content/22_2_Archives.aspx (Accessed: 25 June 2022).

Safari, A., et al. (2019). Natural gas: A transition fuel for sustainable energy system transformation?, Energy Science and Engineering, 7(4), pp. 1075–1094. doi: 10.1002/ese3.380.

Mobile Payment Usage: Systematic Literature Review

Veenu Madan[a] and Manjit Kour[b]

[a]Chandigarh University, India
[b]Chandigarh University, India
E-mail: veenu87madaan@gmail.com; manjuz_99@yahoo.com

Abstract

Based on the data set collected from 2018 to 2022 from the Scopus database, this study represents a systematic literature review of mobile payment usage. Advancements in information technology have resulted in a fundamental redesign of the financial services value chain, resulting in a new business model known as Fintech. Fintech reaches the unbanked people through its technologies. Mobile receipts and payments use cellular and other communication networks to pay for products, services, or invoices with the help of portable electronic devices (such as a cell phone, a smartphone, a handheld PC, mobile telecommunications networks, or near-field communication technology). In this study, Descriptive analysis is done by exporting the data set into an excel sheet and using a pivot table option to show the no of the paper published year-wise and citation analysis based on title and author.

Keywords: Mobile payment, Systematic review, Usage, Digital payment

Introduction

Advancements in information technology have resulted in a fundamental redesign of the financial services value chain, resulting in a new business model known as Fintech. Fintech is regarded as one of the technologies that will disrupt the banking industry and enable businesses to compete effectively in the industrial revolution 4.0 with modern and advanced technologies (Putritama, 2019). Mobile receipt and payments make use of cellular and other communication networks to pay for products, services, or invoices with the help of portable electronic devices (such as a cell phone, a smartphone, a handheld PC, mobile telecommunications networks, or near-field communication technology) (Dahlberg et al., 2006). Digital payments, particularly mobile payments, received a considerable boost after the government abolished 500 and 1000 notes in circulation in Nov2016. People still prefer to use cashless payments when cash again enters the market. Due to the Spread of Coronavirus, cards and money withdrawals from ATMs. are decreasing. People became health-conscious, and to prevent the transmission of viruses, they moved to cashless payments using their mobile phones and app-based payment methods. Now applications have become super applications with funding methods, payment features, bill payment, e-commerce, travel, and financial services, and thoroughly enter into the daily life of users. Our study's primary goal is to assess m-payment studies published between 2018 to 2022. The following research questions are addressed in this study:

(1) To determine the dynamics of the generation of research literature? (2) To determine the most prolific and significant authors in a field? (3) To determine the most impactful Journal in a specific

field? (4) To analyze the prominent publishing houses for researchers in the field of study? (5) To determine citation analysis based on title and author?

Research Methodology

We devised a search technique to find pertinent literature for this comprehensive search. The Scopus database was targeted for this search technique, and the phrases "mobile payment," "digital payment," "e-payment," and "M-wallet" were employed. Journal articles and review papers published in English were included in all searches, covering the database's creation through 2022. The selection criteria were based on the PRISMA statement (Moher et al., 2009). The investigation mainly focused on mapping existing literature on mobile payment in social science, business, and economics. A total of 117 records were extracted at this stage. Some papers are not included because of out of scope, and finally, 102 articles are included in the study.

Descriptive Analysis

The descriptive analysis described what had happened in the past. It is done by exporting Scopus data into an excel sheet, and research is done by using pivot tables, drawing various tables and charts, and analyzing the data based on no of papers published, citations per author, and famous journals for publications.

Table 1. No. of a paper published

Year	No of Papers
2018	9
2019	16
2020	38
2021	32
2022	7
Grand Total	102

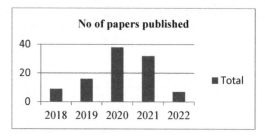

Fig. 1. No of papers published
Source: Author creation

Table 1 and Fig. 1 depict the researchers' interest is increasing in studying mobile payments year-wise. Still, none of the papers published is high volume, so there is an excellent scope for future research in mobile payments in the Indian scenario.

Most Popular Journal

Journals define the area and scope of publication of research articles; through analysis of publication in a journal, a researcher can select the Journal which is well suited to their theme and area and also reveal the acceptance and publication of the paper in a particular theme.

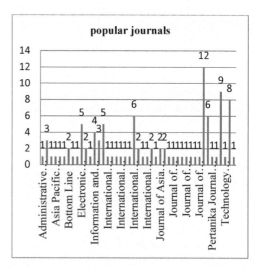

Fig. 2. Popular journals for publication
Source: Author creation

Figure 2 shows the most popular journals for publication in the Journal of retailing and consumer service. After that, the International Journal of information management, the Journal of theoretical and applied electronic commerce, technological forecasting, and social change and technology in society.

Citation Analysis

A citation can be defined as "a reference to a book, article web page or other published or unpublished item that includes sufficient details to uniquely identified that item" It helps to establish a connection between two documents. When one author cites another, a connection is established between them. Citation analysis uses Citations in scholarly work to develop relationships with other works or other researchers. This citation analysis is the process whereby the impact or "quality" of an article is assessed by counting the number of times other authors mention it in their work (Nicolseian et al., 2007).

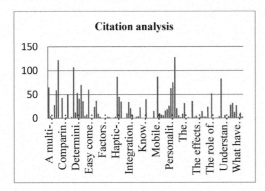

Fig. 3. Citation analysis
Source: Author Creation

The highest citation 128 receiving title is "predicting the determinants of mobile payment acceptance: A hybrid SEM neural network approach" after that, 122 Citations by antecedents of trust and continuance intention in mobile payment platform moderating effect of gender.

Author-Wise Citation

Table 2. Top 5 Authors Based on Citation

Author	Citation
Liébana-Camarillas F., Marinkovic V., Ramos de Luna I., Kalinic Z.	128
Shao Z., Zhang L., Li X., Guo Y.	122
Singh N., Sinha N., Liébana-Cabanillas F. J.	107
de Luna I. R., Liébana-Cabanillas F., Sánchez-Fernández J., Muñoz-Leiva F.	87
Karjaluoto H., Shaikh A. A., Saarijärvi H., Saraniemi S.	87
Patil P., Tamilmani K., Rana N. P., Raghavan V.	83

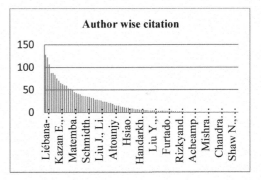

Fig. 4. Author wise citation
Source: Author creation

Figure 4 depicts the citation received by the author, and the highest citation received by Liébana et al. (2019) is 128 and, Shao et al. (2019) got 122 citations, Singh et al. (2021) received 107 citations.

Most Prominent Publishers

A publisher helps writers gain recognition and gives them the chance to showcase their originality. They are crucial as intermediaries between writers and their intended audience. Researchers must get acquainted with publishing houses.

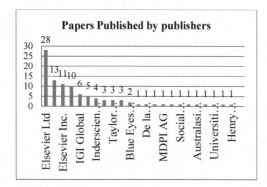

Fig. 5. Papers published by publishers
Source: Author creation

Figure 5 shows the most prominent publishing houses, and analyzing the table, we found that Elsevier ltd published 28 papers and Emerald group holding ltd published 13 articles. It gives a clear idea to researchers' most trusted publishing houses, and they can save themselves from fake publishers.

Limitations of the Study

This review paper has the limitation that it extracts data only from the Scopus database. It may be possible that some documents are left in my study, which provides a more clear idea for further research. So by extracting data from other databases, web of science, j store, and research gate, we can get more papers for systematic analysis and find out research gaps accurately.

Conclusions and Future Research

In essence, this paper dealt with systematic literature on mobile payment usage. After the 2008 financial crisis role of Fintech gained momentum. Fintech payment serves the need of unbanked and small entrepreneurs by using artificial intelligence and machine learning to collect and analyze credit data scores. After demonetization, Indian consumers shifted from cash to cashless transactions using UPI apps and mobile wallets. In the Indian scenario, mobile phone penetration and internet usage are increasing, and payment applications

are also downloaded on their smartphones; people still prefer to keep cash in their pockets despite various digital payment initiatives by the government of India. They perceived the multiple types of risk and insecurity while making payments through wireless devices, so it is necessary to study the factors which affect their intention and actual usage of mobile payment. In the future, we can research the changing structure of incumbents due to the fintech payment and lending system. The role of government in regulating Fintech and creating trust and security among consumers is also a good area for further research.

References

Dahlberg, T., Mallat, N., Ondrus, J., and Zmijewska, A. (2006). Mobile Payment Market and Research - Past, Present, and Future. Presentation at Helsinki Mobility Roundtable, Helsinki, Finland, June 1–2, 2002 (January 2006), 1–16.

de Luna, I. R., Liébana-Cabanillas, F., Sánchez-Fernández, J., and Muñoz-Leiva, F. (2019). Mobile payment is not all the same: The adoption of mobile payment systems depending on the technology applied. Technological Forecasting and Social Change, 146, 931–944.

Karjaluoto, H., Shaikh, A. A., Saarijärvi, H., and Saraniemi, S. (2019). How perceived value drives the use of mobile financial services apps. International Journal of Information Management, 47, 252–261.

Liébana-Cabanillas, F., Molinillo, S., and Ruiz-Montañez, M. (2019). To use or not to use, that is the question: Analysis of the determining factors for using NFC mobile payment systems in public transportation. Technological Forecasting and Social Change, 139, 266–276.

Moher, D., Liberati, A., Tetzlaff, J., Altman, D. G., and PRISMA Group*. (2009). Preferred reporting items for systematic reviews and meta-analyses: the PRISMA statement. Annals of internal medicine, 151(4), 264–269.

Patil, P., Tamilmani, K., Rana, N. P., and Raghavan, V. (2020). Understanding consumer adoption of mobile payment in India: Extending Meta-UTAUT model with personal innovativeness, anxiety, trust, and grievance redressal. International Journal of Information Management, 54, 102144.

Putritama, A. (2019). The Mobile Payment Fintech Continuance Usage Intention in Indonesia. JurnalEconomia, 15(2), 243–258. https://doi.org/10.21831/economia.v15i2.26403

Shao, Z., Zhang, L., Li, X., and Guo, Y. (2019). Antecedents of trust and continuance intention in mobile payment platforms: The moderating effect of gender. Electronic Commerce Research and Applications, 33, 100823.

Singh, N., Sinha, N., and Liébana-Cabanillas, F. J. (2020). Determining factors in the adoption and recommendation of mobile wallet services in India: Analysis of the effect of innovativeness, stress to use and social influence. International Journal of Information Management, 50, 191–205.

Bibliometric Analysis of the Impact of Blockchain Technology on the Tourism Industry

Garima Anand,,a *Anish Shandilya,a Varuna Gupta,a and Vandana Mehndiratta*a

aChrist (Deemed to be University), India
E-mail: *Garima.anand@christuniversity.in

Abstract

The tourism sector is one of the world's fastest-expanding industries. Because of the benefits, it provides to individuals and organizations, the tourism sector has attracted a lot of attention throughout the years. But because of its poor and obsolete data management techniques, this industry is in desperate need of reform. Blockchain technology is one method for managing and exploring data relevant to the tourism industry. This study used bibliometric methods to analyze the impact of blockchain technology on the tourism sector from 2017 to 2022. The publications were extracted from the dimensions database, and the VOS viewer software was used to visualize research patterns. The findings provided valuable information on the publication year, authors, author's country, author's organizational affiliations, publishing journals, etc. Based on the findings of this analysis, researchers may be able to design their studies better and add more insights into their empirical studies.

Keywords: Tourism, Hospitality, Blockchain, Technology, Bibliometric, VOS viewer

Introduction

Tourism is a vital economic force and industry in many nations. The tourism sector has been an attraction for the past few years. Blockchain technology is critical to improving and enhancing data management in this sector (Anand et al., 2022). Amidst the rapidly growing number of published studies and research results concerning the impact of blockchain on the tourism sector, little quantitative analysis of blockchain and tourism research has been conducted in terms of the effect of articles, authors and their institutional affiliations, and so on. This article focuses on the bibliometric analysis of the impact of blockchain on the tourism sector to reach the intended goals.

Bibliometrics is a quantifiable analytic approach that use statistical and mathematical approaches to examine the interdependence and effect of publications in a specific field of interest. This method may be used to effectively identify significant studies, authors, affiliated journals, organizations, and countries across time, as well as offer a visual representation of massive amounts of academic literature. Bibliometric analysis has also been used to examine blockchain technology research trends. The primary goal of this study is to conduct a bibliometric analysis of the literature produced in this field from 2017 to 2022 to give insight on several aspects of the literature. Such information can have significant ramifications for empirical investigations in this field.

The following are the study's contributions:

1. How has paper publication distribution developed over the years?

2. In which journals have the majority of these studies been published?
3. Which country/institution has had the largest influence on the publication?
4. Which keywords are used more frequently?

The following is how the article is structured: The methodology section discusses the bibliometric analysis technique, the results section helps to identify the most important sections of the interpretation, the discussion section brings these findings together and emphasizes topics for future research, and finally, the conclusion section summarises the study's primary conclusions (Sweileh, 2020).

Methodology

The first step in performing the bibliometric analysis is to choose available databases based on their appropriateness and the relevance of using one or the other. The Dimension database includes extensive publishing data and is the most generally acknowledged and utilized database for scientific publication analysis. The Dimension Database was queried for all data using the following search terms in titles, abstracts, and keywords: (Blockchain) AND (Tourism OR hospitality) AND Impact. The field of research was restricted to "Tourism." This study looked at publications published between 2017 and 2022; a total of 190 were identified. Few publications were excluded because they did not meet the inclusion criteria; hence, 102 publications were eventually included in the final analysis. All data obtained from Dimensions were entered into VOSviewer v.1.6.11, a popular tool for analyzing and visualizing interactions between authors, nations, co-citations, and article titles (Lee et al., 2020). This tool uses its mechanism to categorize data and assign a color for every cluster. It was chosen for its easiness of use and capacity to interface with a diverse variety of publications. The findings of the analysis are detailed in the section that follows.

Results

From the analysis, it can be identified that there has been a considerable rise in the publication of these

kinds of articles, and this study can steer toward potential research needs. This article identified the document type, research areas, primary authors, countries with the most publications, study areas with the most publications, and most prominent journals covering this area of research (Ali et al., 2018).

Document and Source Types: In total, 102 publications met the selection criteria in the document type. A total of 6 types were found in these 102 documents. The most frequent publication type is an article (70), accounting for 68.6% of total publications. Chapters (18) are in the second position with a proportion of 17.64%. Other document types consisting proceedings (7), edited book (4), preprint (2), and monograph (1). Table 1. shows the frequency of various document types (Hassan et al., 2021).

Table 1. Document and source types

Document Type	Frequency	% (N=102)
Articles	70	68.6
Chapters	18	17.64
Proceedings	7	6.8
Edited Book	4	3.9
Preprint	2	1.96
Monograph	1	0.93

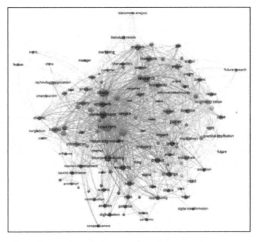

Fig. 1. Keyword Analysis

Evolution of Publications Over Years: As shown in Table 2, a total of 102 publications have been published in the past six years. The first publication revolving around this area of research was published in 2017. The worldwide year-wise distribution of publishing output indicates a rising pattern. It increased from 8 papers in 2018 to 30 articles in 2021. The number of yearly publications is likely to grow even more after 2022 (Sidhu et al., 2020).

Table 2. Publications over years

Year of Publication	Number of Publications
2017	1
2018	8
2019	20
2020	29
2021	30
2022	19

Keyword Analysis: Keywords, being one of the most relevant pieces of publications, are adequate vocabulary in searching for subjects in abstracting and indexing databases. Keyword research is helpful to its diverse readership. VOSviewer is used to do a keyword analysis for this purpose (Hasbullah et al., 2021). A total of 133 words were discovered and categorized into 5 clusters based on a minimum frequency of 5 times per word. As indicated in Fig. 1, the most commonly used keywords were "tourism" 74 times, "study" 48 times, and "technology" 45 times.

Analysis of Authorship: The publications were further examined to obtain author information. The 102 publications were contributed by 277 various authors. The majority of articles, 31.37 percent, are authored by two authors, for a count of 32 articles. This is followed by 19 articles published by a single author, accounting for 18.62 percent of the total, and articles written by three authors, accounting for 17.64 percent of the total. Articles with four or more authors accounted for 32.35 percent or 33 articles (Patrick et al., 2020). Table 3 includes the nine most prolific authors in this field of study. Tarik Dogru of Florida State University in the United States authored the most publications (8), followed by Azizul Hassan of the Tourism Society in London (6 articles).

Table 3. Top 9 Contributing Authors

Author	Document	Citations
Dogru, Tarik	8	130
Hassan, Azizul	6	0
Valeri, Marco	4	238
Treiblmaier, Horst	4	130
Mody, Makarand	4	70
Suess, Courtney	4	70
Baggio, Rodolfo	3	187
Onder, Irem	3	137
Rahman, Muhammad Khalilur	3	0

Analysis of the Author's Main Affiliation and Countries: This article retrieval analysis comprises the collection of publications from 34 countries and 159 institutions. This section discusses listings of countries with at least five publications (Viana-Lora et al., 2022). According to Fig. 2, the United States published the most articles, 13 in total, followed by Bangladesh with nine publications. Table 4 displays the top eight organizational affiliations of the 102 publications identified in this analysis. Florida State University had the most publications, with eight, followed by Modul University, which had seven publications

Table 4. Top 8 Organizational Affiliations

Organization	Documents	Citations
Florida State University	8	331
Modul University Vienna	7	177
University Niccolo Cusano	4	238
Tomsk Polytechnic University	3	187
Texas A&M University	3	70
Boston University	3	70
Griffith University	3	29
Univeristy Malayasia Kelantan	3	0

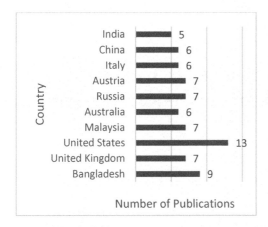

Fig. 2. Country Analysis

Discussion

The present bibliometric analysis of studies investigating the impact of blockchain technology on the tourism sector published in the past six years shows that the overall number of articles increased substantially during this period. The United States has the highest proportion of articles in this category, followed by Bangladesh and the United Kingdom. Furthermore, a network analysis based on keyword co-occurrence identified two categories of domains to focus on: Tourism and Technology. Author and organizational affiliation analysis showed that several research organizations from non-Asian countries published many studies on this research area. However, the authors with the most citations were from the United States and the United Kingdom. The latest findings may be beneficial for organizing and prioritizing future research on blockchain technology in the tourism industry.

Conclusion

Based on the bibliometric analysis, a significant perspective is given related to the impact of blockchain technology on Tourism sector between 2017 and 2022. The publications were retrieved from the Dimension database. In total 102 journal articles were analyzed systematically to determine the evolution and patterns of publishing, the top most prolific organizations, the top cited articles,

countries, author's keywords, and bibliographic couplings of sources. According to the findings, yearly publications increased dramatically in 2020. The emergence of blockchain technology in tourism sector was demonstrated by the involvement of 34 countries and 159 organizations. The United States produced the most publications, with 13 papers receiving 530 citations. Next, Florida State University appeared to be the most productive organization. These insights can be used to assess the strengths and limitations of this field's future advancement.

References

Anand, G., Prajeeth, A., Gautam, B., Rahul, and Monika (2022). Data Communication Technologies and Internet of Things, Lecture Notes on Data Engineering and Communications Technologies, 101, 709, doi: https://doi.org/10.1007/978-981-16-7610-9_53

Sweileh, W. M. (2020). Global Health, 16, 44, doi: https://doi.org/10.1186/s12992-02000576-1

Lee, I. S., Lee, H., Chen, Y. H., and Chae, Y. (2020). J Pain Res., 13, 367, doi: https://doi.org/ 10.2147/ JPR.S235047

Ali, P. M. N., Malik, B. A., and Raza, A. (2018). Annals of Library and Information Studies, 65, 217.

Hassan, R., Chhabra, M., Shahza, A., Fox, D., Diana and Hasan, S. (2021). Journal of International Women's Studies, 22, 1. Available at SSRN: https://ssrn.com/abstract=4011499

Sidhu, A., Singh, H., Virdi, S., and Kumar, R. (2020). Journal of Content Community and Communication, 12, 20, doi: https://doi.org/10.31620/JCCC.12.20/04

Hasbullah, N., Sulaiman, Z., Mas'od, A., and Ahmad, S. (2021). Turkish Journal of Computer and Mathematics Education (TURCOMAT), 12, 1292, DOI: https://doi.org/ 10.17762/turcomat.v12i5.1796

Patrick, Z., and Hee, O. (2020). International Journal of Academic Research in Business and Social Sciences, 10 (5), 770, doi: https://doi.org/10.6007/IJARBSS/v10-i5/7248

Viana-Lora, A., and Nel-lo-Andreu, M. G. (2022). Humanit Soc Sci Commun, 9, 173, doi: https://doi.org/10.1057/s41599-02201194-5

Analysis of Marketing Strategy & Customer Satisfaction Over Select Mobile Service Providers

Gurpreet Singh Matharou,[*,a] *Simran Kaur,*[b] *and Pramod Gupta*[c]

[a,b]Associate Professor, Manav Rachna International Institute of Research and Studies,
Faridabad, Haryana India
[c]Professor, Lords University, Alwar, Rajasthan
E-mail: *gsmatharou1@gmail.com

Abstract

The paper addresses the analysis of critical factors influencing the selection of mobile service providers (MSP) by the public as per the preferences based on socioeconomic conditions. Broadly, seven factors comprising cost, communication, service, network, customer care, innovativeness, and advertisement were studied and analyzed using a developed rotated factor matrix followed by in-depth analysis through factor loadings, communality (h^2), and Cronbach's alpha (α). The cost has a maximum influence on the selection of MSPs followed by communication and innovativeness with α values of 0.817, 0.753, and 0.749. The variable, call rates have a maximum h^2 value of 0.936 which shows the maximum influence on MSPs selection followed by unlimited calls with a h^2 of 0.897, roaming charges (0.873), ISD charges (0.859), and customer-oriented promotion (0.825). Out of the three MSPs (Jio, Airtel, and VI) under study, Jio was a winner with the best offering and customer support to the customers.

Keywords: Mobile Service Providers, Factor loading, Communality, Cronbach alpha

Introduction

Telecom Services have been appreciated across the globe as an essential medium for the socio-economic progress of a state. Telecommunication is the backbone required for the accelerated expansion and modernization of different regions of the economy. India is the world's second-leading media transmissions market with a support base of 1.16 billion and has shown steady progress in recent times. The Indian convenient economy is growing speedily and will strengthen India's Gross Domestic Product appreciably. The Government of India illustrates the relevance of a world-class telecommunication framework and intelligence as indicator of the accelerated socio-economic stride of the nation. In 2019, India surpassed the US to turn into the second-leading market in the volume of mobile application downloads. India is also the second-largest broadcast communication market. The genuine support base stood at 1189.15 million in September 2021. The specific volume of web supports appeared at 794.88 million in September 2021. Gross revenue of the telecom sector prevailed at Rs. 64,801 crores (US$ 8.74 billion) in the prior quarter of FY22. The complete remote knowledge usage in India developed into 16.54% quarterly to turn up at 32,397 PB in the primary quarter of FY22. The engagement of 3G and 4G knowledge handling to the ultimate size of inaccessible knowledge applications was 1.78% and 97.74%, respectively. It has been observed that the last five years, with an increment in mobile phone usage and a considerable decline in call and data cost, will add around 450 million users by 2025. Until then, around 21 million skilled manpower with well-versed knowledge in the 5G technology, and other related technologies such as the Internet of Things (IoT), Artificial Intelligence (AI), and cloud computing shall be required. The country has undergone tremendous changes recently and has experienced higher economic, industrial, and technological growth rates. Communication has become integral to any business' growth, success, and efficiency. Mobile service providers (MSP) are considerably helping connect people across the planet. Increasing infrastructure facilities for the telecommunication industry have created more and more competition in the telecommunication sectors. Upadhyaya and Sharma (2012) have made a comparative study between BSNL and Airtel MSPs, and suggested a few improvements to capitalize on the ever-expanding mobile market. (Shah, 2013) focused their study on Samsung handsets to analyze the effect of independent variables (features, price, etc.) on the dependent variable, customer satisfaction (CS). They made the analysis using the correlation between the dependent and independent variables. (Refai et al., 2014) emphasized trust as a golden word for achieving the best CS in the mobile phone market. They also suggested some future scopes in the related areas. Kumar and Soundararajan (2016) emphasized adopting scientific strategies as a key to surviving in today's MSPs competition. They also stressed the importance of broadband in solving most of the technical issues faced during wireless communications. Mahalaxmi and Kumar (2017) analyzed the service quality and CS offered by JIO MSP in the local Trichy region. The in-depth analysis put forward the Reliance on company penetration marketing strategy and its capability of assessing the customer's voice. (Hashem et al., 2017) analyzed the quality of service offered by the Jordanian telecommunication sector as per customer perception. The success keywords described in the paper include reliability, responsibility, assuring policies, empathy, and tangible networks. Selelo and Lekobane (2017) studied the relationship between quality (service) and CS in the mobile sector. They compared the

factors, levels, and their effect on CS through mathematical models. (Mannan et al., 2017) in-depth analyzed the factors causing a lack in CS, customer intention to switch to another MSPs, and the cost associated with changing and also suggested a check sheet to analyze the alternatives. The region of the study was Bangladesh. Tandon et al. (2020) examined the importance of artificial intelligence in the corporate sector. It can be summarized from the above critical papers that if the MSPs cannot compete with other service providers In the market, it will lead to the closure of the company, or amalgamation or absorption will occur.

Research Methodology

The present study aims to assess the respondents' socio-economic conditions and preference over particular MSPs and examine the quality parameters designed by the selected MSPs to influence their customers. This research adopts a multi-stage, random sampling method, taking the Jaipur district as a study region. The inputs of 600 respondents were finalized out of the 660 surveys, out of which 60 were found inappropriate. This study is only restricted to selected mobile service providers like JIO, AIRTEL, and VI. Since the study is based on the random sampling method instead of the census method, the findings cannot be generalized. After an exhaustive literature review, the seven factors (cost, communication, service, network, customer care, innovativeness, and advertisement) were identified after a detailed literature review. Using multiple regression analysis, the impact of factors influencing the selection of MSPs is being made.

Respondents Profile

Table 1 shows that out of 600 respondents, a maximum of (45.78%) of respondents are male and Jio users, while a minimum of (25.37%) of respondents are female and VI users. A maximum (30%) of respondents are in the age group of 20–30 years. 33% of Airtel users reside in urban areas, and 43% reside in rural areas. Around 77% of respondents had opted for pre-paid connections, Around 69% use 4G connections in urban setups, whereas 25% opted for 2G, 3G setups. 52% of respondents prefer unlimited call data (UCD), only 18% prefer unlimited data (UD), followed by 17% full talk time (FTT), and the rest 13% unlimited outgoing calls (UOC).

Table 1. Respondents' Profile

		Jio		Airtel		VI		Total	
		No	%	No	%	No	%	No	%
Gender	Male	152	45.7	112	33.7	68	20.5	332	55.3
	Female	108	40.3	92	34.3	68	25.3	268	44.7
Age (in years)	Below 20	36	33.3	44	40.7	28	25.9	108	18
	20 - 30	88	48.8	56	31.1	36	20	180	30
	30 - 40	92	65.7	28	20	20	14.2	140	23.3
	40 and above	44	25.5	76	44.1	52	30.2	172	28.7
Residential Area	Urban	108	52.9	68	33.3	28	13.7	204	34
	Rural	64	28.5	96	42.8	64	28.5	224	37.3
	Semi-urban	88	51.1	40	23.2	44	25.5	172	28.7
Type of Plan	Post-paid	36	25.7	92	65.7	12	8.57	140	23.3
	Pre-paid	224	48.7	112	24.3	124	26.9	460	76.7
Time Spent on social media	Up to 1 hr.	4	50	0	0	4	50	8	1.3
	1 to 3 hr.	40	58.8	20	29.4	8	11.7	68	11.3
	3 to 5 hr.	80	47.6	52	30.9	36	21.4	168	28
	Above 5 hr.	136	38.2	132	37.0	88	24.7	356	59.3
Network Type	2G	0	0	0	0	36	100	36	6
	3G	0	0	20	13.1	100	65.8	152	25.3
	4G	260	63.1	184	44.6	0	0	412	68.7
Preferable Offer	FTT	64	62.7	18	17.6	20	19.6	102	17
	UOC	12	15	64	80	4	5	80	13.3
	UCD	180	58.1	90	29.0	40	12.9	310	51.7
	UD	4	3.7	32	29.6	72	66.7	108	18

*FTT: Full Talk Time;

UOC: Unlimited Outgoing Calls;

UCD: Unlimited calls+data;

UD: Unlimited data

Rotated Factor Matrix

In this research, factor analysis was undertaken to elicit individual factors. The variables, which empower each component based on the strength and orientation of factors bundled in persuading the choice of service providers were determined. Factor loadings reveal how considerably an element defines a variable. Table 2 illustrates the rotated factor matrix for the variables of the components, prompting the choice of service providers. Loadings near –1 or 1 point out that the component greatly affects the variable. Loadings adjacent to Zero reveal that the component bears a low influence on the variable. Some variables may show large loadings of multiple factors (Watkins, 2018).

Table 2. Rotated factor matrix for the variables of the components prompting the choice of service providers

S. N	VARIABLE	FACTORS						
		F1	F2	F3	F4	F5	F6	F7
1	Unlimited calls	0.923	0.096	0.007	0.132	0.038	0.104	0.079
2	Roaming charges	0.827	–0.365	0	0.157	0.139	0.107	0.002
3	ISD charges	0.736	–0.523	0.021	0.123	–0.044	0.129	0.096
4	Falling tariffs	0.68	0.437	0.086	–0.161	–0.308	–0.149	–0.007
5	Call rates	0.212	0.911	0.038	–0.165	–0.114	–0.138	–0.018
6	Promotional offers	–0.31	0.783	–0.008	–0.095	0.071	0.147	0.074
7	Call forwarding/ waiting	–0.13	0.782	0.01	–0.007	0.229	0.157	0.102
8	Reliability	–0.06	0.014	0.838	0.105	0.078	–0.1	–0.034
9	Responsiveness	–0.05	–0.074	0.707	–0.132	0.069	–0.01	–0.066
10	Assurance	0.037	0.105	0.698	–0.171	–0.112	0.117	–0.053
11	Technology use	0.112	–0.026	0.683	0.113	0.15	–0.247	0.306
12	Product range	0.116	0.066	0.564	0.084	0.175	–0.192	0.419
13	Good coverage	–0.06	–0.035	–0.136	0.817	–0.208	0.107	–0.058
14	Fast connectivity	0.387	–0.237	–0.032	0.69	–0.095	–0.159	–0.108
15	Signal clarity	0.149	0.027	–0.223	–0.64	0.037	0.451	–0.061
16	Brand image	0.3	–0.075	–0.136	0.599	0.186	–0.182	0.058
17	Empathy	0.244	–0.191	0.326	0.466	–0.454	0.317	0.04
18	Customer complaint management	–0.11	0.056	0.205	–0.097	0.738	0.221	0.026
19	Complaint resolution	–0.06	0.201	0.05	–0.159	0.633	0.036	–0.168
20	Sincere in solving problem	0.3	–0.174	0.026	0.109	0.0555	0.034	0.084
21	Customer oriented promotion	0.146	–0.031	–0.14	–0.013	0.07	0.882	–0.022
22	CRM platform	–0.04	0.212	–0.044	–0.3	0.273	0.611	0.093
23	Catchy advertisement	–0.04	–0.073	0.044	0.015	–0.044	0.093	0.848
24	Media selection	0.12	0.186	0.003	–0.088	–0.053	–0.036	0.841

F1: Cost	F2: Communication	F3: Service Factor	F4: Network	F5: Customer Care	F6: Innovativeness	F7: Advertisement

Results and Analysis

It is apparent from Table 2 that all twenty-four descriptions have been translated into seven components. The rotated factor matrix, for the factors prompting the selection of service providers with established alternative names, is explained further in Table 3. In Table 3, the factor loadings marked were further taken into consideration from table 2. The loadings selected are near +1 and have a major influence on the variables. In Principal Component Analysis (PCA) and Factor Analysis, a variable's communality is an appropriate method for forecasting the variable's importance. More precisely, it states what percentage of the variable's variance is a result of either the PC or the interrelationships between each variable and individual factor (Matharou and Bhuyan, 2020, 2021). Communality (h^2) specifies the extent of variance in every variable and can be calculated as per the function $\sum_{0}^{n} f_n$. Where f_n is the factor loading for the nth factor for a corresponding variable. The communality of variable unlimited calls is thus the summation of all squares of the factor loadings of seven factors, which can be mathematically shown as Unlimited calls

$$h^2 = \sum \begin{array}{c} (0.923)^2 + (0.096)^2 + (0.007)^2 + (0.132)^2 + (0.038)^2 \\ + (0.104)^2 + (0.079)^2 \end{array}$$

Likewise, h^2 for all other variables are being calculated and shown in Table 3. The Cronbach Alpha (α) is a projection of internal uniformity and shows a relation between a group of set of items. It is acknowledged to be a method of scale-reliability measurement and internal consistency. It is calculated as per equation 1. where k is the corresponding highest value in the variable, s^2y is the variance along the y axis and $\sum s_{i^2}$ is the variance along the x-axis. The present study aims to assess the respondents' socio-economic conditions and preference over particular MSPs and examine the quality parameters designed by the selected MSPs to influence their customers.

$$\alpha = \left(\frac{k}{k\text{-}1} \right) \left(\frac{s^2y\text{-}\sum s_{i^2}}{s^2y} \right) \ldots\ldots\ldots\ldots(1)$$

After the rotated factor matrix analysis, the following decision was made. Unlimited calls have got maximum factor loading of 0.923 on first-factor cost. Likewise, call rates (0.911) on communication, reliability (0.838) on service factor, good coverage (0.817) on good coverage, customer complaint management (0.738) on customer care, customer-oriented promotion (0.882) on innovativeness and catchy advertisement (0.848) on the advertisement. If the communality is small, this indicates that the variable suggests little prevailing with the alternative variables and is possibly a spot for expulsion. It can be seen from the table that one of the factors cost variables, fallings tariffs have the least h^2 value in comparison to unlimited calls, roaming charges, and ISD charges.

Table 3. Variables Factor loading, Communality, and Cronbach's Alpha

S.N	Factor	Variables	Factor Loading	Communality h^2	Cronbach's Alpha
1		Unlimited calls	0.923	0.898	
2	Cost	Roaming charges	0.827	0.873	0.817
3		ISD charges	0.736	0.859	
4		Falling tariffs	0.68	0.803	
5		Call rates	0.911	0.935	
6	Communication	Promotional offers	0.783	0.753	0.753
7		Call forwarding/waiting	0.782	0.717	

8		Reliability	0.838	0.735	
9		Responsiveness	0.707	0.534	
10	Service Factor	Assurance	0.698	0.558	0.722
11		Technology use	0.683	0.67	
12		Product range	0.564	0.586	
13		Good coverage	0.817	0.749	
14		Fast connectivity	0.69	0.729	
15	Network	Signal clarity	−0.64	0.691	0.695
16		Brand image	0.599	0.544	
17		Empathy	0.466	0.727	
18		Customer complaint management	0.738	0.661	
19	Customer Care	Complaint resolution	0.633	0.502	0.716
20		Sincere in solving problem	0.555	0.449	
21	Innovativeness	Customer-oriented promotion	0.882	0.825	0.749
22		CRM platform	0.611	0.595	
23	Advertisement	Catchy advertisement	0.848	0.739	0.685
24		Media Selection	0.841	0.767	

Likewise, the same relation can be seen with call forwarding/ waiting on communication, the responsiveness on service factor, the brand image on the network, sincerity in solving the problem of customer care, CRM platform on innovativeness, and media selection on the advertisement.

The α shows a relation between a group of the set of items. More the value of α more will be the significance of the influence of the group items. The α value of >0.7 is considered to be significant (Taber, 2018). It can be seen through α analysis, that the factor cost with variables (unlimited calls, roaming charges, ISD charges, and falling tariffs) has the maximum value of 0.817, which shows the cost to be the most influential factor on which the selection of MSPs depends. Followed by cost, is the factor communication with variables (call rates, promotional offers, and call forwarding waiting) and has an α value of 0.753. It can be seen that call rates and promotional offers has a positive influence on customer buying behavior. Similarly, Innovativeness as a factor has the third largest α value (0.749). The related variables are customer-oriented promotion and CRM platform. This shows that customer-oriented promotion along with a suitable CRM platform will lead to innovativeness in the working style of the company and hence attracts customer towards it. Service factor and Customer care shares the fourth spot with α value of (0.722 and 0.716). The less value of two factors in comparison to the before-said factors shows the MSPs service offered is up to the mark and hence fewer concerns amongst the respondents. In addition, since fewer concerns about service less is interaction with customer care support. Maximum respondents reported very less interaction with customer support and very less service-related issues. The fifth spot can be equally shared amongst network and advertisement factors. Both factors have α values of (0.695 and 0.685)

respectively. Maximum respondents have reported of less issues with the selected MSPs and the buying behavior has less influence on media type and catchy advertisement.

Conclusions

The present study aims to assess the respondents' socio-economic conditions and preference over particular MSPs (Jio, Airtel, and VI) and examine the quality parameters designed by the selected MSPs to influence their customers. This research adopts a multi-stage, random sampling method, taking the Jaipur district as a study region. Some of the critical discoveries of the research are being outlined.

1. The cost has a maximum influence on the selection of MSPs followed by communication, and innovativeness with α values of 0.817, 0.753, and 0.749. Internal consistency was observed amongst the group of variables related to the factors.

2. The variable (unlimited calls) has got maximum factor loading of 0.923 on cost. Likewise, call rates (0.911) on communication, reliability (0.838) on service factor, good coverage (0.817) on good coverage, customer complaint management (0.738) on customer care, customer-oriented promotion (0.882) on innovativeness and catchy advertisement (0.848) on the advertisement.

3. The variable call rates have maximum h^2 value of 0.936 which shows maximum influence on MSPs selection followed by unlimited calls with h^2 of 0.897 followed by roaming charges (0.873), ISD charges (0.859), and customer-oriented promotion (0.825).

4. Jio was the winner of Airtel and VI. The Airtel and VI companies should make the customer care services operational round the clock to cater to the customers' needs. The service providers should also upgrade their services to address the customers' needs across the world.

5. Poor connectivity or network issues should be addressed utilizing using the latest technology. Disconnection while talking is another problem encountered by most customers. This can be overcome with high-technology backups.

References

Hashem, T., Farah, J., and Hamdan, F. (2017). Measuring Service Quality Level in The Jordanian Telecommunication Sector from its customers' perspective using the SERVPERF Scale. European Journal of Business and Social Sciences, 5(12), 15–27.

Kumar, A. S. N., and Soundararajan, S. D. (2016). A Study on Customer Attitude Perception towards Branded Broad Band. International Journal of Interdisciplinary Research in Arts and Humanitie, 1(1), 133–136.

Mahalaxmi, K. R., and Kumar, S. N. (2017). A study on service quality and its impact on customer's preferences and satisfaction towards Reliance JIO in trichy region. International Journal of Advanced Education and Research, 2(3), 35–41.

Mannan, M., Mohiuddin, Md. F., Chowdhury, N., and Sarker, P. (2017). Customer satisfaction, switching intentions, perceived switching costs and perceived alternative attractiveness in Bangladesh mobile telecommunications market. South Asian Journal of Business Studies, 6(2), 142–160. https://doi.org/10.1108/SAJBS-06-2016-0049.

Matharou, G. S., and Bhuyan, B. K. (2020). Parametric Optimization of EDM Processes for Aluminum Hybrid Metal Matrix Composite using GRA-PCA Approach. International Journal of Mechanical and Production Engineering Research and Development 10, 367–378. https://doi.org/10.24247/ijmperdjun202034

Matharou, G. S., and Bhuyan, B. K. (2021). Modelling and combined effect analysis of electric discharge Machining process using response surface methodology, Materials Today: Proceedings, 146(15), pp.6638-6643. https://doi.org/10.1016/j.matpr.2021.04.103.

Refai, A., Nayef, A., and Mohammed, N. A. B. (2014). The Influence of the Trust on Customer Satisfaction in Mobile Phone Market: An Empirical Investigation of the Mobile Phone Market. International Journal of Management Research and Reviews, 4(9), 847–860.

Selelo, G. B., and Lekobane, K. R. (2017). Effects of Service Quality on Customers Satisfaction on Botswana's Mobile Telecommunications Industry. Archives of Business Research, 5(3), 212–228.

Shah, N. P. (2013). Customer Satisfaction of Samsung Mobile Handset users. Voice of Research, 2(3), 76–79.

Taber, K. S. (2018). The Use of Cronbach's Alpha When Developing and Reporting Research Instruments in Science Education. Res Sci Educ 48, 1273–1296. https://doi.org/10.1007/s11165–016-9602–2

Tandon, N., Kaur, S., and Matharou, G. S. (2020). Contemporary Trends in Education Transformation Using Artificial Intelligence, in Transforming Management Using Artificial Intelligence Techniques. CRC Press, pp. 89–103. Available at: https://doi.org/10.1201/9781003032410–7

Upadhyaya, R. C., and Sharma, V. (2012). Customer Satisfaction with Network Performance of BSNL and AIRTEL Operating In Gwalior Division (M.P.). IOSR Journal of Business and Management, 4(3), 18–21.

Watkins, M. W. (2018). Exploratory Factor Analysis: A Guide to Best Practice. Journal of Black Psychology, 44, 219–246https://doi.org/10.1177/0095798418771807

Awareness of Platform Skills Amongst the Faculty of Business Management, Pune, India

Prathamesh Nadkarni,,a *Priya Singh,b and Padmashri Chandakc*

aMIT Art Design & Technology University, Pune, India
bMIT Art Design & Technology University, Pune, India
cIndira Institute of Management PGDM, Pune, India
E-mail: *prathameshnadkarni21@gmail.com

Abstract

When it comes to enhancing a Post Graduate Student's employability skill in today's world of education, faculties need to be very much aware of the requirement of the corporates from the classroom learning. Student-Teacher cognitive approach to learning approach and participation in class is dramatically improving in day-to-day business school sessions. The different certification modules, leadership development modules, and communication skills training are inclusive in every Faculty Training FDP (Faculty Development Program). The most crucial aspect, however, has remained untouched to this day: post-graduate business management students' and Faculties are Platform Skills Training. Classroom learning is a process a student goes through in his tenure of education for a particular course. Faculty's platform skills in the field of management studies must be reviewed and developed. In terms of providing value to the audience, the distribution of knowledge or content is equally as crucial as the knowledge itself.

Keywords: Platform Skills, Faculty of Business Management, Presentation Skills, Faculty, Training, Trainer

Introduction

When it comes to enhancing a Post Graduate Student's employability skill in today's world of education, faculties need to be very much aware of the requirement of the corporates from the classroom learning.

Student-Teacher cognitive approach to learning approach and participation in class is dramatically improving in day-to-day business school sessions. The different certification modules, leadership development modules, and communication skills training are inclusive in every Faculty Training FDP (Faculty Development Program). The most crucial aspect, however, has remained untouched to this day: post-graduate business management students' and Faculties are Platform Skills Training.

Students have been observed being reviewed and promoted for campus selections depending on their exam and certification scores. However, Platform skills remain untouched as the Faculties and the Universities are unaware of the same, but the Corporates use it as an ultimate criterion to select he student. As discussed above, Classroom Content Delivery, Confidence, Voice Projection, Speaking Rate, and Facial Expression are the top 5 Elements Faculties need to get trained on to percolate it to the students. 16 Elements of the Platform Skills *(Qualifications Pack - Occupational Standards for Training and Assessment- IT/ITeS SSC NASCOM): -*

1. Volume
2. Pitch
3. Rate
4. Pause
5. Pronunciations
6. Filler Words
7. Attitude
8. Confidence
9. Controlling nervousness
10. Posture
11. Facial Expressions
12. Gesture
13. Body Movements
14. Personal Appearance
15. Voice
16. Speed
17. Classroom Content Delivery Principle

In numerous Training Programs done by the trainers, platform skills have demonstrated promising outcomes in effective learning and understanding of the audience's information.

Theoretical View for Platform Skills

Platform skills are presentation behaviors that a trainer employs to effectively transmit content. To get their messages across, both presenters and trainers must display great platform skills. This is not to be confused with skills that guarantee participation (which, in general, only trainers utilize). One intriguing area where training and presentation collide is platform skills.

Practical view for Platform Skills

- At the start of the session, communicate the topic, goal, and relevance of the session to the participants.
- Nervousness is effectively managed so that it does not distract the participants.
- Uses humor, analogies, examples, metaphors, stories, and non-lecture or PowerPoint delivery methods.
- In a "full-frontal" body position, he faces the participants the majority of the time.
- With vigor, summarizes and concludes the class.

As a result, on similar lines, Classroom learning is a process a student goes through in his tenure of education for a particular course. Faculty's platform skills in the field of management studies must be reviewed and developed. In terms of providing value to the audience, the distribution of knowledge or content is equally as crucial as the knowledge itself. Platform skills should be part of Faculty Training Programs.

Theoretical and practical knowledge is one of the most significant things a Post Graduate Student should have. Is this, however, sufficient knowledge for a student to operate as a trainer, instructor, or corporate employee in the corporate world?

Hence, the Inclusion of Platform skills training should be part of the curriculum and should be implemented by Post Graduate Students. It should be the part of Academic Calendar as the implementation can be a strict process to follow.

Review of Literature

- Sheri Staak (January 6, 2015) *"Tune In to Wow Leadership:10 Lessons Learned from America's Favorite Shows"*

This book gives an in-depth insight into the platform skills required for a trainer to conduct the session. It is a fascinating book that can help a researcher to explain the process to sharpen the platform skills of Post Graduate Students

- Consumer Dummies (April 2017) *"Career Development All-in-One For Dummies"* Chapter 2

This book helps a researcher to understand the evaluation rubrics of the teaching Platform skills and also the techniques required to enhance the platform skills of a Post Graduate Student.

- Kimberly Devlin (31st March 2017) *"Facilitation Skills Training"*

It helps a researcher to understand how Platform skills lay a foundation for any session. Workshops for reference are designed in the book that gives a fair idea about the enhancement criteria of Platform skills.

- Susan Meyer. (1992). *Cultivating Reflection-in-Action in Trainer Development.* Volume: 3 issue: 4, page(s): 16–31 Issue published: January 1, 1992

In this article, researcher has significantly put light upon how a trainer can impart professional values and attitudes in others by using the plat form skills.

- Donald A. Schön, (1989). *Educating the reflective practitioner in* the Journal of Continuing Education in the Health Professions. 9(2):115–116Publication Date: April 1989

In this article, according to the researcher, certain professional pieces of knowledge cannot be taught and can only be transmitted through the utilization of platform skills.

Ashok Ajmera (2016). *Share the wisdom* explains the importance of feedback in the very session a trainer has to take from the audience. Platform skills assessment is one of the important parameters that need to be evaluated.

Materials and Methods

Research Methodology

Research Design

Researchers used a survey research design to learn more about the current state of platform skills awareness among faculty members of business management institutions. The Instructor's Guide for the Platform Skills Development Model can be finalized based on the analysis of the responses.

Sample Size – 43

Sample Unit - Every Business Management Institute's Teaching Faculty in the Pune Region of India.

Sampling Technique – Researchers used *purposive sampling* in selecting the faculties of the institutions based on the Soft skills training provided for students as well as faculty members, and it is included in their academic calendars.

Data Collection Tool

As a data collection tool, a questionnaire was used. The responses were collected via the internet. The date from the samples was collected using a Google Form. The responses were collected using email as a medium. 43 business management faculties from 9 different institutions within respective universities responded to the questionnaire.

Structure of the Questionnaire

Total No of Questions – 15

Open-ended Questions – 7

Close-ended Questions – 8

Quantitative Analysis - Quantitative analysis of the Opinions are analyzed.

Types of Questions Asked in Questionnaire

Question No. 1 & 2	Multiple Choice Questions	Nominal Scale
Question No. 3 & 4	Open Ended Questions	Nominal Scale
Question No. 5	Close Ended Question	Nominal Scale
Question No. 6 7 8 & 9	Open Ended Questions	Nominal Scale
Question No. 10	Dichotomous Question	Nominal Scale
Question No. 11	Rating Scale Questions	Ordinal Scale
Question No. 12 & 13	Ranking Questions	Ordinal Scale
Question No. 14	Dichotomous Question	Nominal Scale
Question No. 15	Open Ended Questions	Nominal Scale

Theory Behind the Research

As a result of the Platform Skills Model, participants will be able to grasp an event and then alter it using the cycle of Concrete Experience, Reflection Observations, Abstract Conceptualization, and Active Experimentation. Hence, experiential research theory is the foundation of the research theory. *(Based on David Kolb's Experiential learning theory model in 1984).*

Finding & Interpretation

- Only 11.62 % of the faculties could define Platform Skills Right. Only 74.41% of Faculties have attended Platform skills workshops till date.
- The majority of faculties claim to incorporate delivery skills, pitch rates, and public speaking skills in their curriculum. However, based on the answers to Questions 1 and 3, we can deduce that the faculty assumed Platform skills were the same as communication skills.
- As shown in the table above the responses of Faculties were organized into five main categories: "Qualities," "Communication Skills," Domain Knowledge Skills," "Technical Skills" and "Platform Skill." Using Axial Coding the theme that arose from the responses was that "Qualities and communication skills were identified as the most important aspects that students should possess to be employable by the faculties. Also, a good deal of domain knowledge and technical skills are required.
- The responses of Faculties were divided into synonymized terms, as shown in the table above. The analysis that emerged from the responses, using ground theory, was that "the found lacunas could give clarity of the required training of Platform Skills that need to be given to business management students while performing on stage as well." This question was posed to get faculty feedback on the presenter's gap that students encounter while performing on stage.
- This Analysis could help us to find the cruel 5 Elements out of 16 platform skills that would be the base for the Model. As per the Average Scores, Top 5 Platform Skills elements are: - Attitude (4.56), Eye Contact (4.53), Classroom Content Delivery Principles (4.46), Confidence (4.40) & Voice Projection (4.40).

Recommendations

- For percolation of the training and effective classroom management, faculties should be made aware of Platform skills through pieces of training, guest sessions, activities, and other means and should be the part of Faculty Development Program. Also should differentiate the Communication Skills and Platform Skills.
- Platform Skills development should be the part of Academic Calendar.
- Platform Skills development Model should be designed for the top 5 elements of Platform skills with the concern of Faculties and Industry Experts and Platform Skills Development Model Instructor's Booklet should be designed (which is in process).
- After attending a Platform Skills Development Program, an expert team should assess the Pre and Post Changes in the faculties.

Conclusion

As per the objectives are satisfactorily tested with the respondents mentioned above, I can conclude this Paper by stating that Faculties of Business management lack knowledge about Platform Skills and are not aware of the term. As Platform Skills is not part of the curriculum, adverse effect of the same is seen on the Placements of the Students. Once the Platform skills development Model is designed for the students, Instructor's guide would be designed for the same. Faculties will also be given primary training on how to conduct the Model. Long-term goal is to apply the same for the inclusion of the Model in the National Education Policy.

References

Adult Learning - Volume 3, Number 4, Jan 01, 1992. SAGE Journals. (n.d.). https://journals.sagepub.com/toc/alxa/3/4.

When a Trainer is a Presenter: 5 Top Platform Skills: Guila Muir. Guila Muir | Developing trainers, presenters and facilitators to make a difference. (2015, May 9). https://www.guilamuir.com/when-a-trainer-is-a-presenter-five-top-platform-skills/.

Flipgrid. Flipgrid. (n.d.). https://flipgrid.com/

(n.d.). TSSC Telecom Sector Skill Council. https://www.tsscindia.com/media/2566/trainer-platform-skills.pdf

Manolis, C., Burns, D. J., Assudani, R., and Chinta, R. (2013). Assessing experiential learning styles: A methodological reconstruction and validation of the Kolb Learning Style Inventory. Learning and Individual Differences, 23, 44–52. https://doi.org/10.1016/j.lindif.2012.10.009

Yeoman, K. H., and Zamorski, B. (2008, May 31). Investigating the impact on skill development of an undergraduate scientific research skills course. Bioscience Education e-Journal. Retrieved August 20, 2022, from https://eric.ed.gov/?id=EJ835795

A research on blogging as a platform to enhance language skills. (n.d.). Retrieved August 20, 2022, from https://www.researchgate.net/publication/248607376_A_research_on_blogging_as_a_platform_to_enhance_language_skills

Semiotic Analysis in Select Advertisements on Nature and Sustainability: Green or Social?

R. Sathvika[*,a] *and V. Rajasekaran*[b]

[a]Research Associate, School of Social Sciences and Languages, Vellore Institute of Technology, Chennai, India
[b]Assistant Professor Senior, School of Social Sciences and Languages, Vellore Institute of Technology, Chennai, India
E-mail: [*]sathvikasanju@gmail.com

Abstract

The paper aims to examine the impact of sustainability advertising on Tupperware products, it seeks attention to know the different effects of environmental and social sustainability. Descriptive content analysis provides insight through the theories of semiotics in environmental advertising messages. The theoretical analysis of Saussure and Barthes helps in investigating the sign language system. The study is based on two select advertisements for Tupperware products. Thus, the findings suggest focusing on sustainable environmental symbolic aspects conveys the effectiveness of advertising. Future studies can be carried out using the same analytical process to gain deep knowledge in semiotics through sustainable advertisements.

Keywords: Advertisement, Sustainability, Semiotics, Nature, Saussure, and Barthes

Introduction

Our Universe has faced increasing levels of environmental issues, namely pollution, ocean pollution, climate change, air pollution, extinction of animals, deforestation, and many to name as well. After the pandemic period, people are more cautious and concerned about the environmental problems in our country. People assume that the significant role of green advertising is the way it communicates pro-environmental measures or tackles environmental sensitivity issues. It is common to see images used in sustainable advertisements. The semiotic signs and symbols are portrayed in such a way. Recent green advertising research pays focused attention to the effectiveness of advertising about environmental problems and the way it conveys to the people (Song and Luximon, 2019). Furthermore, advertising has various visual techniques that play a vital role in semiotics analysis.

Similarly, Thushari and Senevirathna (2020) examined a study that focuses on the aquatic ecosystem and how plastic pollutants contaminate the coastal and marine environments. Microplastics play a crucial role in the aquatic ecosystem. Plastic pollutants are spread in different forms of the aquatic ecosystem that destroy the life of the organisms in aquatic bodies. The author also gives intense caution and is aware of the sources of plastic pollutants, the small organisms try and challenge its survival through the chaos created in the marine ecosystem. Contrastingly, on the other side Song and Luximon (2019) have done an exploratory study with concern for environmental sustainability through a green advertising perspective. The author explains the effectiveness of green and sustainable advertising and people's perceptions that concerns environmental change. The study also supports the author's study that advertising taglines act as signs and symbols that could enhance people's awareness.

Research Objectives

The objective of the study is to examine the semiotic techniques present in environmental advertising. Also, to explore the content in the ad that can bring out the change in sustainable advertising. Further, to investigate the perceptions of green and sustainability in select advertisements.

Research Questions

- What are the perceptions of green and sustainable advertising in select advertisements?
- How do semiotic techniques convey the meaning of the image portrayed in environmental advertisements?
- How does the content in the ads create a change in sustainable advertising?

Rationale to Choose Tupperware Products That Support Environmental Sustainability

"The fruits of life fall in the hands of those who climb the tree and pick them." Says Earl Tupper who is the inventor of Tupperware products. In the year 1946, Tupperware products planted the seed into the world and it's been more than 70 years still it is lively across the world. Tupperware operations were started in India from the year 1996. It then launched Aqua safe bottles in 2009 which created a nationwide sensation. The tagline #tupperwareforlife states the sustainability of the product as well. As the product, cares for the planet and about our people and they have taken its commitment to provide a sustainable solution to single-use plastic pollution. These products are reusable and durable, and they also provide a lifetime warranty. The Tupperware brand has three strategies to maintain the brand's environmental and social sustainability changing lives, Living smart, and Acting responsibly. There was another big reason as this company has collaborated with the United Nations Sustainable Developmental Goals (SDGs) approved in 2015 and set 17 global priorities towards achieving peace and prosperity by 2030.

Significance of the Study

The present study explores the semiotic terminologies such as signifier and signified present in the select Tupperware advertisement. It further illustrates the sustainable pictures, and messages and creates awareness for the people. The benefit of knowing the sign language system where one can interpret different meanings from a single sign or symbol in any advertisement. The commoner will enhance awareness and caution through sustainable environment advertisements. Furthermore, it is effective for the audience to know the sustainability of the environment and the eco-system of our planet. This environmentally sustainable advertising will create an impact in the minds of the readers to develop a better environment to protect our planet Earth.

Analytical Framework

The present study comes under qualitative research that can be further tuned into the methodology of descriptive content analysis which helps to uncover the meanings present in terms of signs and symbols, taglines, captions, color, or background picturized in environmentally sustainable advertisements. The theories of Ferdinand de Saussure and Roland Barthes' in semiotics can be imposed to enhance our knowledge of a sign language system. The proposed theory was discussed in the paper by the researcher attempts to explain, investigate, and analyze the meanings of each sign or symbol used in the advertisement. We can broadly predict the verbal and non-verbal language used in the advertisement. The non-verbal language connecting the visual signs or the visual symbolic representations plays a vital role in understanding the denotative and connotative meanings present in the advertisement as suggested by the theory of Roland Barthes (1977) in his book Rhetoric of the image. Saussure (1916) admits that any linguistic sign can have only a signifier and a signified.

Analysis

Marine pollution is a problem that has never been settled in recent times. Our ocean is being flooded with lots of chemicals and trashes and most of them plastic litter. The sustainable development goals in advertising depend on the product for which it is being advertised. This was in the wake of green advertising, where the advertiser's used the environment as a medium tool for marketing. Meanwhile, it is believed that sustainable advertising endorses the wider knowledge of a better future and creates awareness in society. The Marine and coastal environment plays a major role and is a highly productive zone that combines different kinds of subsystems (Thushari, 2020). The marine environment is the vast area coverage in a water body that covers 71 percent of the earth's coverage proposed by the author Thushari (2020) in his study. In particular, research stated by national geographic encounters that common types of marine debris include plastic items like shopping bags, beverage bottles, food wrappers, and bottle caps. The single-use plastic can take thousands of years to decompose. This trash seems to be dangerous for both humans and animals. The small bits of broken plastic termed "microplastic" are been consumed by the small organisms living in the ocean. These microplastics are less than five millimeters when consumed by small organisms the toxic chemicals are part of their tissues. Microplastic pollution enters the food chain of these organisms and they also become part of our food when we consume fish as seafood.

The Tupperware product uses sustainable advertising as a tool to enhance awareness. The Tupperware product is known for its environmental sustainability as they use recycled plastics and wastes to make the best out of it. Other wastes which cannot be used for recycling purposes are sent to make other products out of it like flower pots. The semiotics sign language executes signifier and signified that is used to depict the image in advertising. The denotative and connotative meanings can be identified through Barthes' theory of semiotics. Few attributes were

encountered in analyzing the image portrayed in the advertisement. The detail of the attributes used is illustrated in the table below.

Table 1. Attributes used for advertisements

Attributes used in the Select Advertisements	
Attribute	Description
Color	Dark shades- deep blue, deep ocean
Space	Background takes up the whole product
Font	Brand's tagline is bold and highlighted, and descriptions and social messages are in the running text
Product display	In the center of the ad, a sea turtle image on the top of the product
Position/way of projecting	Center positioning
Light	Dark mode, night lighting
Real or cartoon-based picture	2D image and product picture
Identity	Tupperware water bottle, ocean pollutants, sustainable terms
Creativity	Sustainable awareness is created using ocean pollutants and sustainable words
Visual and verbal connect	Picturization with taglines

The researcher utilized Tupperware product - water bottle two of its advertisements that focuses on environmental sustainability are retrieved from the Outlook magazine issued in the year 2019. As shown in figure 1, we can observe that ocean picturization along with sea turtles trying hard to survive in the midst of ocean pollutants. The oceanic picture can be divided into signs and symbols in terms of signifiers and signified and the denotative and connotative meanings displayed in the advertisement. From the advertisement image, it is observed the signifier or the denotative meaning is the ocean pollutants created by humans which are destroying the living organisms in aquatic bodies. Firstly, the ocean is the signifier and the iconic symbol. The secondary sign is the Tupperware bottle on top of the product image of a sea turtle is been imposed. The signified that can be witnessed in the advertisement is the marine pollution from the trashes created by humans. More in particular why a sea turtle image has been imposed on top of the product because the signified explores in detail that sea turtles are the ones who consume single-use plastic often imagining it as jellyfish. As the single-use plastic has the forms, shapes, and texture same as the jellyfish the poor creature is not able to find the difference between the two of it. These meanings are considered to be the connotative ones and the signified element from the advertisement. Also, once they consume it they find it difficult the growth or development of their lifespan. They tend to suffer in their internal organs and rupture intestinal blockages leaving the turtles unable to feed and that ends up in starvation. The aftermath of consuming the microplastic if they survive makes them unnatural and puts a stoppage for their growth that further results in slow reproduction.

The set-up seems to be night mode which gives the denotative meaning of the background whereas, the connotative meaning is the color of the background seems to be dark deep blue ocean color, night light, and the background is surrounded by single-use plastic pollutants. The tagline of the brand Tupperware is #tupperwareforlife, which intense the sustainability of the product as well as creates awareness among the people. The caption which is a verbal sign or the iconic sign signified from a semiotics perspective is highlighted in the bolded text "To Protect & Preserve our Earth" which creates a necessary awareness about environmental sustainability. As Tupperware products are known for their durability and sustainability since the products are re-useable with a lifetime warranty. The description explains the sustainable advertising in the advertisement given by Tupperware. It further has a caption that "confidence becomes you," where the product Tupperware gives confidence about environmental sustainability as they do not harm the environment. As Tupperware products are not made of single-use plastic, plastic can be reusable in the future.

Table 2. Semiotics Attributes of Advertisement

Sign	Signifier	Signified
Background	Ocean pollutants	Pollution created by humans
Animal	Sea turtle	Eats plastic solid waste "marine debris"
Object	Water bottle	positioning focused
Text	Tagline & caption	Sustainable products and awareness created

The second advertisement for the same product Tupperware water bottle is entirely contrasting because the select advertisement does not have any image nor background or any set-up in particular but the verbal connection that comprises exclusively sustainable words given in the shape of the water bottle. These sustainable words as verbal language are considered to be the signifier and the iconic symbol of the entire advertisement. The words used in the advertisements that depict environmental sustainability are "recyclable, reusable, durable, lifetime warranty, eco-friendly, green earth, less waste, safe-to-use, non-toxic, eco-friendly, chose wisely, and rethink" and are given designed in the shape of the bottle. The signifier or the denotative meanings are the verbal connection that is illustrated and seems to be iconic in the advertisement. The signified is a bottle full of sustainable words that have been brought into advertising that fills up the shape of a bottle. The particular advertisement where Saussure induces the meaning of arbitrary can be analyzed in the select advertisement. Although, the arbitrariness present in the randomly selected pictures or the signs and symbols depicted in any advertisement when is it seen in sustainable advertising it is very much effective from the reader's perspective. The signifier is that Tupperware products always contribute to a greener environment. The connotative meaning lies in the sustainable words projected in the shape of a water bottle with a blue color bottle cap to give it a complete image of the bottle. The connotative meaning or the iconic signified is how it justifies the environmental sustainability from the product's taglines when you purchase Tupperware you can be confident and promising that the product stays for a lifetime. Further, it contributes to a greener environment for the betterment of the future.

Conclusion

The marine and coastal ecosystems have to be considered as high rate of risk that can harm our planet Earth. Plastic pollutants cause an ecological imbalance in the marine ecosystem where innocent small organisms are been affected by high-risk factors of their life that might even lead to destroying their life of it. The semiotic theoretical analysis opens the way to understanding the systematic representation of both sign and signifier of sustainable advertisement that serves the purpose of creating awareness to the commoner and the caution to act at the right time.

Finally, the study reveals the overall current scenario in marine ecosystems and the single-use plastic pollutants that eradicate the small organisms in the ocean. Sustainable advertising highlights the awareness established from the Tupperware products of green advertising whilst the semiotics sign language plays the unity role in analyzing furthermore for attaining clarity in the sign language and the sustainable advertisements.

References

Barthes, R. (1977). Rhetoric of the image. New York: Hill and Wang.

Cummins, S., Reilly, T. M., Carlson, L., Grove, S. J., and Dorsch, M. J. (2014). Investigating the Portrayal and Influence of Sustainability Claims in an Environmental Advertising Context. Journal of Macromarketing. 34(3), 332–348. https://doi.org/10.1177/0276146713518944

De Saussure, F. (1916). Nature of the linguistic sign. Course in general linguistics, 1, 65–70.

Roman, L., Schuyler, Q., Wilcox, C., and Hardesty, B. D. (2021). Plastic pollution is killing marine megafauna, but how do we prioritize policies to reduce mortality?. Conservation Letters, 14(2), e12781.

Song, Y., and Luximon, Y. (2019). Design for sustainability: The effect of lettering case on environmental concern from a green advertising perspective. Sustainability, 11(5), 1333. https://doi.org/10.3390/su11051333

Thushari, G. G. N., and Senevirathna, J. D. M. (2020). Plastic pollution in the marine environment. Heliyon, 6(8), e04709.

Factors Affecting Organic Food Purchase Intention: A Signaling Theory Perspective

Dipandi Mishra,,a *Alok Tewari,*b *and Amarendra Pratap Singh*a

aIndira Gandhi National Tribal University, Amarkantak, India. bDr. A. P. J. Abdul Kalam Technical University, Lucknow.
E-Mail: *dipandimishra2013@gmail.com

Abstract

The current paper is directed toward understanding how the Signaling theory could work as a theoretical model in understanding consumers' purchase intention toward organic foods. A descriptive quantitative study was administered with the contribution of 228 valid responses. These responses were gathered by distributing an online questionnaires to organic food consumers. Structural equation modeling (SEM) was applied for testing the theoretical model. The impact on brand image identification, sensory attributes, and brand credibility as signalers has been tested on the purchase intention of consumers of organic food. User Generated Content (UGC) has been used to extend the signaling theory used in the current study. The results show that sensory attributes, brand credibility, and UGC strongly correlate with intention to purchase organic food. In contrast, the brand image does not strongly impact organic food purchase intention.

Keywords: Signaling Theory, Organic Food, Brand Credibility, Brand Image identification, Sensory appeal, User Generated Content, Purchase Intention

Introduction

Recently, individuals have become aware of their environmental conditions and pro-environmental movements. This has led people to monitor their consumption patterns; hence, a new market segment is known as the "green market" has evolved, also known as the green consumers' market (Peattie, 2010). Organic food is a prominent part of the green market. U.S. Agriculture Department states "organic foods" as the foods produced using farming techniques that help preserve biodiversity, safeguard natural resources, and use only approved substances. Therefore, pesticides are not used in the production of Organic Foods (Seufert et al., 2017). No growth hormones or antibiotics should be fed to organic foods obtained from animals like eggs, meat, and dairy products (Seufert et al., 2017). So, organic foods are environmentally safe.

Consumers understand the message of marketers when it is believable and relevant. When signals are powerful, the skepticism of consumers is conquered regarding products. In business, marketing is an ingrown domain for implementing signaling theory (Dunham, 2011). There have been studies in the past which have contributed to the advancement of consumer behavior by incorporating an understanding of signaling theory (Griskevicius et al., 2010). Therefore, understanding the signals' forms and the reliable methods from which these could be transmitted brings clarity, appropriateness, and efficiency to the mind of consumers regarding the products (Dunham, 2011).

The current study, therefore, uses sensory appeal, brand credibility, and brand image as signals in the mind of consumers to purchase organic foods. User-generated content is used to extend this signaling concept.

Literature Review

Signaling Theory

The signaling theory is a system that removes confusion when two parties share information (Spence, 2002). The functionality of signaling theory is that signalers are insiders (brand, sensory appeal, costliness, etc.) (Dunham, 2011) who have access to information/signals regarding any product (Kirmani and Rao, 2000). It is then determined whether to transmit this knowledge (say about organic food) to outsiders who are receivers of this information (consumers in case of marketing) who require this information related to the product in which they are interested and would prefer to acquire this information (Spence, 2002).

Brand Credibility

Brand credibility indicates the trustworthiness of a brand (Gong W, 2020). Brand credibility has been described as the believability of the product information enclosed in a brand, which helps consumers recognize that the brand has the expertise (e.g., ability) and trustworthiness (e.g., intent) to constantly offer what has been pledged (Martin-Consuegra et al., 2018). Thereby, consumers identify the brand's credibility by engaging in an appraisal mechanism that outlines the cumulative credibility of a brand (Jun, 2020). According to signaling theory, the proficiency and sureness echoed by the brand are responsible for creating its credibility in consumers' view. This phenomenon involves the aggregated influences attached to the brand's past and present communicated commitments and marketing strategies that symbolize credibility to the consumers (Spry et al., 2011). And hence the more consumers are aware of brand credibility, the more their inclination toward purchasing organic foods which they consider healthy. Therefore, we postulate:

H1: Brand Credibility impacts the Organic Food purchase intention.

User Generated Content

It is a form of unpaid marketing where users promote the brand instead of the brand itself (Ngah et al., 2021). Firms typically ask users to rate their service encounters after a service is consumed as it helps in service improvements and can create positive or negative user-generated content which is a better measure of user satisfaction (Kar, 2020). Mathur et al. (2021) have reported user-generated content to be a strong predictor of purchase intention. Hence, the current study hypothesizes:

H2: User Generated Content affects the consumers' intention to buy Organic Food.

Sensory Appeal

The appeal to receiver psychology, through either manipulating evolved preferences or by reducing skepticism, is the crux of signaling theory (Dunham, 2011). Within the business realm, attention to evolved preferences that capitalize upon sensory bias can guide both product design and marketing efforts (Saad, 2006). appeals to receiver psychology through the five senses may equip marketers with tools derived from evolutionary psychology to better position products and attract consumers (Dunham, 2011). Organic foods have vision, smell, and taste better than contemporary food that uses insecticides and pesticides for their production. In addition to the health benefit, Thøgersen et al. (2015) highlight the importance of flavor in the attitude of consumers that buy organic products. Roitner Schobesberger, Darnhofer, Somsook, and Vogl (2008) performed research in Bangkok, which concluded that together with consideration for a healthier option, organic food is attracting consumers due to it being trendy and "tastier." Sensory Appeal (taste, appearance, etc.) has a positive relationship with the organic food purchase intention (Thøgersen et al., 2015; Lee and Yun, 2015). Therefore, our study postulates:

H3: Sensory Appeal presented by Organic Food impacts the consumers' intention to purchase.

Brand Image Identification

In the context of food marketing, ample research has been carried out that affirms the brand image drives the consumers' inclination towards purchasing organic food (Jaeger and Weber, 2020) genetically modified food, and bio-fortified food (Adeyeye and Idowu-Adebayo, 2019). The literature suggested that a favorable brand image helps food products gain a competitive advantage and purchase intention (Hien et al., 2020).

H4: Brand Image Identification impacts the Organic Food Purchase Intention.

Organic Food Purchase Intention

Ajzen (2011) reported that the most vital factors that decide consumers' actual purchase of a particular product are the consumers' intention. Spears and Singh (2004) refer to consumers' purchase intention as "an individual's conscious plan to make an effort to purchase a product." Organic food purchase intention is defined as the health-conscious consumers' readiness and willingness to purchase organic food products articulated by the consumers for the friendly environment and health benefits (Dagher and Itani, 2014).

Methodology

Details of the Questionnaire Distributed

The development of the questionnaire for the current study was through sourcing items from

applicable literature. Brand Credibility was measured using four items from Erdem and Swait (2004). Purchase intention was assessed using four items from Tewari et al. (2022a,b). User Generated Content was measured utilizing six items from Kim and Johnson (2016). Sensory Appeal was measured with the help of three items from Lee and Yun (2015). Brand Image was assessed using three items from Davis et al. (2009).

Data Collection

To approach the required population of the study purposive sampling technique was employed. Consumers who were familiar with green organic foods were selected to respond to the questionnaire. All the items were measured on a 7-point Likert scale in which "1" was assigned with strong disagreement and "7" was assigned strong agreement with the measured item.

The web-based questionnaire was distributed through online mode to 280 consumers, out of which 241 responses were received. After eliminating incomplete responses, a final sample of 228 was used for further analysis. Our study had 20 items therefore a minimum sample of 200 was required according to the criterion of 10 responses per item (Kline, 2011). Thus, the size of sample 228 was sufficient for this study

Table 1. Factor Loading, Reliability, and Validity

Constructs	Indicators			Std. Factor Loading	CR	AVE	Cronbach's alpha (α)
Brand Credibility (bc)	bc1	<---	BC	0.76			
	bc2	<---	BC	0.862			
	bc3	<---	BC	0.842	0.879	0.647	0.876
	bc4	<---	BC	0.745			
User Generated Content (ugc)	ugc1	<---	UGC	0.741			
	ugc2	<---	UGC	0.737			
	ugc3	<---	UGC	0.749	0.896	0.59	0.895
	ugc4	<---	UGC	0.817			
	ugc5	<---	UGC	0.783			
	ugc6	<---	UGC	0.779			

Sensory Appeal (sa)	sa1	<---	SA	0.799			
	sa2	<---	SA	0.652	0.819	0.604	0.811
	sa3	<---	SA	0.866			
Brand Image (bi)	bi1	<---	BI	0.736			
	bi2	<---	BI	0.756	0.794	0.563	0.793
	bi3	<---	BI	0.759			
Purchase Intention (pi)	pi1	<---	PI	0.872			
	pi2	<---	PI	0.77	0.891	0.673	0.89
	pi3	<---	PI	0.854			
	pi4	<---	PI	0.78			

Data Analysis

AMOS 23 and SPSS 23 software were employed for data analysis. The normality of data was tested using Skewness and Kurtosis indices which were inside the range of ±3 and ±10, respectively, implying that the data set was typically distributed (Kline, 2011). To check the internal consistency of the measurement items, Cronbach's alpha values were calculated. The alpha values of all the constructs were above 0.7 thresholds which is recommended (Nunnally and Bernstein, 1994).

Measurement Model

The measurement model displayed an excellent fit through confirmatory factor analysis with values CMIN = 262.833; CMIN/df = 1.643; DF = 160; CFI = 0.958; SRMR = 0.049; RMSEA = 0.053. The estimates for the subscales of reliability and validity exceeded the cutoff criteria. For all the

constructs, factor loadings were discovered above the 0.6 mark. The values of composite reliability were above 0.7. Good convergent validity was also found with the help of AVE values above 0.5. A comparison was conducted between the square root of AVE and factor correlations to find discriminant validity (see Table 1). The obtained value of the former exceeded the latter, reflecting adequate discriminant validity (Fornell and Larcker, 1981).

Structural Model

The model fit was found to be good with the values CMIN = 262.833; CMIN/df = 1.643; DF = 160; CFI = 0.958; SRMR = 0.049; RMSEA = 0.053. These fit statistics were found with the help of CFA executed on a structural model.

Table 2. The path coefficients of the structural equation model

Paths	Path Coefficients (β)	t-value	p-value	Hypotheses
BC → PI	0.356	5.421	***	H1-Supported
UGC → PI	0.39	6.103	***	H2 -Supported
SA → PI	0.203	3.069	0.002	H3-Supported
BI → PI	0.122	1.748	0.081	H4-Not Supported

***p<0.001

Findings and Discussions

Hypothesis H1, H2 & H3 were supported (see table 2). The consumers were influenced by the smell, vibrant colors, and flavor/ taste of the organic foods. They found that organic foods have

a different overall appeal than conventional food which directs them to consider organic food as a better food choice and hence they are encouraged to buy these foods.

Brand credibility is important for consumers. They feel when it comes to foods which are impacting their health, they can only trust credible brands, and therefore when they have brand choices that are more authentic and credible, they are willing to purchase organic foods. This proves that brand credibility helps create positive signals in consumers' minds regarding organic food.

User-generated content through positive reviews, e-wom, usage experience, etc., of organic foods helps consumers to reach the required organic food. It is helpful for consumers to learn about nearby availability, quality, prices, experience, and pleasant and unpleasant aspects of organic foods. This influences consumers to create an understanding of organic foods, which ultimately helps them to intend to purchase organic food.

The brand image does not impact consumers to buy organic foods. This might be because when it comes to purchasing organic foods, the brand's image does not matter to them. Instead, they consider brand credibility influencing them to buy organic food.

Implications

The current study has tried to contribute to marketing experts and academicians. The study will help marketers to formulate strategies that could function as signals for consumers while making purchasing decisions regarding organic foods. Academicians and scholars can further explore and extend signaling theory to understand consumers' purchase intentions.

This study is one of the latest and original contributions toward organic foods.

Conclusions

In the overall evaluation of the model, the result showed that brand credibility and sensory Appeal could be used as tools to increase the purchase intention of organic foods. UGC has the most vital relationship with purchase intention ($\beta = 0.39$). Firms should also improve UGC as it works as feedback in signal theory. Brand credibility also has a strong relationship with purchase intention ($\beta = 0.356$). Sensory Appeal impacts purchase intention ($\beta = 0.203$). Although, brand image is insignificant in affecting the purchase intention of organic foods ($\beta = 0.122$).

References

Ajzen, I. (2011). The theory of planned behaviour: Reactions and reflections. Psychology & health, 26(9), 1113–1127.

Dagher, G. K., and Itani, O. (2014). Factors influencing green purchasing behaviour: Empirical evidence from the Lebanese consumers. Journal of Consumer Behaviour, 13(3), 188–195.

Davis, D. F., Golicic, S. L., and Marquardt, A. (2009). Measuring brand equity for logistics services. The International Journal of Logistics Management.

Dunham, B. (2011). The role for signaling theory and receiver psychology in marketing. In Evolutionary psychology in the business sciences (pp. 225–256). Springer, Berlin, Heidelberg.

Erdem, T., and Swait, J. (2004). Brand credibility, brand consideration, and choice. Journal of consumer research, 31(1), 191–198.

Fornell, C., and Larcker, D. F. (1981). Structural equation models with unobservable variables and measurement error: Algebra and statistics.

Gong, W. (2020). Effects of parasocial interaction, brand credibility and product involvement on celebrity endorsement on microblog. Asia Pacific Journal of Marketing and Logistics.

Griskevicius, V., Tybur, J. M., and Van den Bergh, B. (2010). Going green to be seen: status, reputation, and conspicuous conservation. Journal of personality and social psychology, 98(3), 392.

Hien, N., Phuong, N., Tran, T. V., and Thang, L. (2020). The effect of country-of-origin image on purchase intention: The mediating role of brand image and brand evaluation. Management science letters, 10(6), 1205–1212.

Jaeger, A. K., and Weber, A. (2020). Can you believe it? The effects of benefit type versus construal level on advertisement credibility and purchase intention for organic food. Journal of Cleaner Production, 257, 120543.

Jun, S. H. (2020). The effects of perceived risk, brand credibility and past experience on purchase intention in the Airbnb context. Sustainability, 12(12), 5212.

Kar, Arpan Kumar (2020). What Affects Usage Satisfaction in Mobile Payments? Modelling User Generated Content to Develop the Digital Service Usage Satisfaction Model. doi:10.1007/s10796-020-10045-0

Kim, A. J., and Johnson, K. K. (2016). Power of consumers using social media: Examining the influences of brand-related user-generated content on Facebook. Computers in human behavior, 58, 98–108.

Kirmani, A., and Rao, A. R. (2000). No pain, no gain: A critical review of the literature on signaling unobservable product quality. Journal of marketing, 64(2), 66–79.

Kline, R. B. (2011). Convergence of structural equation modeling and multilevel modeling.

Lee, H. J., and Yun, Z. S. (2015). Consumers' perceptions of organic food attributes and cognitive and affective attitudes as determinants of their purchase intentions toward organic food. Food quality and preference, 39, 259–267.

Nunnally, J. C., and Bernstein, I. (1994). Elements of statistical description and estimation. Psychometric theory, 3(127).

Peattie, K. (2010). Green consumption: behavior and norms. Annual review of environment and resources, 35(1), 195–228.

Roitner-Schobesberger, B., Darnhofer, I., Somsook, S., and Vogl, C. R. (2008). Consumer perceptions of organic foods in Bangkok, Thailand. Food Policy, 33(2), 112–121.

Saad, G. (2006). Applying evolutionary psychology in understanding the Darwinian roots of consumption phenomena. Managerial and Decision Economics, 27(2-3), 189–201.

Seufert, V., Ramankutty, N., and Mayerhofer, T. (2017). What is this thing called organic?–How organic farming is codified in regulations. Food Policy, 68, 10–20.

Spears, N., and Singh, S. N. (2004). Measuring attitude toward the brand and purchase intentions. Journal of current issues & research in advertising, 26(2), 53–66.

Spence, M. (2002). Signaling in retrospect and the informational structure of markets. American economic review, 92(3), 434–459.

Spry, A., Pappu, R., and Cornwell, T. B. (2011). Celebrity endorsement, brand credibility and brand equity. European journal of marketing.

Tewari, A., Mathur, S., Srivastava, S., and Gangwar, D. (2022a). Examining the role of receptivity to green communication, altruism and openness to change on young consumers' intention to purchase green apparel: A multi analytical approach. Journal of Retailing and Consumer Services, 66, 102938.

Tewari, A., Srivastava, S., Gangwar, D., and Verma, V. C. (2022b). Young consumers' purchase intention toward organic food: exploring the role of mindfulness. British Food Journal, 124(1), 78–98.

Thøgersen, J., Barcellos, M. D., Perin, M. G., and Zhou, Y. (2015). Consumer buying motives and attitudes towards organic food in two emerging markets: China and Brazil. International Marketing Review, 32(3/4), 389–413.

Towards Sustainable Business: Review of Sentiment Analysis to Promote Business and Well-Being

Mausumi Goswami[*,a] *and Arpit Sharma*[a]

[a]CHRIST (Deemed to be University), Bangalore, India
E-mail: *mausumi.goswami@christuniversity.in

Abstract

Sustainability in business is expected considering the growth in the long run. Sustainable development goals are important for our sustainability on this planet. In case of a business, it is essential to ensure sustainable processes and sustainability of the existing customers. Sustainable customers can in turn contribute to improving the process by providing constructive suggestions to the business. This paper is an attempt to review sentiment analysis techniques to improve the customer experience of a business.

Keywords: Sentiment Analysis, Business Process, User experiences, User Reviews, Sustainability, SDG goals

Introduction

Sentiment Analysis is about analyzing the user reviews about a brand. The user reviews can be positive, negative or neutral. Sentiment analysis provides thorough information correlative to public views, to research different types of tweets and reviews of various domains. Sentiment analysis may provide thorough information related to user experiences, public views, and opinions collected through web mining-based research from differing types of tweets and reviews of various domains. Concerning the latest industrial developments contemporary development sentiment analysis using deep learning may be utilized. There is extensive use of social media on the web e.g., reviews, forum discussion, blogs, Twitter, Facebook, and other social networks. The solution to many commonly asked questions is obtained through mining the enormous amount of knowledge set of various social networking sites. Studying and investigating various procedures of the business concern to achieve some new business insight is an important objective. To achieve this objective various machine learning and natural-language processing-based approaches had been used in past but presently the researchers are preparing to use sentiment analysis using deep learning. This research attempts to investigate the essential knowledge based on the application of Deep learning and informative study about the current use of sentiment analysis techniques. Eventually, it gives an outline of deep learning before expanding on to a detailed search of its current applications in sentiment analysis. We believe that, just like deep learning research and applications evolve, there would be more compelling transfer learning development for sentiment classification in the nearest future (Zhang et al., 2018). Deep Learning and CNN models are very popular models to predict sentiments related to financial transactions and CNN also wants to demolish data processing solution to advance toward the available sentiment. Deep Learning models are altered to boost sentiment analysis efficacy on Chart analysis. It is also reported that the convolutional neural network outperforms logistic regression in sentiment analysis. Major Intention of describing (Sohangir et al., 2018) results is to find the right thought for the relevant sites which are to be explored. In this volume, collaborative filtering is described as solving problems. Authors in Liu et al. (2012) survey highlighted the importance and repercussions of sentiment analysis concerns in sentiment evaluation. Based on two comparisons within forty-seven studies, it is concluded that the structure of sentiment review and the obstacles of sentiment classification are largely determined by the subject. The optimum obstacles for evaluation sentiment ratings are dictated by subject structure and review design. Hussein et al. (2018), in this publication, a different deep convolutional neural network is introduced for trend analysis of textual information. The method has been applied to two different corpora: the Stanford Sentiment Treebank (SSTb) and the Stanford Twitter Sentiment corpus (STS) (Dos Santos et al., 2014). Deep learning algorithm for sentiment analysis of tweets is demonstrated. Main contribution of this research is a structured technique for instantiating the deep convolution network's characteristic weights. It is anticipated to use the deep learning technique in the future for even more Information Retrieval applications, such as learning to rank for Microblogging retrieval and answers reranking (Severyn et al., 2015). The neutrality of a phrase is governed not only by the content but also by its concerned aspect. For element sentiment classification, an Adulation Long Short Attention Span Network is proposed. When unique features of a sentence are taken as received input, the proposed technique can concentrate on different segments (Wang et al., 2016). In this dissertation, an interesting alternative to a specific target integral part of mood research is demonstrated. A long selective memory (LSTM) network with a modular learning algorithm is explained. Later part of training of a deep neural network for emotion classification leverages common sense knowledge of notion variables. Category of Research Questions: In this section, various categories of research questions related to Sentiment Analysis are explored. A few categories are mentioned below:

Category 1: Objectives of Natural Language Processing and Sentiment Analysis, Category 2: Challenges of Sentiment Analysis, Category 3: Applications of Sentiment analysis, Category 4: Algorithms and techniques utilized in Sentiment Analysis. All these categories are demonstrated in the next paragraph in detail.

Category 1: The specific and unique objective of natural language processing is to make computer devices like human beings so that they can perceive, emphasize and decipher different text and sentences like the human being. But we are focusing on the principle of sentiment analysis and sentiment analysis is an approach that comes under the natural language processing domain so the effective objective of sentiment analysis is to determine the outcome in the form of a review like positive feedback and negative feedback based on the variety of data. Now let us discuss what are the major features of sentiment analysis the first feature of sentiment analysis is the speed in which there is any computerization system and it can access a large amount of data so based on the speed it can obtain the results or patterns. The second feature is the efficiency with which the system generates the outcomes then it is in some value format and if the value is good then it can be considered that the predicted result is good based on the different types of data. The last but most important feature is material extrication in which the machine or system can divide and define certain attributes by the derivation of the data from different domains.

Category 2: There are certain challenges in sentiment analysis some of them are very major challenges let us understand what are they so the first challenge is Tone, basically the tone is to delineate the spoken text, and to overcome this challenge there are some tone identifier algorithms and some best software also. The other challenge is polarity in which there are some review values that are very congruent to other review values for example hate and the other congruent word is not so bad, so to resolve this there are various good sentiment analysis tools that can easily observe these types of similar values. The last challenge is Negations means some different types of sentences are very complex and confusing in nature and it is very difficult for the machine learning method plan to decode and abstract the meaning from those sentences so to anatomize this problem there are best approaches in which they can convert the negative sentence into the positive one.

Category 3: There is major application of sentiment analysis in data science exists in the real life lets understand all the applications one by one, the first application of sentiment analysis is Uber which is a taxi service mainly behind this concept by the use of sentiment analysis the main dissolution of the impulsion regarding the prices corresponding to various location can be measured. In some cases, in uber the acceptable contextual search can be applied. Then the second most important application of the sentiment analysis idea is the daily news so many thoughts and opinions can be highlighted based on various topics from different domains.

Category 4: In sentiment analysis, there are various approaches and algorithms are which are being described in an architecture diagram but before describing all the algorithms there are some popular algorithms for sentiment analysis LSTM (long short-term memory), Machine learning algorithms such as Support vector machine, Naive Bayes Classifier and the decision tree algorithm. Below there is the description of the architecture diagram of the different approaches of sentiment analysis in which different categories can be divided and under that category different subparts are also shown-

T: Traditional Approaches

M: Machine Learning Approaches

L: Lexicon Approaches

H: Hybrid Approaches

M1:Semi-Supervised Learning

M2: Unsupervised Learning

M3: Supervised Learning

S1: Support vector machine

N1: Neural network

NB: Naïve Bayes

DT: Decision Tree

LC: Linear Classifiers

RC: Rule-Based Classifier

PC: Probabilistic Classifier

S: Statistic Based

S': Semantic-Based

BN: Bayesian Network

ME: Maximum Entropy

DL: Deep Learning

DBN: Deep belief network

CNN: Convolutional neural network

RNN: Recursive neural network

DNN: Deep neural network

RNN: Recurrent Neural network

HNN: Hybrid neural network

Methodology: The proposed methodology is mentioned below:

1. Collect opinions from social media related to business models, Services and Processes, User experiences, Reviews
2. Perform Feature Engineering.
3. Perform exploratory data Analysis.
4. Perform sentiment analysis using AI models like RNN Lstm, Fast AI
5. Analyze the dataset to identify if there exists any significant input related to the sustainability of business Processes, Services, and sustainable development goals
6. Comparative studies of the Models and analysis of the result
7. Validation of the results
8. Visualization using charts and graphs
9. Impact Analysis
10. Reports Preparation

The methodology is being implemented with the deep learning algorithm. The initial dataset contains details about 100 opinions. The advantages of business sustainability are demonstrated in Fig. 2. Figure 3 shows Connection between Business, Business sustainability, and SDG goals. The techniques related to sentiment analysis are demonstrated in Fig. 1.

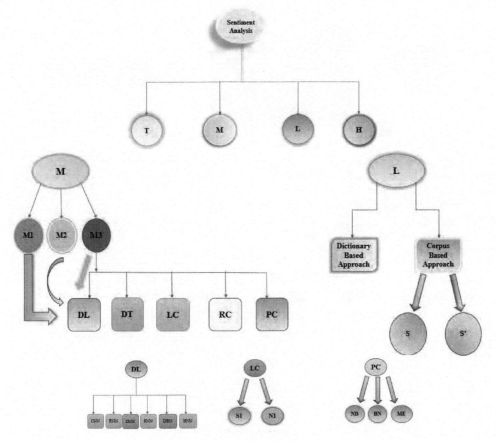

Fig. 1. Architecture Diagram of Different Approaches of Sentiment Analysis. The different

Fig. 2. Diagram Advantages of Business sustainability

Fig. 3. Connection between Business, Business sustainability and SDG goals

Conclusion: This paper tries to propose a framework that can enhance user experiences of various products and services of any business through sentiment analysis techniques. This is a part of ongoing research. The detailed analysis will be included in future work.

References

Zhang, L., Wang, S., and Liu, B. (2018). Deep learning for sentiment analysis: A survey. Wiley Interdisciplinary Reviews: Data Mining and Knowledge Discovery, 8(4), e1253.

Sohangir, S., Wang, D., Pomeranets, A., and Khoshgoftaar, T. M. (2018). Big Data: Deep Learning for financial sentiment analysis. Journal of Big Data, 5(1), 1–25.

Liu, B., and Zhang, L. (2012). A survey of opinion mining and sentiment analysis. In Mining text data (pp. 415–463). Springer, Boston, MA.

Hussein, D. M. E. D. M. (2018). A survey on sentiment analysis challenges. Journal of King Saud University-Engineering Sciences, 30(4), 330–338.

Dos Santos, C., and Gatti, M. (2014, August). Deep convolutional neural networks for sentiment analysis of short texts. In Proceedings conference on Computational Linguistics: Technical Papers, pp. 69–78.

Severyn, A., and Moschitti, A. (2015, August). Twitter sentiment analysis with deep convolutional neural networks. In Proceedings of the 38th international ACM SIGIR conference on research and development in Information Retrieval.

Wang, Y., Huang, M., Zhu, X., and Zhao, L. (2016, November). Attention-based LSTM for aspect-level sentiment classification. In Proceedings of the 2016 conference on empirical methods in natural language processing, pp. 606–615.

Ma, Y., Peng, H., and Cambria, E. (2018, April). Targeted aspect-based sentiment analysis via embedding commonsense knowledge into an attentive LSTM. In Thirty-second AAAI conference on artificial intelligence.

Soleymani, M., Garcia, D., Jou, B., Schuller, B., Chang, S. F., and Pantic, M. (2017). A survey of multimodal sentiment analysis. Image and Vision Computing, 65, 3–14.

Tan, S., and Zhang, J. (2008). An empirical study of sentiment analysis for chinese documents. Expert Systems with applications, 34(4), 2622–2629.

Measuring the Impact of Attitudes of Rural Customers Towards Digital Banking Products After COVID-19: Based on the UTAUT Model

B. Lavanya[*a] and Dr. A. Dunstan Rajkumar[b]

[a]Vellore Institute of Technology, Vellore, India
[b]Vellore Institute of Technology, Vellore, India
E-mail[*]: laya.lavanya1986@gmail.com

Abstract

The rural population plays a vital role in the upliftment of any society. In India, there are 6,40,867 villages with an estimated population of 83.3 crores out of 121 crores as per the 2011 census. The epidemic forced a new model. Amidst COVID-19, the eminence of digital banking products and their features has been realized. This study is based on the "Unified Theory of Acceptance and Use of Technology (UTAUT) model" and incorporates four constructs: performance expectation, effort expectation, facilitating condition, and social influence. Along with the four constructs, the security measure is also included. SPSS and AMOS were used as software. The study employed Structural Equation Modeling (SEM) to measure customers' attitudes with a sample size of 540 rural, digital-using customers. The research work was carried out for 8 months from October 2021 to May 2022. The result shows that there exists some positive relationship among the constructs.

Keywords: Digital banking, UTAUT, Behavioural Intention, Behavioural Usage, Performance Expectancy

Introduction

Technology has led to global digitization and e-banking. With digital inventions in banking, people can transact on their traditional accounts using a personal computer or smartphone with an internet connection for cash withdrawals, account transfers, utility bill payments, reviewing and printing statements, and similar transactions (Mohamed Merhia, et.al., 2019). During and after the COVID-19 epidemic, digital financial products are vital. The Indian economy supports about 60% of the rural Indian people, making it the predominant society. It is important to understand and use a wide range of digital banking services to figure out if a digital economy can work. (Nenad Tomic et.al., 2022). Hence, this study measures the impact of the attitude of rural customers toward digital banking products after the post-Covid phenomena. The objectives of this study are to examine the variables that measure the impact of rural customers on digital banking products and to identify the degree of positive relationship that exists among these factors.

Unified Theory of Acceptance and Use of Technology (UTAUT)

The Technology Acceptance Model (TAM), Combined Technology Acceptance Model (C-TAM), Theory of Planned Behaviour (TPB), Motivational Model (MM), Model of PC

Utilization (MPCU), Innovation Diffusion Theory (IDT), Social Cognitive Theory (SCT), and Theory of Reasoned Action (TRA) are all theories that are included in the UTAUT model (Bilai Eneizan et.al., 2022; Veera Bhatiaselvi, 2015). The UTAUT model's strength is that it contains the majority of the elements described as potential barriers to electronic payment adoption in its most basic version. Additional components can be simply added to the basic model.

Literature Review and Hypotheses

Many ideas and models have been created over the last few years to explain and identify the elements that drive the adoption of e-banking. This research is reliant on the "Unified Theory of Acceptance and use of Technology (UTAUT)" paradigm, which employs the four fundamental notions of performance expectation, effort expectation, facilitating conditions, and social influence. Based on the theoretical backdrop previously explored, the security feature is an additional construct that has been introduced in this study.

Performance Expectancy

Performance expectations are one of the main factors influencing the inclination to adopt new technologies (Manisha Sharma and Sujeet K. Sharma, 2019). This factor measures the Performance Expectancy of digital banking products in the rural population against their Behavioural Intention and use.

Hence the following hypotheses have been formulated.

H1: Performance Expectancy propounds a positive relationship toward Behavioural Intention.

H7: Performance Expectancy propounds a positive relationship toward Behavioural Usage

Effort Expectancy

It is stated as a person's perception of the difficulties in implementing new technology. The easier it is to use an information system, the greater the pace of technology adoption (Ike Kusdyah Rachmawati, et al., 2020). This factor measures the Effort Expectancy of digital

banking products in the rural population against their Behavioural Intention and Usage. Hence the following hypotheses have been formulated.

H2: Effort Expectancy propounds a positive relationship toward Behavioural Intention.

H8: Effort Expectancy propounds a positive relationship toward Behavioural Usage.

Facilitating Condition

The degree to which a person feels that the current organizational and technological infrastructure supports the usage of the system is characterized as facilitating condition (Venkatesh et.al., 2003). This factor measures the Facilitating Condition of digital banking products in the rural population against their Behavioural Intention and Usage. Hence the following hypotheses have been formulated.

H3: Facilitating Condition propounds a positive relationship toward Behavioural Intention

H9: Facilitating Condition propounds a positive relationship toward Behavioural Usage.

Social Influence

External environmental factors that a person encounters and changes their behavior and perception of certain activities and views of family, friends, relatives, or co-workers are referred to as social influence (Tan and Hirnissa, 2021). Social norms are influenced by two sorts of influences: informational which means gaining knowledge and normative which means expectations of others to gain knowledge (Daka, and Phiri, 2019). This factor measures the Social Influence of digital banking products on the rural population. Hence the following hypotheses have been formulated.

H4: Social Influence propounds a positive relationship toward Behavioural Intention

H10: Social Influence propounds a positive relationship toward Behavioural Usage.

Security Measure

Security measure is defined as measures taken as a precaution against espionage. In addition to performance expectations, effort expectations, social impact, and facilitative conditions, the security measure factor is introduced in this research. This measures how secure the customers feel about the security of their transactions. Hence the following hypotheses have been formulated.

H5: Security Measure propounds a positive relationship toward Behavioural Intention

H11: Security Measure propounds a positive relationship toward Behavioural Usage

Behavioural Intention and Usage

Behavioral intention is a measure of user confidence in using the Banking Application in the future. It is defined as the actual conditions under which systems and technology are used (Khurana and Jain, 2019).

H6: Behavioural Intention propounds a positive relationship toward Behavioural Usage.

Methodology

A sample of 540 rural customers residing in and around Vellore District, Tamil Nadu, India is selected for the study. The sample unit is the customers who are digital users and the cluster sampling method is used in the study. The district is split up into five taluks, comprising 285 villages in total. The sample size has been selected among the 285 villages. The research lasted eight months, from October 2021 to May 2022. Only the demographic profile is excluded from the data collection, which is done with the aid of the hypothesis and a five-point Likert scale ranging from strongly disagree to strongly agree. SPSS 3.0 and AMOS 3.0 are used for statistical analysis. Exploratory Factor Analysis is used to verify the provided model by testing the measurement model and model fit. To test the hypotheses, structural equation modeling is used.

Results and Discussion

Demographic Profile

Around 600 Questionnaires were dispersed among the rural population of the Vellore district. There were 540 completed responses, 38 with missing data, and 22 with no response. As a result, 540 replies were used for further investigation. The data represents that the male and female populations accounted for 53% and 47% respectively. Most of the people polled were 36-45 years old (34%) followed by those between the ages of 26-35 years old (28%). Regarding education, the majority of respondents have a bachelor's degree (30%) and diplomas (22%). Regarding occupation, majorities of the respondents are private employees 34% followed by government employees 23%. Regarding marital status, 52% of them are married and 48% of them are unmarried. Regarding the type of account, most of them use savings accounts 37%. The plurality of respondents (20%) has accounts with Canara bank. The majority of the respondents are using digital banking products for payment of bills (28%), checking of statements of accounts (22 %), and withdrawals (20%).

Validity and Reliability

Table 1 shows the validity of the variables. The Average Variance Extracted (AVE) is computed to measure the convergent validity. If the AVE is more than 0.5, the variables are accepted. Here, it is more than 0.5 for all the variables. Composite Reliability is computed for assessing the significance of an item. If the CR value is more than 0.7, the variables are accepted.

Table 1. Validity

	CR	AVE	MSV	MaxR (H)	PE	EE	FC	BU	SEC	BI	SI
PE	0.902	0.699	0.005	0.920	0.836						
EE	0.833	0.558	0.016	0.853	−0.010	0.747					
FC	0.901	0.753	0.183	0.904	−0.003	−0.080	0.868				
BU	0.887	0.723	0.118	0.892	−0.033	0.075	−0.055	0.850			
SEC	0.882	0.718	0.143	0.984	−0.011	0.128***	0.201***	−0.001	0.847		
BI	0.878	0.707	0.118	0.901	0.072	0.010	−0.043	0.344***	−0.034	0.841	
SI	0.819	0.602	0.183	0.828	−0.019	−0.035	0.428***	−0.069	0.379***	−0.027	0.776

Exploratory Factor Analysis (EFA)

EFA was used to determine the factors that measure the impact of rural customers on digital banking products. Before continuing, the EFA must be checked against the KMO and Bartlett's Test of Sphericity value. KMO is 0.737, which is more than the suggested value of 0.50, and Barlett's Test of Sphericity score is substantial. Communalities for all the items are above 0.40, therefore, no item is removed from the study. The number of factors extracted is seven and the total variance explained is 77.747. As a result, the variable chosen for the investigation explains 78 percent of the model. From table 2 it is presumed that all of the factor loadings were greater than 0.7. In the first factor, all the variables depict performance expectations, in the second factor, effort expectations, in the third factor, facilitating conditions, in the fourth factor, social influence, in the fifth factor, security measures, in the sixth factor, behavioral intentions, and in the seventh, behavioral usage of the customers.

Cronbach's Alpha is used to assess the reliability of the variables, and all the reliability values are above 0.722, indicating adequate reliability.

Measurement Model

The proposed structural model's fit indices are tested using Confirmatory Factor Analysis. GFI, TLI, CFI, NFI, IFI, RMR, RFI, Chi-square/Df, and RMSEA include some of the most often used fit indices. The following figure indicates the measurement model for the before and after modification and table 3 indicates the measurement model fit indices. After modification, there is a perfect positive relationship between error variables e8 and e5, as shown by the After-modification figure. This indicates that there is a positive linkage between clear instructions and actual presence in the bank.

Provided that the model produced satisfactory fit indices, it is employed to test the hypothesis.

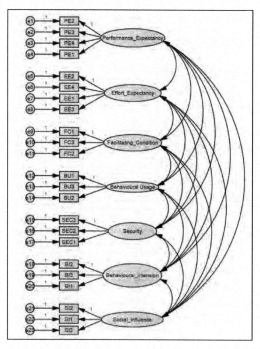

Fig. 1. Before Modification

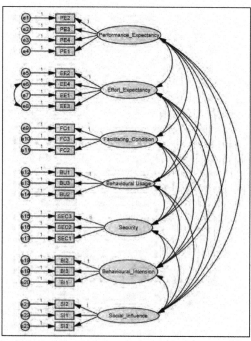

Fig. 2. After Modification

Note: PE denotes Performance Expectancy, EE denotes Effort Expectancy, FC denotes Facilitating Condition, SI denotes Social Influence, SEC denotes Security, BI denotes Behavioural Intention, and BU denotes Behavioural Usage.

Table 2. Matrix of Rotated Component

Variables	Component							Reliability
	Factor1	Factor2	Factor3	Factor4	Factor5	Factor6	Factor7	
Timely work is possible	.918							0.88 Performance Expectancy
Met the expectations during Covid	.902							
User-friendly digital work	.877							
All transactions possible in covid	.805							
Use all digital products in covid		.837						0.816 Effort Expectancy
No difficulties while utilizing		.825						
Avoids physical presence		.820						
Easily understandable procedures		.777						
24/7 customer support			.917					0.890 Facilitating function
Technical team support			.909					
Someone is there to help			.843					
Family & friends' connection				.854				0.826 Social Influence
Many recommend digital products				.848				
Much influenced by pandemic				.776				
Protected from digital threats					.915			0.880 Security Measure
Account information is secured					.871			
Transactions are highly secured					.855			
Intention to refer to others						.927		0.876 Behavioural Intention
Intend to use all digital apps						.876		
Intend to use after a pandemic						.824		
Usage is simple							.892	0.883 Behavioural Usage
Simultaneous Usage							.887	
Anywhere and anytime use							.871	
Extraction Method: Principal Component Analysis Rotation Method: Varimax with Kaizer Normalization								
a. Rotation converged in 5 Iterations								

Table 3. Measurement Model Fit Indices

Fit statistic	Cut-off	Before Modification	After Modification
x^2	-	726.658	519.119
x^2 Significance	$p \leq 0.05$	0.000	0.000
x^2/df	≤ 2-5.0	3.477	2.496
GFI	≥ 0.90	0.90	0.926
AGFI	> 0.90	0.869	0.902
NFI	≥ 0.90	0.903	0.931
RFI	≥ 0.90	0.882	0.916
IFI	≥ 0.90	0.929	0.957
CFI	≥ 0.95	0.928	0.957
TLI	≥ 0.90	0.913	0.948
RMSEA	≤ 0.08	0.068	0.053
RMR	≤ 0.05	0.042	0.037
RFI	≥ 0.90	0.882	0.916

Testing of Hypothesis Using Structural Equation Modeling, the relationship between the construct and variable can be discovered. SEM has the benefit of simultaneously evaluating a large number of variables. The Hypothesis testing is explained in Table 4. Five external factors (i.e., Performance Expectancy; Effort Expectancy; Facilitating Condition; Social Influence; Security Measure) are the fundamental precursors of Behavioral Intention and Usage. The findings indicate that all these factors affect Behavioural Intention and the hypotheses based on these external factors are supported. Behavioural Intention and Behavioural Usage are also impacted, as revealed by hypothesis 6. Hypotheses 7, 8, 9, 10, and 11 are not supported, indicating that there is no correlation between behavioral usage and performance, effort expectancy, facilitation condition, social influence, and security.

Table 4. SEM Output (Hypothesis Tested)

Testing of Hypotheses			Estimate	S. E	C.R	P	Result
Behavioural_Intention	<---	Performance_Expectancy	.078	.051	3.528	.027	S
Behavioural_Intention	<---	Effort_Expectancy	.022	.048	4.460	.045	S
Behavioural_Intention	<---	Facilitating_Condition	.035	.053	3.660	.009	S
Behavioural_Intention	<---	Security	.027	.046	3.593	.013	S
Behavioural_Intention	<---	Social_Influence	.003	.087	4.033	.003	S
Behavioural_Usage	<---	Behavioural_Intention	.287	.039	7.360	***	S
Behavioural_Usage	<---	Performance_Expectancy	.053	.041	1.309	.191	NS
Behavioural_Usage	<---	Effort_Expectancy	.025	.038	.649	.517	NS
Behavioural_Usage	<---	Facilitating_Condition	.014	.042	.333	.739	NS
Behavioural_Usage	<---	Security	.026	.036	.711	.477	NS
Behavioural_Usage	<---	Social_Influence	.077	.069	1.123	.261	NS

*** The arrows indicate that there is a direct link between the dependent and independent variables, which is significant at 5%; S-Supported; NS-Not supported

Conclusion

During this pandemic crisis, customers have seen that they understand the significance of using digital banking solutions. Even after COVID-19, its impact has been seen throughout the country, especially in rural areas. Those customers who were not familiar with digital products before and after the pandemic got somewhat familiar with these digital banking products. This study concludes that the impact of attitudes of rural customers toward digital banking products after Covid-19 based on the UTAUT model has statistically proved that there exists a positive relationship among the constructs with Behavioural intention. The measurement model indices show there is a good model fit. Hence it is concluded that the customers are giving more importance to the Behavioural intention of digital banking products than their usage.

References

Bilal Eneizan, et al., (2022). Older adult's acceptance of online shopping (Digital Marketing): Extended UTAUT model with Covid-19 fear. Journal of Theoretical and Applied Information Technology, 100(7).

Daka, Gladys Chikondi, and Phiri, Jackson (2019). Factors Driving the Adoption of E-banking Services based on the UTAUT model. International Journal of Business and Management. 14(6).

Ike Kusdyah Rachmawati et al., (2020). Analysis of Mobile Banking Use with Acceptance and Use of Technology (UTAUT). International Journal of Scientific and Technology Research. 9(08).

Manisha Sharma., Sujeet K. Sharma. (2019). Theoretical Framework for Digital Payments in Rural India: Integrating UTAUT and Empowerment Theory. International Conference on Transfer and Diffusion of IT (TDIT).

Merhia, Mohamed, Honea, Kate, and Tarhinib, Ali (2019). A cross-cultural study of the intention to use mobile banking between Lebanese and British Consumers: Extending UTAUT2 with security, privacy, and trust. Technology in Society.

Tomic, Nenad, Kalinic, Zoran, and Todorovic, Violeta (2022). Using the UTAUT model to analyze user intention to accept electronic payment systems in Serbia. Portuguese Economic Journal.

Khurana, Sunayna and Jain, Dipti (2019). Applying and Extending UTAUT2 Model of Adoption of New Technology in the context of M-shopping Fashion apps. International Journal of Innovative Technology and Exploring Engineering, 8.

Tan, M.C. Zariyawati and Hirnissa, M. T. (2021). The application of UTAUT theory to determine factors influencing Malaysian Adoption of Mobile banking. International Journal of Academic Research in Business and Social Sciences, 11(1), 146–159.

Bhatiaselvi, Veera (2015). An extended UTAUT model to explain the adoption of mobile banking. The Information Development, 16.

Venkatesh, V., Morris, M. G., Davis, G. B., and Davis, F. D. (2003). User acceptance information Technology: Toward a unified view. MIS, 425.

The Intrinsic Importance of Brand Mascots

Dhanshree Vishwakarma[a] and Sambit Kumar Pradhan[b]

[a,b]Unitedworld Institute of Design, Karnavati University, Gandhinagar, India.
E-mail: sambitkumar@karnavatiuniversity.edu.in

Abstract

One of the reasons many old brands such as Parel-G, Amul, and McDonalds are still recognized by the masses is their mascots. Especially in areas of relatively lower literacy, products are often identified through the symbols and images on the packaging. Brand mascots can become one of the best ways to build a brand identity and use it for advertisement. People might not recognize your logo as much as they recognize a mascot. This paper will talk about what a brand mascot is, how a company can design an effective mascot, and use of mascots in branding, web, and user interface design through case studies. The aim of writing this paper is to illustrate how fictional characters can be used to enhance a brand's identity and to find out if mascots will survive the changing trends.

Keywords: Brand mascots, Branding, UX-UI, Relatability, Recall Value

Introduction

We live in a world where there is huge competition between brands, every brand wants to grab customers' attention and stand out from its competitors. Brand identity is the personality of a company that shows a brand's value, creates a connection between customers and products or services served by the company, and helps in brand recognition. It can be a symbol, logo, tagline, name, or any character.

Developing a brand mascot is a great way to develop a positive brand image that can have an emotional connection with its customers. The brand mascot is a fictional storyteller who personifies your brand. It can be a humanized character or any animated figure used to communicate the company's ideologies, act as a spokesperson, and become the voice, tone, and visual representation of your brand.

Having a mascot as part of your branding may seem like just an aesthetic choice, but it's more powerful than it looks. An efficiently designed mascot is like an asset for a brand because it helps the brand to be recognized for years. For example, "Amul girl," who was created in the year 1966 as a rival to the "Polson Butter girl" more than half a century after its creation, is still as relevant and loved today as she was earlier.

With the involvement of time and technology, use of mascots also evolved from sports to branding now it is also used in web and app design too.

Literature Review

Branding is a marketing practice used by companies to create their brand identity. A brand is defined as an identifying symbol, mark, logo, name, word, and/or sentence that companies use to distinguish their product from others (Will Kenton, 2020). Apart from making your memorable branding also helps the company to distinguish your company from competitors.

Brand represents who you are as a business and what is the vision of your company.

Mascot is a character that personifies a brand, it can be any fictional character, animal, animated character, an inanimate object whose purpose is to communicate company's values and convey message on behalf of the company. Having a mascot as a branding strategy is a great idea because it adds personality and emotion to your brand, helps a brand to stand out and become recognizable, helps in grabbing attention of the customers, and provides brand flexibility (Simon, 2016).

History of brand mascots

The term mascot originates from the French word *"mascotte"* which means lucky charm. Traditionally animals were used as mascots which served as a source of entertainment for the public and for keeping their excitement level high. It was thought that animals were used to bring a different feel to the game i.e., entertainment as well as fear. Over time mascots also evolved from predatory animals to two-dimensional fantasy mascots, to three-dimensional mascots. People started to physically interact with mascots when American puppetry allowed people to have both visual and physical interaction with mascots. After this brands realized the potential and opportunities of using three-dimensional mascots and gradually other companies started creating their brand mascots which gave their customers something that they had a connection with (Startups in Canada, 2018).

Can Brand Mascots Survive the Changing Trend?

According to an article, Brand mascot cannot go out of fashion. This article quotes examples of successful Indian brand mascots that are still recognized by people and it also has interviews of a few marketers, brand experts, and advertising veterans held by the best media info where the speakers spoke about why brands are nowadays not keen on mascots.

KV Sridhar and Anil S Nair spoke about the beginning state of brand mascots. People used to buy products by seeing the symbols and images on the packaging it might be because of the literacy rate of India was not so good so brand mascots worked as a brand identity that distinguishes one brand from another.

B K Rao, Saurabh Uboweja, Anil S. Nair, and RS Sodhi spoke about Increasing clutter, a fragmented audience. Brand mascots used as logos are like assets and are a part of the brand identity. But nowadays very less brands are using mascots as compared to previous years.

There was a question asked to the speakers that are macots dying trend? Rao and Bijoor felt that mascots are going to make a comeback in a big way soon. Rao told about the popularity of Parle-G girl in the rural areas of India. As stated by Rao Parle-Gis still known as the "Babuwa-wala biscuit" there are also examples of ZooZoo brand mascot of Vodafone and Amul girl that have made an impact on its audience and managed to stay relevant and adapt the changing trends. [Roshni Nair, 2018]

Mascots that re-defined Indian advertisements

Fig. 1. Maharaja, brand mascot of Air India

Fig. 2. Amul Girl, Mascot for Amul

Fig. 3. Parle-G girl, brand mascot for Parle-G biscuits

Fig. 4. Gattu, brand mascot for Asian Paints

Fig. 5. The Devil, brand mascot for Onida TVs

Fig. 6. Bholu, the guard, brand mascot for Indian Railways

Fig. 7. Zoozoosl, brand mascots for Vodafone

Fig. 8. Nirma girl, brand mascot for Nirma detergent

Fig. 9. Rabbit Puppet, brand mascot for Lijjat papad

How to Design an Effective Mascot for Your Brand?

An effective mascot should be designed with proper user research; analyzing needs and expectations of the target audience alongside a thorough study of market environment. Features of an effective brand mascot are (Tubik Studio, 2019):

- Memorable
- Recognizable
- Original
- Representing a consistent character
- Flexible to adapt and adjust
- Applicable for diverse tasks
- Looking good in different sizes and resolutions
- Stylistically harmonic
- Lively and user-friendly.

Mascots in branding

A unique brand mascot can make a company look attractive and trustworthy to its potential customers. It helps in building a strong brand identity of company and its products and it also differentiates a product from its competitors. It is one of the most effective ways to promote a brand and engage customers. It adds humor, fun and friendliness to a brand.

Mascots are helpful for companies that are looking for umbrella branding and not for a company that produces too many sub-brands. Developing mascots for each sub brand would be costly and time-consuming told by RS Sodhi, Managing Director, Gujarat Co-operative Milk Marketing Federation (GCMMF) (Roshni Nair, 2018).

Mascot used in logos can create a strong brand identity, spokesperson and can act as a brand ambassadors also. It can be effective across industries; it depends upon the company how they use it and what message they want to deliver to its audience (Mascot logos, n.d.)

Case study of Andre

Fig. 10. Mascot in Logo designing for a landscape company, Andre)

Here is a case study that shows that mascots can be an effective tool that can be used as one of the marketing strategies to create brand's identity., andre, it is a landscape company. Their initial logo was based on letter mark, then they decided to use a mascot as logo which is a friendly hummingbird combined with a leaf shape and it made a foundation for the further development of cooperate identity (Tubik Studio, 2019).

Mascots in UI

Mascots are not only useful in the advertising and marketing industry but also in designing web and mobile interfaces. They bring life to the interface, grab users' attention and add humanized features to the interface. It is also a way to avoid extra copy on screen and it saves precious space which can be used for placing other elements. [Tubik Studio, 2019]

Designing Emotional Interfaces for Boring Apps

An app can become boring if the products are not attractive, exciting, and repetitive. So, to bring some excitement designer should design such an interface that can create an emotional connection with the user. One of the ways to do so is using mascot in UI for example, Tunnel Bear and Remem Bear.

Tunnel bear, a VPN service that uses a bear as its mascot, keeps its users busy and makes an emotional connection with the user. unlike other VPN services, Tunnel Bear evokes endearment.

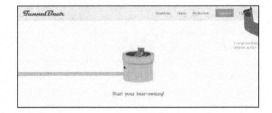

Fig. 11. *TunnelBear*, mascot used by VPN service to overcome boredom

In the same way, RemeBear which is a password manager also uses a bear as its mascot which is illustrated and used in such a manner that it gives the user a friendly feel. The mascot appears when the user completes the first stage of signing up there is an illustration of a friendly bear hug (Alice, 2018).

Fig. 12. *RemeBear,* Screen that shows an emotional connection that the mascots create with the users

Case Study on Toonie, the Cheerful Bird

Fig. 13. Toonie the cheerful bird for an alarm app

Toonie the cheerful bird is a mascot that is designed for an alarm app. It works as a communicator between the user and interface. It has a great impact on establishing the voice and tone of the product. The aim of using Toonie as the mascot was to help users interact with app and to spread the voice and tone of the product. It tells the users about news, rewards, errors, and also adds fun and color to monotonous lives. Toonie is created in such a manner that it humanizes the app makes its users feel like they are interacting with a real person and tells the users about all the features of the app. The app has used the mascot to fill the space effectively, if the user has not set any alarm then the user can see a funny illustration of fallen bird.

Fig. 14. Toonie alarm app screen that shows that alarm is not set

This app showed the efficiency and diversity of how mascots can be useful in designing an interactive user interface. The funny, interactive, cheerful, and tone and voice chosen for the character establish an emotional background of perception and user experience that works for wide target audience (Tubik Studio, 2019).

Mind Map

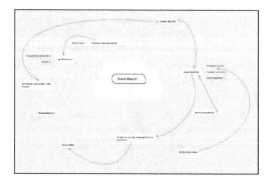

Fig. 15. Mindmap summarizing what are brand mascots

The mind map is the summary of what is a brand mascot and what are its characteristics. It tells that brand mascots are fictional characters that create brand's personality and identity that acts as company's spokesperson to convey the message that the company wants to deliver to its customers. It portrays the company and what are its ideologies. It helps the brand to make an emotional connection with the audience which leads to recognition and memorability of the brand and its product.

Conclusion

While going through the whole concept of brand mascot, I got to know how to design and where to use a mascot for building a strong brand identity, I also came to know that brand mascots can never go out of trend because of the brand identity that the brand has developed which has created a real connection with the audience, as a result, people will recognize your brand just by looking at your mascot.

As we saw above mascots are not only effective in branding but also in web and user interface design. It all depends upon how intelligently the company chooses its character that can become a visual identity of the company and also the spokesperson and how much the company knows about its consumers and their interest.

References

Angie Maxin (2018). "Mascots in branding," Bam Mascots with Impact, web blog post, 29th June, viewed on 22nd July 2020, https://www.bammascots.com/blog/brand-mascots-and-their-positive-impact-on-consumer-purchasing-behaviour-part-2#:~:text=Strengthening%20brand%20identity,that%20makes%20you%20st and%20out

Alice (2018). Designing emotional interface for boring apps, Smashing Magazine, viewed on 17th July 2020, https://www.smashingmagazine.com/2018/04/designing-emotional-interfaces-boring-apps/

Mascot Logos (n.d.). Mascots in Branding, Tailor Brands, web blog post, viewed on, 17th July 2020, tailorbrands.com/blog/mascot-logos

Nair, Roshni (2018), "Can brand mascots survive the changing trend?" "Mascots in Branding," Best media info, viewed on 16th, July, 2020, https://bestmediainfo.com/2018/06/in-depth-are-brand-mascots-a-dying-trend/

Simon (2016). "What is mascot and benefits of brand mascots," Marvellous, web blog post, September, viewed on 20th, July, 2020, https://wearemarvellous.com/character-mascot-design-branding/

Startups in Canada (2018). "History of brand mascot," Startups in Canada, web blog post, 22nd August, viewed on 14th, July, 2020, http://startupsincanada.com/brand-mascots/

Tubik (2019). "How to design effective mascot for your brand?," "Mascots in branding," "Case study on Andre," "Mascots in UI," "Case study on Toonie Alarm"; viewed on 15th July 2020, https://uxplanet.org/the-power-of-mascots-in-branding-and-ui-design-5973d12be955

Will Kenton (2020,s brand definition, Investopedia, viewed on 23rd July 2020 https://www.investopedia.com/terms/b/brand.asp

Borikar, H., Bhatt, V., and Vora, H. (2022). Investigating The Mediating Role Of Perceived Culture, Role Ambiguity, And Workload On Workplace Stress With Moderating Role Of Education In A Financial Services Organization. Journal of Positive School of Psychology, ISSN, 9233-9246.

Malek, MohammedShakil S., and Bhatt, Viral (2022). Examine the comparison of CSFs for public and private sector's stakeholders: a SEM approach towards PPP in Indian road sector, International Journal of Construction Management, DOI: 10.1080/15623599.2022.2049490

Efficient Market Hypothesis: A Conceptual Framework

Dharmendra Khairajani[*,a]

[a]Karnavati University, Gandhinagar, Gujarat
E-mail: [*]dharmendrak@karnavatiuniversity.edu.in

Abstract

Efficiency is a key notion in the finance industry. The idea of proficiency as it relates to money markets has been explored by academics and economists for a long time, with the efficient market hypothesis (EMH) currently being a key matter of study. The extent to which stock prices and other securities prices include all readily accessible, relevant information is referred to as market efficiency. In the view of well-known author Fama the efficient market hypothesis (EMH) a shareholder cannot outperform the market since stock prices previously take into report all appropriate information. Shareholders who consent to this testimonial are more likely to invest in index funds that emphasize passive portfolio management and follow the performance of the entire market. Academic discussion of stock price behavior has recently been very active. The Random Walk model operates in this manner. The basic premise is that investment strategies based on long-ago performance may not always make higher returns in the future. There are several research initiatives that many academicians and economists from around the world have started. There are 3 types of EMH weak form, semi-strong form, and strong form of EMH.

Keywords: Market Efficiency, EMH, Stock Market, Weak form, Semi Strong Form, and Strong Form

Introduction

People, who dealt with the market while standing beneath a banyan tree established the first stock market of India, which existed for more than 200 years. In 1899, the Bombay stock exchange was formally opened. There have been numerous booms and busts since then. The start of the Second World War in 1939 brought about a roar that was swiftly followed by a depression. December 1946, India had just seven stock exchanges and 1125 listed companies. In December 1985, there were 4344 listed businesses and 14 stock exchanges, which marked a significant change.

The theory has its origins in the 1960s, when the majority of research studies—beginning with (Fama, 1965 and Samuelson)—considered the capital markets to be efficient. Over the ensuing decades, an increasing number of researchers began to refute the idea in all three forms. Along with defining efficient markets, Eugene Fama published an article in 1970 that also distinguished between the three levels of efficiency. Efficient market describes that "a great number of rational, profit-maximizes aggressively compete, everyone try to predict the opportunity available in the market, and where present significant information is freely accessible to all players."

Every country has a central body that is responsible for overseeing the capital market, such as the SEC in the US or SEBI in India. Due to its size, the capital market of India has drawn more than 30 million investors. The two major stock exchanges are the BSE and the NSE. The

SEBI was established to oversee the market. The company was founded on April 12, 1988, and it is based in Bombay. It is the principal regulatory body overseeing the expansion and management of the stock market.

A money market is considered to be competent when all relevant information is appropriately reflected in the prices of securities. Capital market efficiency is important when it comes to investing because choosing winners is useless if the market is proficient. Additionally, wasting time, thus there is no way to profit from it. This is justified by the fact that no investor can earn an abnormal profit because all details are accessible, assimilated, and distributed there and reflects in the price of shares in a fast and accurate manner. As a result, even if undervalued assets with higher or lower predicted returns are available, there is no risk of being undervalued. Information is readily available, assets reflect genuine value, and it is no longer possible to keep investors, brokers, financial institutions, and others in the dark. The contemporary era has been characterized by globalization or the integration of economies and global markets. The conflict scenario between Russia and Ukraine in March 2022, a single event affecting the markets of other countries.

With more mature markets, more open borders for cross-border investment, and the present financial crisis, which resulted from a lack of liquidity, the significance of the developing market is growing. The financial disaster that strikes the United States in 2007 is the finest example of how interconnected economies are. As seen by declining stock prices and changes in the currency index, the crisis spread to economies all over the world as a result of a lack of capital.

Today's stock prices are different from those of the past; they could at any time be high, low, or unchanged. This theory holds that all assets on the market are proficient; their prices are affected by fresh flow of information. No investor could reasonably outperform the market at the current stock prices. A consistent return on investment—defined as a yield that is commensurate to the amount invested—can be expected by an investor who makes investments in an efficient market.

Review of Literature

Numerous academics have looked into the effectiveness at Indian stock market and the international stock market. Numerous research encompassing various time periods and employing various statistical methodologies have been conducted to observe Indian stock market's inadequate type of market competence, but only a small amount of work has been done recently. Studies carried out in developed nations (Bachelier 1900) introduced market efficiency to the finance industry at the beginning of the 20th century by mentioning various theories of speculation. In his words, market prices describe the history, present, and potential occurrence, but frequently explain the refusal of clear indication of price variations. In the end, commodities prices were subject to unpredictable swings. As a result, investors could not profit from unusual returns. This circumstance is referred to as the true view and is the foundation of market competence text. A sizable group of academics has historically subscribed to the notion of random walks in stock-market values, particularly economists and statisticians.

In order to get higher profit with fresh information it is not possible because the market takes into account both historical pricing and information from the public. Semi-Strong variant of EMH (Fama, 1970) holds that no stock market investor can make an abnormal profit by using any fresh knowledge because all secondary data, available information, and public information are exposed in the present securities prices.

To evaluate the market efficiency hypothesis, (Lee et al., 1998) applied the (Bartlett 1954) goodness of fit test to the stock prices of 12 different nations from January 1921 to December 1930. The market efficiency theory is not disproven for the majority of European nations, but it is for non-European nations. This finding supports the current theory that the highly volatile stock price changes of the 1920s may have resulted from the infancy of the globalized market.

The RMW model investigates a study by (Sunil Poshakwale, 2001) India's expanding stock market. A lot of studies have been done on these markets' stock prices' risk and return characteristics as a result of the increased investor interest in emerging economies.

Anamika Sharma (2009) share prices accurately reveal all pertinent details, rendering it hard to consistently produce unexpected stock market returns using information that is readily available to the public. How quickly and precisely share prices respond to every news or incident is a sign of how effective stock markets are. This study evaluates Indian stock exchange utilizing the paradigm of market-adjusted anomalous returns.

Mishra (2011) claimed and made an effort to examine the Indian capital market's market efficiency. He used econometrics of time series to demonstrate the market's increased fluctuation and weak form inefficiency. The market's subpar performance was related to the global fallout from the US subprime mortgage crisis.

Kaur, Manjinder, Dhillon, and Sharanjit (2011) used daily information for the years from the Bombay Stock Exchange 1991 to 2006 to examine efficient market hypothesis at weak level. In this work, parametric and non-parametric test were used.

Sujith Kumar and Sadanand Halageri (2011) projected that Stock market efficiency has been of substantial relevance in financial literature. Researchers have looked at several manifestations of the efficient market hypothesis (EMH). This study examines stock price responses to stock split announcements, a publicly accessible piece of data to assess the semi-strong variant of the EMH. When the price appropriately represents all of the available information, a market is said to be efficient. This implies that prices follow a random walk and that there is no way to produce long-term excess returns. The foundation for capital development is a robust and integrated capital market. The effectiveness of capital markets and financial institutions determines how efficiently capital generation occurs.

The goal of this article, according to Pradipta Kumar Sanyal, Padma Gahan, and Smarajit Sen Gupta (2014), is to assess the effectiveness of the stock market in emerging economies. We assume so as to stock markets in developing nations have become efficient as a result of market integration, quick information transmission, technology advancements around the world, and structural developments of stock markets, particularly in emerging economies. However, the results of previous literature research point to contradictory findings. This prompts a revalidation of the subject of market efficiency or random walk behavior given the expanding market under examination. Examining the random walk behavior of stock markets in 10 emerging economies is the goal of this study. Market efficiency literature reviews have been conducted for both developed and developing economies. The majority of the study's findings point to the absence of random walk behavior in emerging economies and are consistent with earlier findings.

Neeraj Gupta and Deepika Singh Tomar (2014), the different manifestations of several empirical research have been conducted on the validity of the efficient market hypothesis. The weak-form efficiency assumption test is used to determine if the data that was incorporated in previous stock prices are accurately reflected. The weak version of the Efficient Market Hypothesis (EMH) states that current stock prices accurately represent all of the data in the price history sequence. The recent price swings are utterly unanticipated. They are produced by new information and are independent of or dependent on earlier price changes.

According to Navratil et al. (2021), in the days and weeks following the start of the COVID-19 pandemic, United States equity market was inefficient. This is proven by the researcher in this paper during that period of Covid. Researchers develop a flexible trading method for entities that benefit the user by applying a unique approach based on a probability ratio to generalize Merton's optimal portfolio problem. These tactics are demonstrated to have good risk and performance statistics as well as statistically significant profitability during the COVID-19 time period.

Market Performance

In a competitive market, stock prices incorporate and reflect all pertinent information, both recent and recent. It indicates that prices adhere to RWM in a successful market. Since it is widely accessible to everyone, as far as we know, there is no chance for anyone to make a significant profit. The effectiveness of a variety of elements, including goodwill, has an impact on the security markets. The characteristics of the issuing firm, the securities to be exchanged, the extent of the market where the securities will be traded, and the amount of technology used, as well as the qualities of the securities to be traded Analysts may use it to look over trade data.

Efficiency is the capacity of the capital market to function in such a way Prices for those securities are based on demand and supply and realistically reflect additional knowledge. The Market Efficiency Theory implies that information is easily accessible in a market, as shown in Figure 1. That is operating efficiently. Information flows freely because investors and the market cash in on all of the accessible information. An unbiased reflection of this can be found in security pricing. There is no way for investors to exploit the information to generate a higher return than that projected by random portfolios because asset prices have already absorbed all of it. Testing market efficiency entails determining whether all three iterations. One investor cannot surpass another unless the latter is prepared to take excessive risks.

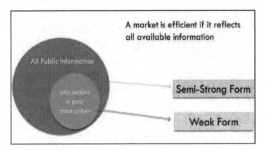

Fig. 1. Types of Efficient Market Hypothesis
Source: Author's construct

Forms of Efficient Market Hypothesis

Numerous researchers have supported Bachelier's findings since the Random Walk Hypothesis was first proposed in 1930, but Eugene F. Fama's Ph.D. thesis from 1960 was the first comprehensive

investigation of the topic, and today the general consensus is that the Random Walk is measured by the EMH. In a competitive marketplace, making price predictions based on fundamental or technical research is useless.

According to this argument, it is challenging for a trader to outperform the stock market by forecasting future security prices using historical data because existing share prices truly represent all info included in past price fluctuations. Therefore, there should be no significant correlation between the values of securities over time for a market to be efficient in this way.

If prices accurately represent all information found in historical price data, a market is said to have Weak efficiency. Technical analysis is ignored because it is useless. In contrast to the random walk hypothesis, it doesn't go into as much detail about the stochastic mechanism that governs price behavior. The weak variety of the EMH is based on the following two hypotheses: (a) that subsequent security price changes are unrelated to one another; and (b) that price changes are evenly distributed among random variables. One can examine the trend to see how effective the market was in the past or search for brief variations in the total return of the market to determine the underlying mechanism generating those rewards, leading to the conclusion that the returns are random. The walking motion In essence, random walk inference suggests that the returns are either higher or lower, or that the process model employed to identify it is flawed and unable to recognize the real mechanism generating returns.

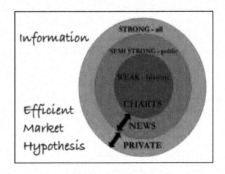

Fig. 2. Three Forms of EMH in diagram
Source: https://tradebrains.in/

Weak Form of Efficient Market Hypothesis

The Weak Form of EMH says with the purpose of the assets or securities' current price reflects prior data. Info is readily available in a modern environment with the click of a mouse. This sort is known as a weak form since security expenses are open to the general public. It implies that nobody ought to be capable of beating the market by using the collective wisdom of everyone. Nobody can employ past data to calculate possible values and produce abnormal gains because current stock prices represent the previous information. There is no correlation between subsequent prices in a market with a weakly formed efficient market.

There is no correlation between successive prices when a market is only half efficient, indicating that no one will continuously make anomalous profits by tracking past price movements. Since it is based on earlier period price trends, this kind of learning is sometimes referred to as technical or chart analysis.

In their dissertation, Ko and Lee (1991) noted that, If the random walk hypothesis holds, then the weak form of the efficient market hypothesis must also hold, but not the other way around. Market efficiency is hence evidence in favor of the model of a stochastic process. The nonstationary model need not be breached in order to display the weak form of market inefficiency.

Semi-Strong Form of Efficient Market Hypothesis

The semi-strong version of EMH is predicated on the idea that current stock prices communicate all previous data about a company or share, as well as all publicly available data like interim and annual reports. To assess the impact of the news on the efficiency of the Indian stock market, a semi-strong form of EMH is employed. To evaluate the result, period technique for mergers and acquisitions the company made around the time of the announcement, a standard risk-adjusted event study is being used. It asserts that all freely released data are already considered in asset values. There is more to relevant records than just past prices; it also includes information from corporate announcements, financial records, the economy, and other sources.

Additionally, it means that relying on "what everyone else knows" should not allow anyone to exceed the market. This shows that a company's financial records are useless for predicting future price fluctuations or generating high investment returns because these are already reflected in the stock prices, which take into account both previous and publicly available data. As a result, nobody is able to profit unusually from fresh info. If prices accurately represent all existing knowledge, both formal and informal, then the market is exceedingly efficient. Prestige information encompasses, but isn't confined to, data that a market maker has access to, information that corporate executives have access to behind the scenes, and information that asset managers devote time and money gathering for their gain.

Strong Form of Efficient Market Hypothesis

According to the strong version of EMH, no investor can anticipate a phenomenal return because share prices reflect all data, both public and confidential. To evaluate the market's efficiency and the value of investing in Indian Capital. For example, expert knowledge quickly becomes a component of the overall price and cannot be used to make extraordinary trading profits. As a result, the security's market value completely and precisely reflects all evidence available, both publicly and privately. It indicates that nobody can profit from the knowledge they have access to, not even the company's management (insider). No one can profit from concealed or insider intelligence since current stock prices reflect all accessible data, whether private or not.

Conclusion: Since about 50 years ago, in the world of finance, the market theory of efficiency prevailed. However, as was noted the study includes numerous problems related to hypothesis, and it presents a fundamentally erroneous notion. The concept is that because the pricing constantly is based on all info that is presently accessible, it still represents the information that is presently available. This concept lies at the heart of the efficient market hypothesis discussion in the wake of the financial crisis. The other major problem is that it is predicated on fundamental neoclassical economics assumptions that are not necessarily true in practice, such as the existence of perfectly competitive markets and rational market players. In conclusion, diverse aspects of asset pricing are explained by the financial economics theory and the trade-off theory. Both, however, have significant limitations given the current situation. This indicates that both fundamental theories must still be used in order to completely comprehend asset pricing. By adopting the overreaction/underreaction steady state, It is important to note that the efficient market hypothesis can be explained using an extension of the behavioral finance theory that is testable. As a result, behavioral finance theory provides a more thorough explanation of asset pricing from a theoretical standpoint.

References

Sharma, Anamika (2009). Impact of Public Announcement of Open Offer on Shareholders Return: An Empirical Test for Efficient Market Hypothesis. The IUP Journal of Applied Finance, 15(11).

Deepika Singh Tomar and Neeraj Gupta (2014). Testing Weak Form Of Efficient Market Hypothesis: A Study On Indian Stock Market, International Journal of Applied Financial Management Perspectives © Pezzottaite Journals, 3(2), April – June 2014, pp1005-55, ISSN (Print):2279-0896, (Online):2279-090X.

Fama, Eugene F. (1970).Papers and Proceedings of the Twenty-Eighth Annual Meeting of the American Finance Association New York, N.Y. December, 28-30, The Journal of Finance, 25(2), pp. 383–417.

Mishra, P. K. (2011).Weak form market efficiency: evidence from emerging and developed world. The Journal of Commerce, 3(2), Hailey College of Commerce, University of the Punjab, Pakistan, ISSN: 2218-8118, 2220-604, pp.26-34.

Sanyal, Pradipta Kumar, Gahan, Padma and Gupta, Smarajit Sen (2014). Market Efficiency in Emerging Economics: An Empirical Analysis, GMJ, 8(1-2), January-December 2014, pp. 22–38

Navratil, Robert, Taylor, Stephen, Vecer, Jan (2021). On equity market inefficiency during the COVID-19 pandemic. International Review of Financial Analysis, 77, 101820. https://doi.org/10.1016/j.irfa.2021.101820

Sujith Kumar, S. H., and Halageri, Sadanand (2011). Review of Management, 1(1), Jan–March 2011, 15 Impact of Stock Split Announcement on Stock Price.

Poshakwale, Sunil (2002).The Random Walk Hypothesis in the Emerging Indian Stock Market, Journal of Business Finance & Accounting, 29(9) & (10), Nov./Dec. 2002, 0306-686X.ß Blackwell Publishers Ltd.

Kaur, M., Dhillon and Sharanjit, S. (2011).Testing Weak Form of Efficiency Hypothesis for Indian Stock Market, Indian Institute of Finance, 25(3).

Website References

https://tradebrains.in/efficient-market-hypothesis

A Study Regarding issues in Public Private Partnership Road & Highway Projects

Pranay Brahmbhatt,[*,a] *Dhiraj Bachwani,*[b] *and MohammedShakil Malek*[c]

[a,b]Parul University of Vadodara, India
[c]F. D. (Mubin) Institute of Engineering and Technology
E-mail: *brahmbhattpranay21@gmail.com

Abstract

India, a developing country, has to learn how to effectively implement the Public Private Partnership model because it may greatly aid in the growth of the economy. The motive of this research is to recognize and comprehend the hurdles that arise in PPP road and highway projects. The author extracted 66 problems from various academic papers and articles. To gauge the extent of the impact on the project, a questionnaire survey was done. Three categories of data from the questionnaire survey were created: high impact, moderate impact, and low impact. According to the analysis, there were 11 difficulties with a high impact on the project, 21 issues with a moderate impact, and 34 issues with a low impact. The author mentioned some suggestions that might be beneficial down the road.

Keywords: PPP Projects, issues, Road & highway Sector

Introduction

"American roads are not good because America is rich, But America is rich because American roads are good"-John F Kennedy. The growth of India in recent years has been a truly amazing development for the global economy. India's economy has historically been consumer-driven, and it has experienced unprecedented development in the previous 20 years, necessitating significant infrastructural improvements. Future growth cannot be sustained by the government alone; hence private sector involvement in such expensive infrastructure projects is essential. Government initiatives have increased the private sector's involvement in India's infrastructure development (Reddy and Sharma, 2017). As the nation progresses on its roads, they constitute a valuable national resource. In India, roads account for 65% of freight and 85% of passenger traffic, and it is predicted that this volume has been increasing at a rate of 7–10% annually. Highways are thought to be more significant than other types of roadways because they link various regions of the nation as well as other nations. Even though they make up only around 2% of the entire network, national highways (NH) handle close to 40% of all traffic. One of the key developments that influence a nation's economic climate and success is regarded to be road development. Along with facilitating human mobility, highways also provide crucial infrastructure for the transportation of products and services, ensuring that demand and supply are met (Nallathiga and Shah, 2014).

According to Indian Government Public Private Partnership is: "A project based on a contract or concession agreement between a government or statutory agency and a private sector business for the delivery of an infrastructure service in exchange for user fees is referred to as a public-private partnership (PPP) project."

1.1 Indian scenario for Public Private Partnerships (PPPs)

Fig. 1. Year Wise PPP Projects in India

Public-private partnerships in India date back to the nineteenth century. Some of the earliest examples of PPP in India in the early twentieth century are the Great Indian Peninsular Railway Company, which ran between Bombay (now Mumbai) and Thana (now Thane) in 1853, the Bombay Tramway Company, which operated tram services in Bombay in 1874, and the power generation and distribution companies in Bombay and Calcutta (now Kolkata). Additionally, private enterprises generated 65% of India's electricity prior to the country's independence in 1947. Following independence, a wave of nationalization swept the nation, putting an end to the private sector's role in building infrastructure. At the time, private companies could only provide infrastructure services as contractors and, in some cases, as operators, especially in important infrastructure sectors like transportation, power, telecommunications, and urban infrastructure (Malek and Bhatt, 2022).

The year-by-year PPP projects in India from 1990 to 2021 are depicted in figure 1 above.

The next figure 2 represents PPP projects in India's road and highway sector broken down by state. With 48 projects, Madhya Pradesh leads the pack of PPP projects for roads and highways. The remaining states are ranked lower in figure 2.

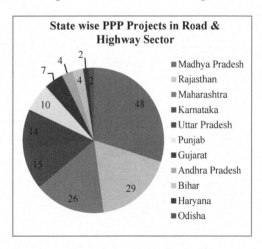

Fig. 2. State Wise PPP Projects in Road and Highway sector of India

Figure 3. Below displays the most recent authority-reported status of the road and highway PPP projects.

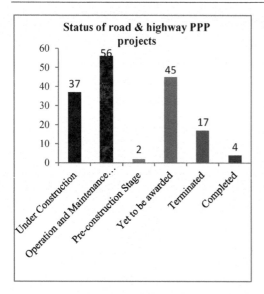

Fig. 3. Status of road & highway PPP projects as per authority

Figure 4 depicts the total number of PPP projects in India in FY21. According to the graph below, 45% of PPP projects in the road and highway sector will be implemented in 2021.

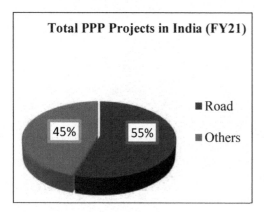

Fig. 4. Total PPP projects in India (2021)

2. Literature Review

The American Society of Civil Engineers, International Journal of Construction Management, International Journal of Project Management, research gate, Transportation Research Proceedia, and many other prestigious journals in the field of civil engineering were studied for this paper's literature review. A few of these journals are shown in figure 5 below.

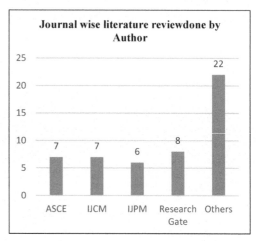

Fig. 5. Journals-wise literature review done by Author

3. Findings from Literature review

The author has outlined the problems encountered in PPP road and highway projects in India based on the literature review. Delay due to land acquisition, a change in the law, a delay in site handover, poor contract management, poor research before tendering the contract, legal costs, and events beyond the parties' control The community's acceptance and support, Technical feasibility of the project, The laws, regulations, and guidelines that have been enacted, Allocating and managing resources efficiently and sufficiently, Having a well-organized public agency and committed employees, proper and efficient employee training, transparency and equity in the procurement process, excessive restrictions on participation, lengthy negotiation delays, The project's environmental impact, financial closure, Client orders for changes, Contractor's poor site management and supervision Client's failure to make progress payments, Difficulties getting work permits, Contractor's insufficient planning and scheduling An accident occurred during construction.alterations to government regulations and laws, a scarcity of high-tech mechanical equipment, The influence of social and cultural factors Inadequate data collection and surveying before design, Contractor's delay in ordering main equipment, confusion over government objectives and evolution criteria, fewer job opportunities The exchange rate, Theft of water, Inadequate PPP experience Partner disagreements, Political unrest and early termination, Technology danger, Raw water of poor quality, funding of PPP projects The threat of climate change, Demand has decreased. Political violence, political and social risks, high financing costs, fluctuating currency exchange rates, a significant cultural divide between foreigners and Indians Excessive participation restrictions, Cost estimation errors, Low traffic demand, cost and time overruns, schedule delays due to unqualified materials rejection Firms with excessive leverage bidding on Projects NPA of a bank (non-productive assets), Disputes over contracts and payments Insurance for a construction company, Hidden department, Consider the economic benefits. alterations to government regulations and laws, Framework for Policy, Approval by the Authority, Eliminating Utility approval, Geological pathogens, Structure demolition, purchasing a Heritage Structure, and Providing land for construction purposes.

Author created a survey form asking respondents to rank the concerns according to their influence on the project in order to determine the impact of the aforementioned issues. There were three options given for each issue: "low impact," "moderate impact," and "high impact." State government organizations and private contractors with expertise in PPP road and highway projects filled out the survey form.

Conclusion: Although the PPP model is effectively used all over the world, it failed miserably in India. The author of this research paper has examined the numerous problems that have arisen in PPP road and highway projects and is aware of how they have affected the projects. 11 concerns that are highly impacted, according to data are: (1) Delay due to land acquisition (2) Change in law (3) Delay in site handover (4) Project feasibility (5) Lengthy delay in negotiation (6) Difficulties in obtaining work permits (7) Effect of social & cultural factors (8) Political violence (9) Cost & time overrun (10) Delay

in getting approval from various government departments (11) Acquiring heritage structure. As indicated in figure 6 below, there were 11 issues with a significant impact on the project, 21 issues with a moderate impact, and 34 issues with a low impact.

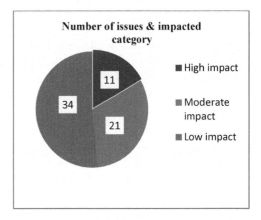

Fig. 6. Issues and their impact on project

Recommendations: Based upon the conclusion, a few significant PPP-related difficulties have been identified that can be fixed by altering PPP policies and by setting up a single point of contact for PPP issues.

References

Reddy, N. S., and Sharma, P. (2017). Why PPP Modeled Infrastructure Projects Failed: A Critical Review with a Special Focus on Road Sector. International Journal of Advanced Engineering, Management and Science, 3(4), p. 239816.

Nallathiga, R., and Shah, M. (2014). Public private partnerships in roads sector in India. In International Conference on "Public Private Partnerships: The Need of the Hour."

Malek, MohammedShakil S., and Bhatt, Viral (2022). Examine the comparison of CSFs for public and private sector's stakeholders: A SEM approach towards PPP in Indian road sector, International Journal of Construction Management, DOI: 10.1080/15623599.2022.2049490

Sarkar, S., and Kovid, R. K. (2015). Framework of Risks Factors and Financing Implications for Road Projects in India: Study of Selected Cases. Pacific Business International, 8(2), pp. 110–122.

Tieva, A., and Junnonen, J. M. (2009). Proactive contracting in Finnish PPP projects. International Journal of Strategic Property Management, 13(3), pp. 219–228.

Ling, F. Y. Y., and Hoi, L. (2006). Risks faced by Singapore firms when undertaking construction projects in India. International journal of project management, 24(3), pp. 261–270.

Sastoque, L. M., Arboleda, C. A., and Ponz, J. L. (2016). A proposal for risk allocation in social infrastructure projects applying PPP in Colombia. Procedia Engineering, 145, pp. 1354–1361.

Patil, S. K., Gupta, A. K., Desai, D. B., and Sajane, A. S. (2013). Causes of delay in Indian transportation infrastructure projects. International Journal of Research in Engineering and Technology, 2(11), pp. 71–80.

Nallathiga, R., and Shah, M. (2014). Public private partnerships in roads sector in India. In International Conference on "Public Private Partnerships: The Need of the Hour.

Rahangdale, P. (2020). Role of Public-Private Partnership in Infrastructural Development Projects: An Indian Perspective. Lex Revolution.

Effect of Lean Principles on Indian Highway Pavements

Pooja Gohil[,a] and MohammedShakil Malek[b]*

[a]Gujarat Technological University, India
[b]F. D. (Mubin) Institute of Engineering & Technology, Gujarat Technological University, India
E-mail: [*]pndave23@gmail.com

Abstract

Indian highway pavements are facing issues with subpar quality, poor performance, bad working conditions, etc. Pre-requisite happens to cut waste, improve quality, and incorporating new technology for highways since they are related to social, economic, and environmental issues. Here, "Pull Systems" one of the Lean Principles was used to prevent overproduction on highway pavement. Owing to time constraints, just two of the most cost-effective materials: bitumen and cement, were subjected to lean approach. On-site monthly consumption rates of both materials were made and were used to determine the cost of dead inventory. Later, lead time stock was added. An analysis led to the discovery that a sizable portion of material inventory could be reduced on-site, freeing up excess funds trapped in dead inventory stock thereby gaining good liquidity. Proposition aiding on-site inventory monitoring was provided. Conclusion disclosed the effect of lean principles on dead inventory cost and the successful elimination of waste in highway pavements.

Keywords: Lean Principles, Highway Pavement, Waste, Pull Systems

Introduction

Lean Construction is currently one of the largest revolutions in the building sector. There is n number of examples that can be used to intricate the concept of Lean thinking in the practical world, whether it is the manufacturing industry or construction such as Ford Production, Toyota Production System, Japanese lean construction, and so on. Lean Construction is a "way to design production systems to minimize waste of materials, time, and effort in order to generate the maximum possible amount of value" (Koskela et al., 2002).

The construction industry, which was once seen as a symbol of industrialization and development but is now criticized for being "ackward" and static, is such a crucial industry that its shortcomings can have significant negative effects. As it is linked to many environmental constraints, these criticisms are now being used as research areas to improve the sector.

Lean Principles: Generally, Lean concept is ruled by five major theories

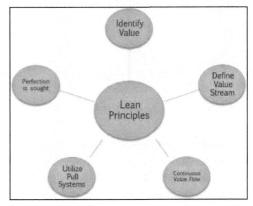

Fig. 1. Lean Principles

1. Identify Value: In building projects, the activities can be divided into three categories depending on value: Value Added; Non-Value Added; and Necessary Waste. Only that value should be taken into consideration; the rest should be discarded.
2. Define Value Stream by identifying each stage in order to identify activities that do not provide value and work toward their abolition.
3. Reduce downtime, defects, inventories, and delays to ensure continuous value flow.
4. Utilize Pull Systems: To supply the finished product or service to the customer at precisely the time that consumer desires it, all components and information must be manufactured and supplied at the appropriate times.
5. Perfection is sought: Lean entails continuously improving by jointly identifying and eliminating wastes to produce the desired results.

Construction Waste: Anything which does not add value to the final product or customer requirement can be considered waste in the construction industry. From a Lean point of view, waste can be categorized into 8 kinds as:

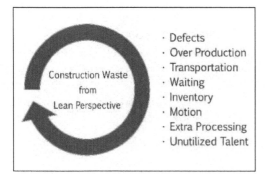

Fig. 2. Construction Waste from Lean Perspective

Pilot Study of Indian Highway Pavement

To eliminate waste brought on by "Overproduction," Lean Principle of "Pull Systems" is used in to project in current work, which considers a section of Indian National Highway.

Details of the Case Study are shown in Table 1.

Table 1. Case Study Details

Project Name	4 Laning of Ahmedabad - Godhra Section of NH-59
Length	118.22 Km
Chainage	4+200 To 122+420 Km
Duration	36 Months
Cost	₹1008.5 Crores
Type of Project	DBFOT- Design, Build, Finance, Operate and Transfer
Type of Contract	EPC- Engineering, Procurement, Construction
Client	NHAI- National Highway Authority of India

Bitumen and cement were the only two significant materials employed in the project that could be applied to the lean concept in the present study due to time constraints. After selecting major materials, various information regarding the same was acquired, including:

Table 2. Details of Materials

S. No	Details of Bitumen	Details of Cement
1	Type: VG30	Grade: M-30
2	Quantity: 19 MT & 19.53 MT	Quantity: 1000-2400 MT
3	Supplier: BPCL (Vadodara)	Supplier: ACC (Vadodara & Ahmedabad)
4	Mode of transport: Tanker	Mode of transport: Open trucks
5	Distance from source: 100km	Distance from source: 90 km From Vadodara;
6	Cost: 30,000 ₹/MT	Cost: 4000 ₹/MT
7	Storage Capacity: 4 Bins of 50 MT each per site	Storage Capacity: 10,000 MT
8	Lead time: 15 days	Lead time: 5 days
9	Procurement: 30 days prior to requirement	Procurement: 15 days prior to requirement
10	Order cycle: 15 days	Order cycle: 30 days
11	Time required for testing: 15 days	
12	Ideal time: 15 days	

Based on the information gathered, the monthly consumption rate for both materials was calculated in metric tons as shown below:

Table 3. Monthly Bitumen Consumption Rate

Month	Opening Stock	Quantity Purchased	Total Quantity	Quantity Consumed	Closing Stock
Mar	70	76	146	81	65
Apr	65	70	135	89	46
May	46	80	126	102	24
June	24	78	102	98	4
July	4	71	75	45	30
Aug	30	0	30	0	30
Sep	30	49	79	17	62
Oct	62	63	125	57	68
Nov	68	74	142	101	41
Dec	41	80	121	97	24
Total	440	641	1081	687	394

Table 4. Monthly Cement Consumption Rate

Month	Opening Stock	Quantity Purchased	Total Quantity	Quantity Consumed	Closing Stock
Jan	3000	1100	4100	2675	1425
Feb	1425	1430	7100	6285	815
Mar	800	1980	2780	2578	202
Apr	202	1750	1952	1286	666
May	666	2160	2826	2143	683
June	683	2370	3053	2987	66
July	66	1080	1146	1093	53
Aug	53	0	53	0	53
Sep	53	1022	1075	987	88
Oct	88	1765	1853	1597	256
Nov	256	1398	1654	1593	61
Dec	61	2300	2361	1879	482
Total	7353	18355	29953	25103	4850

Following analysis of aforesaid data, dead inventory cost and lean cost for both commodities were computed; as anticipated, lean cost was found to be lower than the usual dead inventory cost. The comparison of the two results is shown in the figure below.

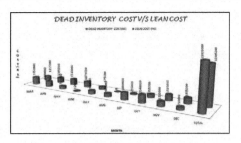

Fig. 3. Lean Application to Bitumen Cost

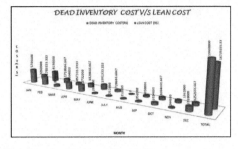

Fig. 4. Lean Application to Cement Cost

Conclusion

On implying lean principles, a sizable portion of material inventory on-site could be reduced, freeing up the excess funds trapped in dead inventory stock.

The investigation shows that on average, 57.35% of dead BITUMEN stock is stacked on-site monthly, costing an average of ₹1,18,20,000 per month (known as dead inventory cost), which if invested properly can generate a good income that is currently being lost by the company. Implication of "Pull Systems" (lean principle), reduces percentage of dead stock to 50.00% and an average cost of dead inventory to ₹1,18,20,000 per month (known as pulled cost).

Same is true for CEMENT, where we can see that on average, 19.32% of dead stock is stacked on-site monthly, costing around ₹1,94,00,000 (known as dead inventory cost), but by implementing "Pull Systems" (lean principle), this percentage can be decreased to 16.66% and the average cost of dead inventory can be abridged to ₹1,67,35,334 each month (known as pulled cost).

Proposition

To improve inventory control on-site, I would recommend the below approach based on the findings.

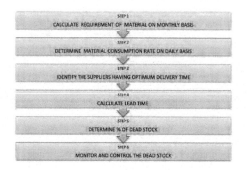

Fig. 5. Proposition

Step 1: First, decide what needs to be done each month. Next, figure out how much material is needed for that work in that month.

Step 2: Using the quantity needed each month, calculate the material's daily consumption rate.

Step 3: Find the providers with the best delivery times in step three.

1. One can learn the delivery time at which a specific provider will supply his material to the site by meeting with several material suppliers.

2. One can choose the supplier from whom the material can be obtained after determining the source who can deliver the material at the ideal delivery time.

Step 4: Calculate the lead time in step four.

1. After choosing the provider based on his ideal delivery window, a safety stock level should be established.

2. We can calculate the lead time by adding the delivery time and the number of days maintained in safety stock.

Step 5: From the computed lead, daily consumption, and monthly requirement, determine the proportion of dead stock.

Step 6: Critically monitor and control the dead stock percentage to ensure that it does not deviate from the fixed derived percentage and maintain inventory effectively without impeding the liquidity of dead stock.

Research Bridge

Though key cost-effective materials are included in study; due to time constraints, study was limited to the application of just one of the lean principles "Pull systems" and that too only on two of the major construction materials, namely Bitumen and Cement. Combination of other lean principles and different construction materials is yet to be explored.

References

Ansell, M. (2007). Lean construction trial on a highway maintenance project. Proceeding IGLC-15 In, 97–107.

Gleeson, F., and Townend, J. (2007). Lean construction in the corporate world of the U.K. construction Industry, University of Manchester, School of Mechanical, Aerospace, Civil and Construction Engineering.

Howell, G. A. (1999). "What is Lean Construction" Lean Construction Institute.

Koskela, L., Lahdenperaa, P., and Tanhuanperaa, V.-P. (1996). "Sounding the potential of lean construction: A case study Proceedings of the 4th Annual Conference of the International Group for Lean Construction, Birmingham, U.K.

Liker, J. (2004). The Toyota Way, McGraw-Hill Publishers, New York, U.S.A.

O'Flaherty, Coleman A. (2002). Highways: The Location, Design, Construction & Maintenance of Road Pavements, Fourth Edition, Butterworth-Heinemann, London.

Pinch, L. (2005). "Lean Construction eliminating Waste" Construction EXECUTIVE (Nov.), 34–37.

Polat, G., and Ballard, G. (2003). Waste in Turkish Construction: Need for Lean Construction Techniques, LCI White Paper, 11, 14–17.

Trivedi, Jyoti, and Rakesh Kumar. (2014). Optimisation of construction resources using lean construction techhnique. International Journal of Engineering Management and Economics 4.3–4, 213–228.

Wolbers, M., Evans, R. J. E., Holmes, M., Pasquier, C. L., and Price, A. D. F. (2005). "Lean construction trial on a highways maintenance project" Proceeding IGLC-15 In., 119–128.

A Critical Review on Cash Flow Management for An Engineering Procurement Construction Sector

Devarsh Patel,[,a] Dhiraj Bachwani,[b] and MohammedShakil Malek[c]*

[a,b]Parul University of Vadodara, India
[c]F. D. (Mubin) Institute of Engineering and Technology
E-mail: [*]pdevarsh1999@gmail.com

Abstract

A Cash flow statement must be provided by businesses, according to International Accounting Standard 7 (IAS 7), which was released by the International Accounting Standards Board (IASB) in 1992 and went into effect in 1994. Since Cashflow's launch, a significant number of studies have been published, and it appears that it is now time for a thorough assessment of the literature. To find out the status of ongoing research on cash flow, this study describes the methodologies used by researchers, and their contributions, and makes recommendations for future research. The primary goal of this study is to explain the state of cash flow research at the moment and to identify areas of cash flow that need more investigation. It covers the proportions of cashflows and working capital. As a result of this study, the state of cash flow research is critically examined, and potential cash flow research areas are suggested.

Keywords: Cashflow, Working capital, Finance management

Introduction

The amount of money that enters and leaves your business over a given period is known as cash flow. Cash flow is important because it enables you to both meet your current financial obligations and make long-term plans. Cash flow, however, is frequently a problem for small businesses. The most crucial resource for a construction company is money. More construction companies fail because they don't have enough money to cover day-to-day expenses than because of poor resource management. claimed that economic problems account for nearly 60% of building contractor failures (Hamburger, 1986). Various forecasting methodologies have been used to better understand the real business environment in the manage cash flow in the construction industry. The degree of automation and the accuracy and detail of cash flow forecasting and management techniques vary. both the process of integrating them and automating the compilation process time and money-related issues Although most methods are not probabilistic, some are there. Cash flows are influenced by a number of variables, including the project's duration, retention requirements, the client's payment due dates, credit agreements with suppliers or vendors, equipment leases, and the subcontractors' payment due dates.

Current Scenario On Business Cash Flow Across India is as follows

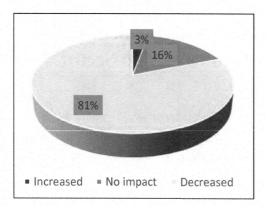

Fig. 1. A survey of cash flow for public and private sector

The poll included participants from India's private and public sectors, as well as multinational organizations. Another aspect is financial planning is a guide to corporate expansion and change (Ceren Oral, 2015). Decisions on financial planning for the four fundamental cases are:-

1. The level of investment in fixed assets.
2. In the planning period, the company's liquidity or working capital requirement level.
3. Debt and equity composition.
4. How to evaluate business decisions.

A positive cash flow indicates that income is greater than expenditures, whereas a negative cash flow indicates the opposite (Sampa Chakrabati, 2022).

The analysis starts with an initial balance and ends with a balance that takes into account all cash receipts and expenses that have been paid over time. Financial reporting frequently makes use of cash flow analysis. A project's cash flow analysis, which involves keeping track of how much money comes in and goes out of the project, is a crucial financial activity. A contractor can estimate future cash flows using (Ali and Shash, 2018) cash flow analysis to determine the project's budget. The timing of these cash flows as well as their amount are both factors that are considered in cash flow analysis. In the construction sector, the majority of cash flow is monitored on a monthly basis.

Research Methodology

This study aims to describe the status of research on cashflow and explore cashflow aspects.

Data was collected from science direct for different keywords. I used the keyword cashflow in Science Direct and received 27 results of research articles from 2015 to 2022.

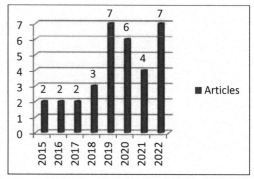

Fig. 2. Illustrates the proportions of cash flow from 2015–2022

I selected the keyword working capital in Science Direct and received 321 results research articles from 2011 to 2022.

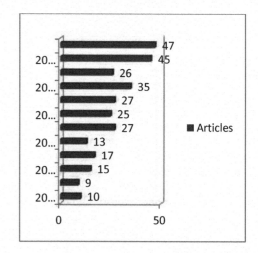

Fig. 3. Illustrates the proportions of working capital from 2011–2022

I selected the keyword finance management in Science Direct and received 37 results from research articles from 2015 to 2022.

Fig. 4. Illustrates the proportions of finance management from 2015–2022

Data was collected from web of science for different keywords I used the keyword cashflow in the Web of Science and received 25 results of research articles from 2015 to 2022.

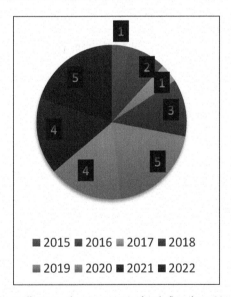

Fig. 5. Illustrates the proportions of cash flow from 2015–2022

I selected the keyword working capital in the Web of Science and received 264 results of research articles from 2011 to 2022.

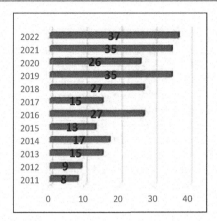

Fig. 6. Illustrates the proportions of working capital from 2011–2022

I selected the keyword finance management in Web of Science and received 28 results of research articles from 2015 to 2022.

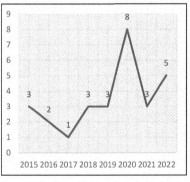

Fig. 7. Illustrates the proportions of finance management from 2015–2022

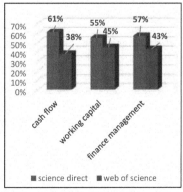

Fig. 8. Comparisons of both the journals

By comparing both the journal results I found out 61% of research articles of cashflow from science direct and 38% from web of science from 2015-2022. For working capital, results showed that 55% of research articles from science direct and 45% from web of science from 2011–2022. For finance management results showed that 57% of research articles were from science direct and 43% from the web of science from 2015-2022. So mainly the higher preference is given to science direct, and the data are much more available.

Risk Factors In Engineering Cash Flow Management

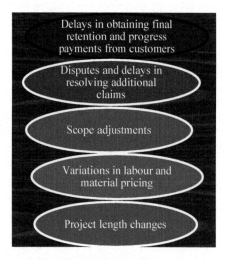

Fig. 9. Risk factors for (CFM)

Future Scope of Research

Cash flow is not yet regarded as the industry standard; nonetheless, it has overcome obstacles such as project management community ignorance and resistance. Academics have generated many ideas for enhancing early Cash flow tools and methods over the years to increase Cash flow in mainstream project management, but there are still several topics that require researchers' attention.

According to future researchers, other cash flow measures (such as working capital from operations, earnings after tax plus depreciation reported cash flow (including cash flow from financing and cash flow from investing), sales, and cash flow ratios could all play an important role in predicting future cash flows.

Conclusion

In this, I used the secondary data collection method for the data collection to critically review the literature on cash flow published in journals and further there is still ongoing research being conducted on the method and although at a slower pace. MS Project, Primavera, and other project management software create "cash flow" reports based on timeframes and resource costs. A thorough review of the software's operating manuals, as well as actual use of the software, found that the cash flows are computed on an accrual basis, with no regard for the cash flows' real timing.

The improvements brought by these studies were compared in order to find out the shortcomings, it also covers the risk factors for cash flow management. It also showed the comparison between two journals to know about proper terms regarding cashflow which are the following:-

Table 1. Comparison of keywords from journals in % for every year

Keywords	Science Direct	Web of Science	From Year
Cashflow	61%	38%	2015–2022
Working Capital	55%	45%	2011–2022
Finance Management	57%	43%	2015–2022

References

Chen Mark T. (2007). ABC of cash Flow Projections, AACE International Transactions PM.02-05.

Fellows, R. F. (1982). Cash Flow and Building Contractors. Quantity Surveyor. September 1982, pp. 121–124.

Hamberger, David H. (1986), Three Perceptions of Project Cost, Project Management Journal, June 1986, pp. 51–58

Kenley, R., and Wilson, O. D. (1986). A Construction Project Cash Flow model—an Idiographic Approach. Construction Management and Economics 4, pp. 213–232.

Malek, MohammedShakil S., and Bhatt, Viral (2022). Examine the comparison of CSFs for public and private sector's stakeholders: a SEM approach towards PPP in Indian road sector, International Journal of Construction Management, DOI: 10.1080/15623599.2022.2049490

Malek, MohammedShakil S., and Gundaliya, P. J. (2020). Negative factors in implementing public–private partnership in Indian road projects, International Journal of Construction Management, DOI: 10.1080/15623599.2020.1857672

Malek, MohammedShakil S., and Gundaliya, Pradip (2021). Value for money factors in Indian public-private partnership road projects: An exploratory approach, Journal of Project Management, 6(1), pp. 23–32. DOI: 10.5267/j.jpm.2020.10.002

Malek, MohammedShakil S., Saiyed, Farhana M., and Bachwani, Dhiraj (2021). Identification, evaluation and allotment of critical risk factors (CRFs) in real estate projects: India as a case study, Journal of Project Management, 6(2), pp. 83–92. DOI: 10.5267/j.jpm.2021.1.002

Ng Ghim Hwee, Tiong, Robert L. K. (2002). Model on cash flow forecasting and risk analysis for contracting firms, International J. of Project Management. 20, pp. 351–363.

Navon, R., Member, L., 1980. Company-level Cash-flow Management.

An Advanced Project Management Technique for the Indian Construction Industry: Critical Chain Project Management

Rahul Bharadiya,[a,] Dhiraj Bachwani,[b] and MohammedShakil Malek[c]*

[a]MTech (Construction Project Management - Pursuing) Parul University of Vadodara, India
[b]Research Scholar, Gujarat Technological University, Ahmedabad, India
[c]F. D. (Mubin) Institute of Engineering and Technology
E-mail[*]: rahulbharadiya555@gmail.com

Abstract

Considering the infrastructure development in India is accelerating at present. According to the Ministry of Statistics and Programme Implementation (MOSPI), As of 1st December 2021, 1679 (Mega - 480 and Major - 1199) infrastructure projects were monitored by the ministry. The majority of infrastructure projects experience time and cost overruns during the implementation of construction work due to improper planning and inadequate project management technique. Generally, the traditional method Critical Path Method (CPM) & Programme Evaluation and Review Technique (PERT) are used for monitoring and controlling the projects. However, these techniques are not feasible for managing infrastructure projects. This paper introduced an advanced technique Critical Chain Project Management (CCPM) for managing large-scale projects. Also, this paper highlighted various problems which affect the time and cost overrun of projects. Future work could be done by adopting Critical Chain Project Management to minimize the constraints.

Keywords: Project Management, Critical Chain Project Management, Programme Evaluation and Review Technique, Critical Path Method, Infrastructure Projects, Uncertainty, Cost Overrun, and Time Overrun

Introduction

"Project management is the application of knowledge, skills, tools, and techniques to project activities to meet project objectives," according to the Project Management Body of Knowledge (PMBOK). The goal of project management is to complete projects on time, within budget, and with the desired quality. Project management, which provides a variety of tools and techniques, provides the ability to complete a project. The project manager is responsible for understanding and balancing the project's scope, planning, schedule, cost, and quality objectives as the project progresses and challenges and changes emerge.

According to Guofeng et al. (2014) and Assaad et al. (2020), the construction project faces a variety of constraints, including time and cost overruns, resource constraints, project complexity, and uncertainty. The constraints and uncertainty in infrastructure projects may lead to schedule delays (Sarkar et al. 2018). Also Narayanan et al. (2018) describe that majority of infrastructure fails to achieve time and cost objectives. Extraordinary constraints and inadequate project management techniques frequently cause construction projects to run behind schedule (Chester and Hendrickson 2005). In the context of the successful completion of construction (Ghanbaripour et al. 2018) investigate critical success factors based on a questionnaire survey.

Indian Infrastructure Projects

The Ministry of Statistics and Program Implementation (MOSPI) (2021–2022) claims Infrastructure projects are classified into two categories: mega projects (In which the project cost is Rs. 1000 Crore and above) and major projects (In which the project cost is Rs. 150 Crore to less than 1000 Crore). As of 1 December 2021, 1679 infrastructure projects (480 mega and 1199 major) will have been started. Figure 1 depicts the proportion of time and cost overruns in infrastructure projects during the year 2015–2021.

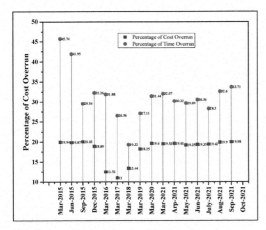

Fig. 1. Time & Cost Overrun in Infrastructure Projects
Source: Compiled from MOSPI by Author

Determination of the Problem by Literature Analysis

The author has reviewed 80 plus research papers and the Ministry of Statistics and Programme Implementation Report to identify the problems in infrastructure projects. The determination of problems is classified into two categories: Problems indemnification from research papers and problems identification from MOSPI Report.

Fig. 2. Major Problems from Literature Review

Figure 2 shows the major problems related to infrastructure projects which are identified from the literature analysis.

Problems Indication from MOSPI Reports

- Delay in ordering the material and equipment.
- Delay in technical approval.
- Inadequate manpower.
- Change in scope.
- Delay in project financing.
- Delay in finalization of detailed engineering.

Frequency of Problem Occurrence in Research Papers

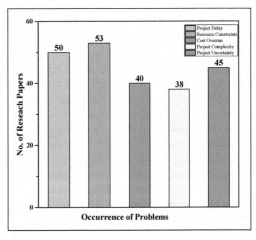

Fig. 3. Frequency of problems occur in Research Papers

Figure 3 shows that Project delay, Resource constraints, Cost overrun, Project complexity, and Project uncertainty are the top most occurred in various research papers.

Project Management Techniques

There are two types of project management techniques: traditional and network methods. The Gantt Bar Chart, Mile Stone Chart, and Line of Balance are examples of traditional methods. The Critical Path Method (CPM), Programme Evaluation and Review Technique (PERT), and Critical Chain Project Management are all part of the network method (CCPM).

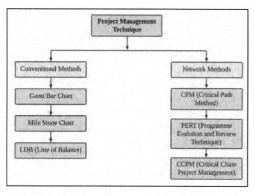

Fig. 4. Project Management Technique

Schedules can be created and managed using well-known and widely used techniques such as CPM and PERT. However, when projects become more complicated and resource demands rise, these approaches may not always be reasonable or effective due to their reliance on specific assumptions that become more problematic with time (Guofeng et al., 2014; Sarkar et al., 2018). According to Petroutsatou (2019) the majority of infrastructure projects continue to base their project scheduling on the more traditional methodologies of CPM and PERT. According to Hall (2012) states that the Critical path is not considered resources and uncertainty in task times. PERT makes it possible to estimate the extent to which the unpredictability of the times required for specific tasks contributes to the overall unpredictability of the duration of the project. However, PERT is heavily dependent on certain

robust statistical assumptions that are challenging to justify for most projects (Hall 2012).

Limitation of Traditional Method

- CPM and PERT method for planning and Scheduling is not feasible in real-life construction because their assumptions are not possible in real-life construction.
- CPM/PERT is not feasible for Risk Management or in Megaprojects.
- CPM/PERT does not consider resource constraints in calculations and allocation.
- CPM/PERT does not provide any safety time (Buffer Time) to eliminate the constraints.
- CPM/PERT does not fulfill the requirement of effective planning and scheduling for infrastructure projects.

Critical Chain Project Management - An Advanced Project Management Technique

Eliyahu Goldratt created the Critical Chain Project Management (CCPM) technique in 1997 in response to the poor performance of traditional project management techniques (CPM & PERT). The critical chain project management technique is used to track infrastructure projects and reduce uncertainty. Izmailov et al. (2016) describe CCPM as an approach to planning and managing projects that takes into account the constraints imposed by available resources. According to Petroutsatou (2019) research conducted on the adoption of critical chain project management techniques, conclude a significant reduction of 30% to 40% of the project's duration. The fact is that CCPM typically results in a far more efficient project schedule than the CPM and PERT.

Critical Chain Project Management is more practical than CPM since it solves problems with "BUFFER" rather than "FLOAT," which is more realistic. Buffers are classified into three types: project buffers, feeding buffers, and resource buffers. The research work of Izmailov et al. (2016) concludes that, when using CCPM, the progress of a project and the accuracy of its planning are typically checked not by the traditional method of earned value analysis, but rather by the percentage of the buffers that have already been used.

Application of Critical Chain Project Management

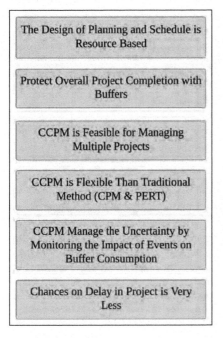

Fig. 5. Application of Circle Chain Project Management

Conclusion

Infrastructure projects in India are rapidly increasing because of Growth in the construction industry. However, these projects cannot achieve their time and cost objectives. The majority of infrastructure project faces problems of time and cost overrun due to improper project management technique. The CPM and the PERT are still used by the vast majority of construction companies. This method is neither feasible nor effective for managing infrastructure projects. The infrastructure project faces numerous constraints, including time and cost overruns, resource constraints, and uncertainty. Also in this paper, major problems were identified from research papers and the Ministry of Statistics and Programme Implementation (MOSPI) Report (2021–2022). Figure 4 shows the top problems (Project Delay, Resource Constraints, Cost Overrun, Project Uncertainty, and

Project Complexity) which are faced by infrastructure projects.

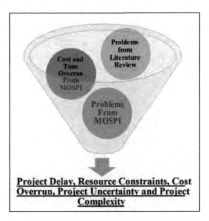

Fig. 6. Filtration of Problems from MOSPI Report, Research Papers, and Time and Cost Overrun data

This paper also introduced an advanced project management approach Critical Chain Project Management (CCPM). The multiple research papers reviewed and identified that CCPM is far more effective and feasible than CPM and PERT. CCPM monitors the project progress based on the buffer consumption. The main of the buffer is to absorb the constraints and complete the project within the time and budgeted cost.

An effective project management technique is critical to the successful completion of a construction project. It also aids in the management of uncertainties and the reduction of constraints.

Future Direction for Research Work

As a future direction for research work, an advanced project management technique for infrastructure projects CCPM can be applied in real-life projects to minimize the duration of projects. Also, CCPM can provide more effective planning and scheduling than traditional techniques (CPM & PERT).

There is planning and scheduling software available for future work, such as Microsoft Project and Primavera P6. As a result, real-life project planning may be more feasible in large-scale projects with a greater number of activities.

References

Sarkar, D., Jha, K. N., and Patel, S. (2021). Critical chain project management for a highway construction project with a focus on theory of constraints. International Journal of Construction Management, 21(2), pp.194–207.

Assaad, R., El-Adaway, I., and Abotaleb, I. (2020, November). Holistic risk management approach for predicting cost and schedule overruns at project completion. In Construction Research Congress 2020: Project Management and Controls, Materials, and Contracts (pp. 583-592). Reston, VA: American Society of Civil Engineers.

Izmailov, A., Korneva, D., and Kozhemiakin, A. (2016). Project management using the buffers of time and resources. Procedia-Social and Behavioral Sciences, 235, pp.189–197.

Ma, G., Wang, A., Li, N., Gu, L., and Ai, Q. (2014). Improved critical chain project management framework for scheduling construction projects. Journal of Construction Engineering and Management, 140(12), p.04014055.

Chester, M., and Hendrickson, C. (2005). Cost impacts, scheduling impacts, and the claims process during construction. Journal of construction engineering and management, 131(1), pp. 102–107.

Petroutsatou, K. (2022). A proposal of project management practices in public institutions through a comparative analyses of critical path method and critical chain. International Journal of Construction Management, 22(2), pp. 242–251.

Ghanbaripour, A. N., Sher, W., and Yousefi, A. (2020). Critical success factors for subway construction projects–main contractors' perspectives. International Journal of Construction Management, 20(3), pp. 177–195.

Narayanan, S., Kure, A. M., and Palaniappan, S. (2019). Study on time and cost overruns in mega infrastructure projects in India. Journal of The Institution of Engineers (India): Series A, 100(1), pp. 139–145.

Hall, N. G. (2012). Project management: Recent developments and research opportunities. Journal of Systems Science and Systems Engineering, 21(2), pp. 129–143.

Malek, MohammedShakil S., and Bhatt, Viral (2022). Examine the comparison of CSFs for public and private sector's stakeholders: a SEM approach towards PPP in Indian road sector, International Journal of Construction Management, DOI: 10.1080/15623599.2022.2049490

Research and Design of Packaging of Fruits in India: A Consumer's Perspective

H. Mehta,[a] R. Bhaiya,[b] P. Patil,[c] A. Parekh,[d] and M. Arunachalam[e,]*

Karnavati University, Gandhinagar, India

E-mail[*]: arunachalam@karnavatiuniversity.edu.in

Abstract

India is the second largest producer of fruit hence, good packaging is an essential component that ensures the safe handling, delivery, and storage of fresh fruits. The demand for fresh fruits has significantly grown owing to the consciousness of the consumer about its numerous health benefits. It is also known that several factors affect the freshness and shelf life of fruits and hence packaging becomes vital to enhance those. However, the packaging is also an important element when we talk about sustainable food consumption. Innovative sustainable packaging reduces fruit loss by preserving fruit quality, satisfies consumer expectations, and at the same time is environmentally conscious. This paper discusses the various developments in fruit packaging techniques along with the recent advances in synthetic and bio-based systems. The main objective of this review is to stimulate conversation around green packaging that is desirable for the consumer.

Keywords: Packaging, innovative, sustainable, consumer, bio-based

Introduction

1.1 Background

Since the 1980s, the international trade in fruits has risen tremendously. The quantity, as well as the varieties of fruits produced, has increased dramatically during this period. Today India is the second largest producer of fruits in the world just after China. There is in general an overall increase in demand for fruits both fresh as well as processed form. India has got climatic conditions that favor agriculture and large land areas to support the same. However, despite all the favorable agro-climatic conditions, India still suffers. Despite high production, fruits are being sold at higher prices in some regions. This is mainly because of irregular distribution. In the process of handling fruits, over 20-40% of the produce is lost many times. Packaging is one of the most important considerations in the fruit market. The losses are incurred not only due to unsafe transportation but also improper storage. The use of proper packaging to transport and market the fruit produced could influence the post-harvest losses incurred. In the beginning, the fruits were only to be transported from the production center to the local mandis but today the relevance of packaging is so much more; today a modern packaging system must be capable of filling up all the gaps from the initial production unit to the final consumer.

In designing fruit packages, one must consider all the aspects; from the physiological properties of fruits to the distribution channel. There are various types of packaging currently being used in the market, for example; polythene films, wooden pallets, mesh bags, foam nets, the recent innovation of modified atmosphere packaging and so much more.

Nevertheless, there is an urgent need to also consider the long-term needs of the ecosystem along with economic growth. Millions of tons of fruits are wasted annually due to spoilage. Thus, let's discuss a little more about how we can design a packaging system for fruits to make it more convenient for the consumer and at the same time help us progress towards a green economy.

1.2 Literature Review

Fruits are living plant-derived products which are usually consumed raw without major processing. Over the last few years, there has been a rise in the consumer demand for high-quality fruits owing to their numerous health benefits. This trend comes with a series of challenges of its own for food business operators as they try constantly to offer fresh produce and better post-harvest solutions to avoid operational losses. On the other hand, when fruit quality is a key concern for the consumer; good packaging is one of the most viable options to extend the shelf lives of fresh fruits. In the past, for example in the year 2012–13, India produced 81.28 million metric tons of fruits under an area of 6.98 million hectares while fruits worth Rs. 2503.75 crores were even exported. Fruit production in India has only seen a rise from thereon; most recently, in the fiscal year 2020, India produced over 100 million metric tons of fruits and is constantly growing. Although India is one of the largest producers of fruits in the world, more than 20% of the total production is often lost due to spoilage at various post-harvest stages. On a more technical side, fruits have high moisture and high equilibrium relative humidity (around 90%) because of which if not packaged properly, they have a high tendency to dry quickly and even wild due to loss of rigidity. As mentioned earlier, fruits are living organisms and respire even after harvesting, hence they need their metabolism to continue to remain fresh. With several advancements in packing technologies, the shelf life of fruits has seen a drastic improvement. However, are these modern methods viable for the consumer as well as the environment? In this paper, published research on the existing packaging of fruits is critically reviewed to utilize the information to design a more sustainable and convenient experience of packaged fruit for the consumer.

What role does packaging play in fresh fruits and their shelf life?

Good packaging and its gas exchange system design can extend the shelf life of the fruit produced in a considerable amount of time. On the topic of fresh produce, let's understand a little in depth how the factors affect the shelf life of fruits first.

Respiration is a major metabolic process that leads to natural aging and subsequent deterioration of fresh fruits. In the process of respiration, oxygen from the surrounding environment is taken in by the fruit and is converted into carbon dioxide, water, and heat (oxidative reduction of respiratory substrates such as carbohydrates, organic acids, etc. takes place). This conversion results in a loss of food nutritional value, poorer flavor, and in turn product quality depletion. Poor ventilation during storage and excessive carbon dioxide build-up can also lead to fermentation and yeast formation. Ethylene is a naturally occurring compound in plant tissues and is responsible for normal ripening of many fruits. The effect of the composition of naturally occurring ethylene in different fruits, exposure to it artificially for early ripening, or even induced ethylene production due to cutting or peeling of fruits can considerably affect the shelf life of the produce.

Storage temperature is the most vital part of understanding the shelf life of fruits and storing them at an optimum temperature and packing can reduce respiration rates, reduce sensitivity to ethylene and even reduce susceptibility to decay. Handling fresh fruits again is an essential aspect as physical damage or even damage by heat or sunlight can drastically affect the fruit's freshness.

So, to summarize, the biggest threat to fruit spoilage is during transport and storage. Fresh fruits get very little attention after being harvested and packed. Many harmful microorganisms affect the fresh produce due to bad processing. Of the potential seriousness of these risks, regulations on fresh fruit are becoming stricter, especially in developed countries. Physical deterioration, such as bruising, softening, and water loss, which can cause the product to shrink can be prevented by packaging. Packaging is an essential component in the food system that assures the safe handling and delivery of fresh produce. It has been around since the early days of agriculture when leaves and animal skins were used as packaging materials during transportation and even storage for later use. Furthermore, packaging can also protect the product from the increase or loss of moisture, dust, and light, especially ultraviolet rays, which can cause deterioration of some photosensitive products. It can also protect the contents of the package from temperature fluctuations during the transportation of refrigerated and frozen foods. Proper packaging can also prevent microbial spoilage caused by bacteria, yeast, and mold. It can also prevent microbial spoilage of stored products by rodents and insects.

Now that we have discussed the various factors we have to keep in mind while understanding packaging and storage of fresh fruits but gone are the days when packaging was solely for shielding organic produce. Today packaging of fruits has evolved and is responsible for comfort, keeping up with the necessities of the buyer, visual communication, protecting the produce from physical or chemical damage, and at the same time must have a limited ecological effect on the environment.

Aim

The study aims to research and design sustainable packaging for fruits keeping in mind an Indian consumer.

Objective

This research thus aims at understanding the issues and challenges faced to plan and implement strategies to develop better packaging for the final consumption stage. The present study is conducted with the following objectives:

- Study and analyze the current packaging adopted by the consumer on the selected varieties of fruits.
- To suggest improvements/modifications in present packaging options or recommend suitable alternatives.

- Packaging options should be easily transportable, easy to assemble, provide adequate ventilation and cost-effective.
- To draw up the final report supported by drawings and prototypes.

Methodology

The present study undertakes a thorough review of the literature available and identifies issues affecting the consumer.

Based on this, we conducted research on Fruit Packaging and generated a questionnaire about the same. The questionnaire was made using Google forms and distributed via online media. It was answered by people of a mixed age group starting from 10. This was done to prove our hypothesis- Edible packaging is an emerging strategy for food quality optimization and the consumer is receptive to it because of an increase in consciousness.

The important questions of contention in the survey were as follows:

- Where do you often buy fruits from? (Local vendor, supermarket, online)
- What material would you prefer for your packaging of food? (cardboard paper, plastic films, foam nets, etc.)
- Do you think there are any problems with the existing food packaging?
- Would you be willing to pay a price difference for packaged fruit?

After this, we designed a new packaging based on the responses from the questionnaire, to meet user needs.

Results

Feedback and data from over 340 people from different backgrounds. The majority responded by saying that they bought their fruits from street vendors while some opted for supermarkets and as low as 5% of people chose online platforms like a big basket for their supplies. 71% of the people that participated in the survey had a positive response upon being asked whether they buy their own groceries while the rest 29% opted otherwise. A surprising 96% of people even chose whole fruit over pre-cut fruit and non-packaged fruit over the ones that come in nearly packed containers. While some of them preferred to save their cut fruit for later, 62% opted to eat it in one go. Another interesting observation was that around 50% of the people were willing to pay a price difference for packaged fruit if necessary. The found data only prompts us towards a single idea that people in India are more prone to packaged fruit if the packaging is simple to understand and sturdy as paper or cardboard and would provide value, or else they prefer their local vendors.

This could be important to note as the majority still prefer whole fruit, possibly because pre-cut fruit may not be able to retain its freshness in existing packaging options. Many even despise the plastics being used for packaging and would want to opt out of it.

Fig. 1. This packaging is ideal for spherical fruits bought in a limited quantity and is possibly can be made of biomaterials

Table 2. Given above are a few suggestions of by-products of popular fruits that can be used to enhance food packaging performance. These materials are thoroughly researched and worked on, helping us achieve a sustainable lifestyle. These emerging technologies are surely going to bring about change in the current packaging trends by reducing plastic consumption and costs of production.

By-Products	Packaging System	Physical and Mechanical Properties
Citrus peel and leaves	Kraft paper + peel:leaf extract (2:0, 2:1, 3:0).	Peel:leaf extract (2:1) increased WVB** and O_2B**.
Grape seed (GSE)	Chitosan and gelatin films with GSE (1% *v/w*) and *Ziziphora clinopodioides* essential oil (ZEO).	1% GSE + 1% ZEO decreased TS*, PF*, PD* and SI**; increased WVB**.
Grape seed (GSE) + Pomegranate peel (PPE)	Surimi edible films with GSE + PPE (0%, 2%, 4% and 6%).	6% PPE improved TS*; 6% GSE increased WVB** and both reduced light transmission.
Mango peel and kernel (MKE)	Edible mango peel coating with MKE (0.078 g/L).	MKE reduced WVB** and film solubility (from 60.24 to 52.56%).
Mango peel extract (MPE)	Fish gelatin film with MPE (1%, 3% and 5%).	MPE improved TS* (from 7.65 to 15.78 MPa) and reduced solubility from 40% to 20%.
Mango kernel starch	Composite film (kernel starch and guar/xanthan gum 10%, 20% and 30%).	The different % of gums increased TS and O_2B**, but decreased the film solubility and WVB*.
Pomegranate peel extract (PPE)	Zein film with PPE (0, 25, 50, and 75 mg/mL of film forming solution).	PPE improved TS* and WVB**, increased film solubility from 6.166% (control) to 18.29% (75 mg PPE).

* Mechanical properties: tensile strength (TS); puncture force (PF); puncture deformation (PD); ** physical properties: swelling index (SI); water vapor barrier (WVB); oxygen and carbon dioxide barrier (O_2B and CO_2B).

Conclusion

Responsible resource utilization may contribute to reduced waste in both food consumption as well as packaging, it also plays a vital role in maximizing food safety and ensuring enhanced quality of fruit. Packaging is an essential element in reducing food spoilage while transporting and storage of fruit; tonnes of fruits get spoiled while transportation and distribution can be minimized with the help of packaging. Currently, packaging trends available in the market are not well adapted for fruit packaging particularly, packaging currently available is primarily focused to be inexpensive. Respiration, Ethylene, Storage, and Handling are the main factors that affect the shelf life of fresh fruit; early spoilage of fruit can be prevented by deriving more functional and sustainable packaging, thus promoting the concept of a green economy.

When food (fruit peels, leftovers) is thrown away, the packaging is also thrown away leading to being a burden to the environment. This paper provides a link between consumer perception of packaging and environmental sustainability. Packaging is a silent salesman for a product, as time has passed consumers have become more demanding.

Still, efficient sustainable packaging is facing hurdles to be adopted by both the manufacturers and the consumer. Nowadays many packaging techniques have influenced fruit packaging as we know it like wooden pallets, corrugated fibreboard, plastic shrink, plastic bags, paper, smart packaging involving advanced technologies, and so on but what is the future? It is something we must consciously think about.

References

National Horticulture board (2013). Cited from: www.nhb.gov.in.

Ali, J., Kapoor, S., and Pradesh, U. (2008). June. Consumers' perception on fruits and vegetables packaging in India. In Annual World Symposium of the International Food and Agribusiness Management Association, Monterey, California, USA.

Smith, J. P., and Ramaswamy, H. S. (1996). Packaging of fruits and vegetables. Processing fruits: Science and Technology. Lancaster, PA: Technomic Publishing Co, pp. 379–427.

Ragaert, P., Verbeke, W., Devlieghere, F., and Debevere, J. (2004). Consumer perception and choice of minimally processed vegetables and packaged fruits. Food quality and preference, 15(3), pp. 259–270.

White, A., and Lockyer, S. (2020). Removing plastic packaging from fresh produce–what's the impact?. Nutrition Bulletin, 45(1), pp. 35–50.

Galgano, F. (2015). Biodegradable packaging and edible coating for fresh-cut fruits and vegetables.

Ščetar, M., Kurek, M., and Galić, K. (2010). Trends in Fruit and Vegetable Packaging - a Review. Hrvatski Časopis za Prehrambenu Tehnologiju Biotehnologiju i Nutricionizam - Croatian Journal of Food Technology, Biotechnology and Nutrition 5.

https://www.researchgate.net/publication/342585172_Sustainable_Use_of_Fruit_and_Vegetable_By Products_to_Enhance_Food_Packaging_Performan

Bala, S., and Kumar, J. (2017). Export Potential And Packaging of Some Important Fruits of India. Journal of Plant Development Sciences, 9(3), pp. 157–164.

Nayik, Gulzar Ahmad, and Khalid Muzaffar (2014). Developments in the packaging of fresh fruits-shelf life perspective: A review. American Journal of Food Science and Nutrition Research, 1(5), 34–39.

Aari Embroidery and Extent of It's Usage in Indian Couture

Pritam Saha[*,a]

[a]Karnavati University, Gandhinagar, India
E-mail: [*]pritamsaha@karnavatiuniversity.edu.in

Abstract

Aari embroidery has been a predominant craft in traditional Indian textiles since the Mughal period of 16th Century India and currently, this forms a major technique of surface design in Indian Couture. However, quite often the technique is not credited and is generalized as hand embroidery. This study looks at the couture collections of 5 of the most prominent Indian Couturiers and maps the usage of Aari embroidery in their collections. The methodology adopted is mainly observational from secondary data. This is a review paper that mainly analyses images of high-end Indian couture garments and visually examines the content of the same with respect to aari embroidery. The constructs of the study will be measured using indicators like the percentage of embroidery in the entire product which will be visually observed and noted by the researcher, followed by observations and discussions on the data.

Keywords: Indian - Couture, Aari Embroidery, Craft, Indian, Fashion

Introduction

Haute couture, literally translated as high dressmaking, is one of the most luxurious segments of fashion since its inception in the late 19th Century. Luxury garments have been one of the most popular forms of opulence amongst the royals and high society people worldwide and the dressmakers behind them have been of utmost importance in royal courts. In the late 1900s, Charles Freddrick Worth known as the father of haute couture was the first person to put a label on a garment, and perception of a dressmaker as a fashion designer was born. This was a key moment in the history of fashion. Today worldwide Haute Couture is an overused term but officially it is passed with extremely stringent certification requirements. These are one-of-a-kind clothing made for a single client. The Chambre Syndicale de la Haute Couture based in Paris, France now selects designer members who go through a qualifying procedure. In a nutshell, the main criteria are that the members must design made-to-order clothing for private clients, with more than one fitting, in an atelier (workshop) with at least fifteen full-time employees to qualify as an official Haute Couture firm. In addition, one of their workplaces must have twenty full-time technical personnel. Finally, Haute Couture firms must display to the public a collection of at least 50 original designs — including day and evening clothing — every season, between January and July. But a lot of fashion designers who are not recognized by this body are also making couture garments of a similar magnitude worldwide, and the treds generated by them are trickling down into the other end of fast fashion.

Indian fashion, on the other hand, is nascent in terms of organized fashion industry with the first fashion week happening in 1999, and the first couture week held in 2008. In this short period, the industry has grown manifold. According to estimates by Fibre 2 Fashion in 2006, India's whole garment market was worth around Rs 20,000 crore. The branded garment industry was almost one-fourth the size of this, or Rs 5,000 crore. Designer clothing, in turn, accounted for around 0.2% of the branded garment industry. The highest sales turnover in the designer clothing market is around Rs 25 crore, with other well-known labels having lower turnovers of Rs. 10–15 crore.

In the current scenario, according to McKinsey's FashionScope statistics, India's clothing market would be valued at $59.3 billion in 2022, ranking sixth in the world after the United Kingdom ($65 billion) and Germany ($63.1 billion). India is becoming a fashion industry focus point, thanks to a fast-developing middle class and an increasingly dominant manufacturing sector. These pressures, along with solid economic foundations and expanding technological sophistication, make India much too crucial for multinational businesses to overlook.

Couture is one of the most prominent segments in the Indian fashion market mainly among high-end designers and local markets have seen a trickle-down version of that. The category has been popularized in almost every corner of the country due to the popularity of fashion designers and celebrities wearing these. An Indian couture garment would mainly be a range of traditional silhouettes like Lehenga sets, saris, etc., with each of the products having a wide range of surface design techniques. Aari embroidery is one of the most prominent forms of embroidery amongst these product categories, some form of it is usually seen as a major part of these couture garments and has been used extensively by most of Indian fashion designers.

Aari embroidery has been a predominant craft in traditional Indian textiles since the Mughal period of 12th Century India and currently, this forms a major technique of surface design in Indian Couture. Once donned by the country's rulers, the patterns would always shine out, on the sheen of silk and the softness of velvet. The art of aari embroidery has gone a long way from

its regal days, with Persian designs entrancing the clothes and commodities producing a sheen of wealth and elegance.

Aari work is created with a pen-like long needle that resembles the shape of a crochet needle, a long needle with a hook at the end. The thread is taken to the upper side and used to fasten the preceding stitch once the aari is pierced through the material. This one-of-a-kind stitch, similar to the cobbler's stitch, is repeated until the desired shape is achieved on the fabric's surface. This work is well-known for its exquisite threadwork, and usage of beads, sequins, and other embellishments, which highlights the essence of hand embroidery.

The purpose of this study is to find out the extent of usage of aari embroidery by the major fashion designers and establish the importance of this technique in the current couture market in India.

Review of Literature: The interaction of the Indian Fashion System with Indian society is inherent and displays itself via the epicenter of Indian culture's main institution of marriage. Marriage in Indian society is about the union of two families, not just the couple, and as a result, has numerous dynamic socioeconomic and occasionally political undercurrents. The traditions and ceremonies of a traditional Indian wedding last many days, with specific indications for a clothing code that is common in most groups and has few changes. In Hindu tradition, the sari represented femininity, hence growing up was marked by a change from a skirt-blouse (Ghagra-choli) to a more mature clothing; the sari (Ranavaade and Karolia, 2016). These weddings have transformed into a platform for wearing the most luxurious clothing by the elites in India wherein most of the traditional crafts are visible. Embroidery is the most prominent royal craft it is predominantly seen in these couture products (Kuldova, 2016).

Fashion forerunners for both the local and international markets owe a great deal to Indian traditional textile techniques as raw material for fashion. Fashion may be described as anything that people accept at a specific period and location. Textile craft has been a key contributor to fashion in India for ages and continues to be so now (Gupta, 2012).

The main style in Indian couture is "royal chic," Kuldova (2016) a fashion that re-imagines current India via emblems of its past glory, which is projected on the imagined canvas of India's future. The trend's defining element is an artistic reinterpretation of aristocratic aesthetics from bygone centuries, from the Rajputs to the Mughal Empire and beyond, allowing the modern commercial elite to reinvent themselves as a neo-aristocratic class within the putative Indian democracy.

Methodology: The paper follows the method of a qualitative review of designer collections and maps the usage of aari work in them.

A sample of 5 designers showcased in the India Couture in July 2022 was selected based on the popularity of designers, which was determined by the number of Instagram followers for these labels.

3 looks were chosen from each of these 5 designers' couture week collections which were

of the highest price as per the online store of their website and mapped for getting certain data points. Certain designers who have a significant number of followers on Instagram, and made to the sample list were omitted as their online stores either did not have or did not reveal the prices of their latest couture collection. Use of aari work by the selected designers was informed through the product descriptions on their websites and behind-the-scenes videos posted by them where the usage of long aari needles was evident, hence inferred that the hand embroidery being used by them is primarily aari work.

The following data points were identified from reviewing the designer's work:

Name of the Designer Label, % of aari work used in the couture collection (this interpretation was done purely through observation based on the researcher's expertise), and Other surface design techniques used.

The data was collected and tabulated and observations were made based on the same. The same is summarised below in "Table 1."

Table 1. Mapping of % of aari embroidery used against selected Indian couture garments, based on interpretation done by observation.

Designer Label	% of Aari used (based on observations of 3 most expensive looks from India Couture Week Collection)	Other Surface Design Techniques Used
Manish Malhotra	0	*Chikan Kari*
Manish Malhotra	0	*Chikan Kari*
Manish Malhotra	0	*Chikan Kari*, Mijwan Thread work
Tarun Tahiliani	90	*Zardozi*
Tarun Tahiliani	0	3D Applique, Cutwork
Tarun Tahiliani	90	*Zardozi*
Anju Modi	60	Not Identified
Anju Modi	80	Not Identified
Anju Modi	50	Not Identified
Amit Aggarwal	70	Fabric Manipulation
Amit Aggarwal	70	Fabric Manipulation
Amit Aggarwal	85	Fabric Manipulation
Rahul Mishra	90	Not Identified
Rahul Mishra	85	Not Identified
Rahul Mishra	80	Not Identified

Based on the above data, it was observed that a total of 57% of these garments are using aari embroidery, which is a significant number for one technique to be used when there is an availability of many to be chosen from.

Observations: It was observed that a significant percentage of high-end couture garments have aari work in them establishing the importance of the craft in today's scenario.

It was observed that the usage of aari needle is done for most of the embroideries done in these couture pieces but in a majority of designer websites the name of the technique is not mentioned in the product description of the same, and it is referred to as simply hand embroidery. It is inferred from their behind-the-scenes videos which clearly show aari embroidery needle being used, thus the embroidery is recognized as aari work.

It was also observed that the aari embroidery needle was used in a very versatile manner by the couture designers, balancing both traditional and modern interpretations of the same.

Discussion: The key point of discussion that emerges out of this study is that of the identity of a recognizable craft, how the same craft of aari embroidery is perceived by the consumers and designers.

The designers as mentioned here showcase very contemporary versions of aari embroidery in the couture week and it can be discussed at this point whether a craft is identified by its aesthetic or technique. As it has been observed in most of the couture collections, it can be discussed that the visual form of the craft is what is getting modernized but the technique remains the same. On the contrary, there are also a lot of collections where the visual aesthetic also draws inspiration from traditional times and is presented in a new manner.

It may also be discussed that the story behind a product or a collection is what justifies the techniques used in them and thus there can be multiple stories, both technical and visual behind the same collection itself.

Conclusion: From the analysis of the observational data it can be seen that aari embroidery, in spite of its wide extent of usage in Indian couture garments, is not always recognized by the designers in their product descriptions. Owing to the lack of this recognition, it can be observed that aari when compared to other forms of embroidery like chikankari has not been able to create a distinct identity of its own in the eyes of the modern-day consumer. This is wider context may contribute to the fact that the aari embroidery artisans other than the ones employed directly by the designers, have not been able to create a brand for themselves and are overly dependent on job work. It can be suggested that the visual popularity but cultural anonymity of this form of embroidery may be carried forward by the aari artisan who directly come up with exclusive aari embroidered products, made and sold by them, thereby leading to their economic growth.

References

Gupta, T. (2012). Textile Crafts and their contribution in India Fashion. The American Sociological Association ASA, 2–16.

Kuldova, T. (2016). Indian "Haute Couture": Ornamentalism and the Aesthetic Economy of Neo-Imperial Atmospheres. Culturologia: The Journal of Culture, 2(2), 1–9.

Ranavaade, V. P., and Karolia, A. The study of the Indian fashion system with special emphasis on women's occasion wear. International Journal of Textile And Fashion Technology (IJTFT), ISSN (P): 2250-2378; ISSN (E): 2319-4510. 6, 19–34.

Aggarwal, A. (2022). Amit Aggarwal. Retrieved 27 September 2022, from https://www.instagram.com/p/ChWHhbBg9b3/

Haute Couture | Fashion A-Z | BoF Education | The Business of Fashion | #BoFEducation. (2022). Retrieved 26 September 2022, from https://www.businessoffashion.com/education/fashion-az/haute-couture

How India's ascent could change the fashion industry. (2019). Retrieved 26 September 2022, from https://www.mckinsey.com/industries/retail/our-insights/how-indias-ascent-could-change-the-fashion-industry

Modi, A. (2022). Anju Modi on Instagram: "The Road Less Travelled Couture 22UnveilingWednesday, 27th July, 2022Stay tuned." Retrieved 27 September 2022, from https://www.instagram.com/p/CgcEdt5JHJ0/?hl=en

Rahul Mishra. (2022). The Process • The Tree of Life [Video]. Retrieved from https://www.instagram.com/p/ChmzpbeB7bx/

Malek, MohammedShakil S., and Bhatt, Viral (2022). Examine the comparison of CSFs for public and private sector's stakeholders: a SEM approach towards PPP in Indian road sector, International Journal of Construction Management, DOI: 10.1080/15623599.2022.2049490

Malek, MohammedShakil S., and Gundaliya, P. J. (2020). Negative factors in implementing public–private partnership in Indian road projects, International Journal of Construction Management, DOI: 10.1080/15623599.2020.1857672

Appendix

Aari - A form of long needle with a hook in an end, using which a type of hand embroidery is done, popularly used in Indian clothing.

Zardozi - A type of hand embroidery, popularly used in Indian clothing, done with a small needle and often combined with a set of metallic embellishments along with threadwork.

Chikankari - A form of hand embroidery using thread and small needle, originating from Lucknow, India.

Knowledge Management – A Sustainable Yet Dynamic Advantage Facilitated by Interpersonal Trust within the Organisation

Dr. Bindiya Gupta [a,*] and *Dr. Bhumika Achhnani* [b]

[a]Associate Professor, UWSB, Karnavati University, Gandhinagar, India
[b]Associate Professor, Faculty of Management Studies, Marwadi University
E-mail: *bindiya@karnavatiuniversity.edu.in

Abstract

It is increasingly being recognized that knowledge-based dynamic capabilities are potentially more viable to generate competitive advantage for a firm, rather than employing knowledge process capabilities or dynamic capabilities separately. The current study unbundles the dynamic capability concept and attempts to elucidate the nature of the associated processes, linking its roots to knowledge management. An exhaustive literature review was conducted to understand the concepts of knowledge management and dynamic capability through the lens of structuration theory. Further, an attempt is made to propose a framework built upon the existing literature, suggesting that sustained competitive advantage can be obtained by building dynamic capabilities with the help of knowledge management which is moderated by the trust within the organizations. This paper thus serves as a preliminary study introducing an integrated conceptual model and making propositions for future empirical research.

Keywords: knowledge management, dynamic capabilities, strategic management, Interpersonal Trust

Introduction

The dynamic capabilities approach seeks to explain the reasons behind the differential success achieved by organizations in sustaining their competitive advantage within the context of dynamic environment (Teece, 1997).

Teece, Pisano, and Shuen (1997) defined dynamic capabilities as "the firm's ability to integrate, build, and reconfigure internal and external competencies to address rapidly changing environments." This definition still holds true, although the speed of change in the environment may be less important now than the level of uncertainty.

The concept of dynamic capabilities stresses the revitalization of organizational resources by reconfiguring them into fresh skills, competencies, and capabilities (Teece, 1997) and enables firms to spontaneously respond to new and recessionary situations. On the other hand, knowledge management emphasizes on providing solutions to managers to generate, preserve, mediate and utilize firm's explicit and tacit knowledge. Knowledge management is a set of strategies, practices, and supporting technologies that help people make better decisions by using and sharing data, information, and knowledge. Owing to the complex, uncertain, and ever-changing environment as faced by the firms, the significance of developing knowledge competencies to ensure organizational success has been acknowledged by researchers (de Rezende et al., 2021). The knowledge competencies lead organizations to a competitive edge over competitors which is gained by building capabilities instead of merely having access to the resources. Thus the way knowledge is managed in the organization impacts the dynamic capabilities of an organization.

Putrika and Syahrizal (2021) studied the impact of interpersonal trust on the knowledge management of team members and its subsequent effect on team performance. Knowledge management is an essential strategic initiative and the most momentous factor of sustainable competitive advantage for firms.

Emphasizing dynamic capabilities, researchers have realized that their nature and origin can be correlated to knowledge (Steininger et. al, 2022). The firm's ability to sense, seize, integrate, mediate and use the knowledge on a regular base holds the key to the firm's abilities and competitive advantage. Similarly, scholars whose primary interest lies in knowledge processes have started to examine conceptual links to dynamic capabilities. It has been investigated how dynamic capabilities can guide knowledge management (Cyfert, 2016). Also, the emergence of complex organizational structures leading to the creation of virtual teams, self-managed teams, cross-national teams, etc. has increased the interest of scholars in studying the role of interpersonal trust in organizations (Zuofa and Ochieng, 2021).

Many papers on Knowledge Management are found to embrace the themes of structuration theory. This paper attempts to understand the potential relationship between knowledge management and dynamic capability through the lens of structuration theory and to establish a theoretical link between their collective impacts on the firm's performance. Here, we put forward a framework that unequivocally integrates the two concepts with trust as a moderating factor and discuss their potential link to a firm's sustained competitive advantage. The study expands on the body of literature already in existence and investigates the moderating impact of trust between knowledge management (and all of its constituent parts) and the dynamic capabilities of an organization for sustainable competitive advantage.

Theoretical Background

Knowledge Management

People are paying more attention to the idea of knowledge management because they are becoming more aware of how important knowledge is to the survival and longevity of a business. The company's ability to adapt to the changing market dynamics is the only thing that will allow it to stay in business. Knowledge management is a growing model that pulls from many different fields and focuses on knowledge in the context of an organization. Knowledge management is mostly about finding and developing knowledge that will help organizations compete in a fast-changing world (Edwards, 2022). The literature on knowledge management also addresses the firm's internal features, such as the types of knowledge and the kinds of activities that influence the applicability of the knowledge management effort. By focusing on the distinction between knowing and knowledge, knowledge management clearly analyses whether knowledge is an asset (a property) or if it is nested in tradition (Munjal et. al, 2021). From this point forward, knowledge can be implicit both in terms of what people, groups, and businesses "have" and what people, groups, and businesses "do" (as practice). However, depending on the knowledge kinds and the tasks involved, both of these types of knowledge have an impact on the performance of organizations. The processes of capturing, meditating, sharing, and utilizing knowledge to create expertise should therefore be carefully and actively considered in knowledge management activities.

In an organization, knowledge can be seen as a set of systematic processes that start with getting information from both outside and inside sources. The next step is for members of the organization to share their knowledge to create new knowledge. Finally, members of the organization process and use the shared knowledge. Consequently, the major goal of knowledge management is still to detect, seize, and reorganize the firm's resources (Becerra-Fernandez, 2001). Knowledge management systems include four crucial components: a technological, human resources, process, and context facet. These components drive any firm's effectiveness.

Nonaka and Takeuchi (1995) looked at how dynamic knowledge management is and how it changes in response to outside influences and individual and group factors when getting and spreading knowledge. They put different kinds of knowledge into two groups: explicit and implicit. Organizations usually collect, organize, and share both tangible and intangible knowledge in a way that we call explicit or tacit knowledge. Since it is derived from one's own experience and may be challenging to transfer, tacit knowledge is typically not simple to impart. Conversely, explicit knowledge can be thought of as the written form of knowledge.

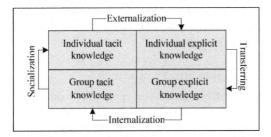

Fig. 1.

Source: adapted from Nonaka and Takeuchi (1995, p. 62)

Structuration Theory

Giddens (1979), a forerunner of the structure theory, asserts that acts of an employee (in the context of an organization) are viewed in relation to the past. While every action creates something new, it also continues the history of something else. The social structure and the actor, who are both crucial elements of each action, interact constantly with one another. As a result, a structure cannot be seen in isolation or as a hindrance to action. The structuration theory postulates a dynamic relationship between these various components of society while acknowledging the interaction between meaning, standards and values, and power.

In earlier studies of knowledge management through the prism of structuration theory, organizations were seen as social systems that replicated social practices through the actions of knowledgeable actors (Glisby and Holden, 2003). Organizations are typically divided into parts that need autonomy as well as specific coordination. Giddens was the first to point out that structure serves as a resource for interaction, allowing people to draw from pre-existing structural resources and rules rather than having to create social reality from scratch. Although the parts of the structure allow people to function within the system, human action will likely be limited by the structures already in place. This implies that structure has the potential to be both limiting and facilitating.

Giddens' structuration theory has been successfully applied by scholars to support the argument that Fig. 2.

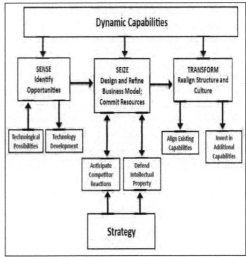

Fig. 2.

Source: Teece, D. J. (2018). Business models and dynamic capabilities. Long Range Planning

Knowledge Management needs to advance beyond crude models of information exchange and develop skills to fully capitalize on social interaction within organizations to gain enduring competitive advantage.

Dynamic Capability

Teece, Pisano, and Shuen (1997) defined the phrase "dynamic capacity" in their seminal study.

It is commonly viewed as an extension of the resource-based view (RBV), which states that each company possesses a unique profile of tangible and intangible resources that contribute to the company's competency and are likely to affect its performance (Amit and Schoemaker,1993). The RBV implied that resources and competencies are the primary sources of competitive advantage. The RBV has been criticized, however, for being a static theory). In addition, the RBV is criticized for failing to understand the origins of heterogeneity and the nature of unraveling mechanisms that permit sustained competitive advantage. Teece et al. (1997) came up with the idea of "dynamic capabilities," which focuses on the fact that what's important for business is "corporate dexterity": the ability to sense and shape opportunities and threats, take advantage of opportunities as they come up, and stay competitive by improving, combining, and rearranging the business's intangible and tangible assets (Refer Fig. 2).

Dynamic capabilities have been important for a long time when it comes to getting a competitive edge. But its importance has grown because the global economy has become more open to sources of invention and innovation, both in terms of where they come from and how they work (Teece, 2018). By interacting with their customers, suppliers, and other parties, firms can improve their skills, and this is likely one of the most important factors that determine their long-term competitive advantage.

Teece (1997) emphasized how important a company's current assets, path dependencies, and organizational routines are. These things lead to organizational learning and help companies do their jobs better and faster. Businesses are very interested in the management of knowledge because it can help them achieve strategic goals like making more money, staying competitive over time, and improving their skills. It can undoubtedly be a tool for developing dynamic capabilities.

Interpersonal Trust

In the context of an organization, trust is crucial for members' performance and well-being during times of crisis (Hungerford and Cleary, 2021), and its presence encourages a culture of information sharing and knowledge transfer. Knowledge sharing was seen (Johan, 2021) as a decision based on two distinct socio-cognitive actions: passing knowledge and accepting knowledge. When selecting whether to share knowledge or not, people's beliefs and values play a significant role; several of these characteristics are connected to trust. One of the main barriers to knowledge sharing among organizational members can be a lack of trust (Muniz et al., 2022).

According to Niedlich et al. (2021) trust primarily consists of four elements: capability, goodwill, behavior, and self-reference. The trust aspect is crucial in developing an organization's dynamic capability. Trust is a dynamic concept that is influenced by both individual and organizational factors.

Discussion and Significance

Knowledge management primarily focuses on the creation, storage, transmission, distribution, and coordination of existing knowledge assets inside the organization to develop and utilize core competencies that result in higher performance. The exchange of ideas to create and develop new learning, enhanced by excellent organizational structures and cultures, and supported by strong knowledge management systems, is an essential element for developing and utilizing core competencies.

The idea of dynamic capabilities offers a crucial theoretical foundation that allows us to comprehend the dynamics of competitiveness in the global era. The proposed model puts a lot of emphasis on how knowledge management affects dynamic capabilities leading to sustainable competitive advantage. It also suggests that dynamic capability is a result of the knowledge management process and is a group response to how the firm and the market are changing. They lead to sensing, seizing, and rearranging based on how the dynamics are changing. The constant cycle of sensing, seizing, and reconfiguring leads to a competitive advantage that lasts.

In today's competitive world, organizations that can change and adapt to the way their environment works have a good chance of staying alive. As a result of environmental shifts, businesses must continuously seek out the most effective strategies with which they can deal with their current situation and gain a competitive edge. In the fiercely competitive modern corporate landscape, knowledge has emerged as a crucial resource for many businesses.

Companies need to continually update their skills and resources to maintain a competitive advantage. To thrive and innovate in today's more competitive and volatile industries, businesses must develop internal processes for learning. Organizational learning, increased corporate performance, and sustainable competitive advantage are all fostered by the dynamic capabilities made possible by knowledge management.

The concept of dynamic capabilities is crucial to the study of competitive dynamics. Despite the obvious status of this topic, additional research into methods that outline dynamic capacities is needed in the current management literature.

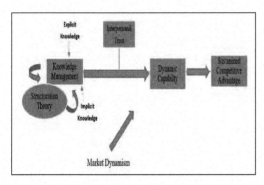

Fig. 3. Proposed framework

The proposed framework posits that the structural elements of an organization set the tone for knowledge management in the organization. Individuals in an organization understand the flow of knowledge based on the structure created for managing the organization even without having any social connections with the organizational members. However, knowledge is not shared freely unless interpersonal trust among the members of the organization exists. Be it explicit knowledge

residing in the previously created documents or the sharing of tacit knowledge residing in the minds of organizational members, it is not conveyed and transferred if the organizational members mistrust each other. Since knowledge is seen to be the currency of the current century, sharing knowledge with others is seen as a threat of losing individual competitiveness in the organization. That is why trust in the proposed framework is mentioned as a moderating variable between knowledge management and the creation of dynamic capabilities.

Organizations face dynamic challenges from the external environment, however in absence of Interpersonal trust, the creation of dynamic capabilities becomes difficult. To sense, seize and reconfigure the existing knowledge employees in the organizations are required to share and discuss the segments of relevant knowledge with each other and come up with strategies to adapt the competitive advantage for dynamic challenges. Only with continuous reconfiguring of capabilities, organizations can target to achieve sustainable competitive advantage. To keep the whole process under perspective, organizational structures need to be created in such a way that employees get enough opportunities to interact with each other without any challenges. Only with such lean structures, employees will be able, to begin with the process of knowledge management which is believed to be imperative for a sustainable competitive advantage.

Scope for Future Research

The proposed model may be tested with empirical research. Till now no study has been undertaken which tests the knowledge management processes for creating dynamic capabilities based on organizational structures with interpersonal trust as a moderating factor.

References

Buzzao, G., and Rizzi, F. (2021). On the conceptualization and measurement of dynamic capabilities for sustainability: Building theory through a systematic literature review. Business Strategy and the Environment, 30(1), 135–175.

Edwards, J. S. (2022). Where knowledge management and information management meet: Research directions. International Journal of Information Management, 63, 102458.

Hungerford, C., and Cleary, M. (2021). "High trust" and "low trust" workplace settings: Implications for our mental health and wellbeing. Issues in Mental Health Nursing, 42(5), 506–514.

Johan, M. (2021). The Effect of Knowledge Sharing and Interpersonal Trust on Innovation: An Empirical Study in Indonesia Higher Education. International Journal of Social and Management Studies, 2(3), 106–122.

Muniz Jr, J., Wintersberger, D., and Hong, J. L. (2022). Worker and manager judgments about factors that facilitate knowledge sharing: Insights from a Brazilian automotive assembly line. Knowledge and Process Management, 29(2), 132–146.

Munjal, S., Andersson, U., Pereira, V., and Budhwar, P. (2021). Exploring reverse knowledge transfer and asset augmentation strategy by developed country MNEs: Case study evidence from the Indian pharmaceutical industry. International Business Review, 30(6), 101882.

Niedlich, S., Kallfaß, A., Pohle, S., and Bormann, I. (2021). A comprehensive view of trust in education: Conclusions from a systematic literature review. Review of Education, 9(1), 124–158.

Putrika, C. S., and Syahrizal, S. (2021). Interpersonal trust and knowledge sharing: the moderating role of leadership support. Human Resource Management Studies, 1(4), 237–245.

Steininger, D. M., Mikalef, P., Pateli, A., and Ortiz-de-Guinea, A. (2022). Dynamic capabilities in information systems research: A critical review, synthesis of current knowledge, and recommendations for future research. Journal of the Association for Information Systems, 23(2), 447-490.

Zuofa, T., and Ochieng, E. G. (2021). Investigating barriers to project delivery using virtual teams. Procedia Computer Science, 181, 1083–1088.

Examining Potential Dangers and Risk Factors in Building Construction Projects

Deep Upadhyaya[*,a] *and MohammedShakil Malek*[b]

[a]Gujarat Technological University, India
[b]Principal, F. D. (Mubin) Institute of Engineering and Technology, Gandhinagar, India
E-mail: [*]deep.upadhyaya@gtu.edu.in

Abstract

Not only is the construction business a huge driver of economic activity, but it is also among the most dangerous in every single country on the planet. To make construction projects safer, it is essential to research the risk variables that are accountable for accidents that take place on construction sites and to have an understanding of the risk factors that play a role in the occurrence of these incidents. This study's objective is to discover the many aspects of danger that have an impact on the safety of construction work in the hopes of reducing the number of accidents that take place in this industry. According to the findings of this investigation, there are a total of thirty-six danger variables, each of which has to be treated with extreme caution in any future building endeavors.

Keywords: Risk, Safety, Health, Building, Construction

Introduction

Any nation's social and economic progress depends on its building sector. Several studies, like the ones below, have verified the relevance and significance of the construction sector to a nation's economy (Abang et al., 2005). There are a variety of reasons why the construction industry has worse accident statistics than the manufacturing industry. In the manufacturing industry, the working environment is often controlled, there is little change in the tools and processes used, and the labor force is steady. As a result, the workforce in the manufacturing environment will be fully aware of the risks and the precautions that should be taken to lessen workplace hazards. Contrarily, the construction industry is characterized by a mobile labor force and a constantly changing work environment. "Work should take place in a safe and healthy working environment, and working conditions should be consistent with worker's human dignity, and that occupational health and safety policies must be implemented both by the government and organizational levels," the International Labour Organization Conference in Geneva in 1985 declared. The International Labour Organization (ILO) and the World Health Organization promote the occupational health of all global enterprises (WHO). "Zero-incident" is the HSE manager's stated objective. Typically, risk identification is defined as a collection of principles, guidelines, and procedures designed to reduce the likelihood of mishaps, injuries, and other unfavourable results from using a product or service (Che et al., 2007). A safety manager will be hired to identify and address system flaws in order to guarantee the security of the construction phase. This technique, which serves an organisational purpose, makes sure that all risks to safety have been accurately recognized, evaluated, and effectively controlled. It is the manager's responsibility to respect social and legal obligations.

Literature Review

Bhattacharjee, S., Gosh, S. (2011): In India, 11,614 to 22,080 persons die in construction. The Indian construction industry alone accounts for 24.20% of the 48,000 occupational accidents that occur yearly in India. According to this estimate, the mortality rate per 1000 employees in the UK, Singapore, UAE, and Taiwan construction industries was 0.02 in 2013, 0.05 in 2012, and 0.125 in 2011.

Examining construction contractors' safety performance was done in 2005 by Ng, S.T. et al., R.M. Factors: In terms of an organization: safety management system required by law. Establishing a safety policy and defining who is responsible for safety. At the project level: Establishing evacuation plans and procedures, providing a safe and healthy work environment, and setting up a safety committee.

Cooper, D. (2010): Qatari construction must improve safety risk management. Companies should train personnel in safety risk management. Factors: More seminars/workshops.

Ismail, K.I., Periaiah, and Che Hassan (2010): The building site has reliable and organised safety measures, including a safety policy, the survey claims. Instruction, safety meetings, OSHA inspections, fall prevention systems, and safety marketing.

Bakshi A., Kumar, K., Rani, Ekta. (2009): The risk of accidents and injuries in construction is increased by cumbersome equipment, dangerous instruments, and poisonous chemicals. Large project owners can take an extra vigorous part in construction safety management at each phase of the project's implementation, together with choosing safe contractors for the construction phase, developing the contracts, and selecting safe workers themselves, as well as creating a culture of safety on the projects through safety training and recognition programs.

Poor top management safety knowledge, a lack of training, and project manager safety awareness are all contributing causes to China's subpar construction safety management, according to Tam, Zing, and Deng (2004). Neglecting safety and operating carelessly.

Identify variables impacting safety performance on construction sites, Sawacha, E. et al. (1999) recognize the parameters: The management discusses safety, providing safety manuals the provision of safety gear. fostering a safe atmosphere assigning a qualified safety representative to the location.

Method to Examine Danger and Risk Factors:

Fig. 1. PDCA steps

1) PLAN: To study Management program concepts, Effective approaches, legal policies, and proper mitigation ideas. OSHA-recognized commonly recognized superior technological practice and BAST evaluation. Safety induction- It's the first step; the HSE department gives a project overview and safety orientation to new employees.

Information should be covered at the official orientation:

- Project safety regulations and rules, as well as the project's policy and mission
- In an emergency, act
- Permission to Work
- Talking about tools, using power tools
- Work at height and electrical work
- Fire safety, handling of materials
- First-aid station, power tool handling
- Environmental conditions and incident reporting
- I.D.s (Individual Defense Equipment)
- Provide Detail to HSE Staff
- Basic Needs and Welfare

Record the information after orientation using the proper format.

2) Do: Before performing an evaluation, the HSE department must review the action and technique description so they can reduce risk.

Table 1. Method Statement content format

Purpose	Why Do You Work?
Scope	where region was the work? Make a move
Reference standard	Drawings approved by the DM and specifications and standards
Control Measure	Make sure every work is in the status of needing permission and approval
Methodology	Observe the defined procedure
Job Responsibility	Who is in charge of the aforementioned work?
Resources	equipment or materials needed

3) Check: Documentation reviews: Examine early team activities, prior project files, and project plans. Brainstorming: The most common identifying method is ideation. The purpose is to identify safety-affecting elements for qualitative and statistical safety assessments. Checklists: Recognition criteria may be created using existing project information and expertise. Using checklists simplifies identifying stability elements. For physical Evidence: Position of wounded personnel, machine, substance, safety feature, wear and tear, cleanliness, climate, illumination, and loudness. Risk Assessment: The HSE manager conducts a risk analysis after getting data from the project specifications. TRA (Task Risk Assessment): Before evaluating risk, the HSE manager determines S×P (Magnitude ×Probabilistic) = Risk grade and Risk rating. Suppose the grading and risk rate is not suitable. In that case, the HSE manager evaluates all the situation and build a complete young evaluation strategy to decrease and reduce the danger in such a manner that the rating and risk level goes down to reasonable standards. The HSE manager determines risk grade and value based on his construction experience.

Table 2. The HSE department of the company prepares a risk matrix

Probability	Consequences				
	Insignificant 1	Minor 2	Moderate 3	Major 4	Catastrophic 5
Rare 1	1	2	3	4	5
Possible 2	2	4	6	8	10
Likely 3	3	6	9	12	15
Often 4	4	8	12	16	20
Frequent 5	5	10	15	20	25

Risk influences: 1-10 – Low, 10-20 – Medium, and 20-25 - High risk

4) Act: To increase construction safety, take these steps. Toolbox Talk: Safety-proof all workers/employees. Work-related. It is done every morning before commencing the job at the designated location and the workers are instructed about the danger related and the safety precautions, they must take to avoid the risk, SOP (Safe operating procedure): Assistant HSE officer gives activity directions, Monthly inspection: The inspector inspects safety gear frequently and tags the record. Every three months, the label color is altered; permit to Work: High-risk activities need HSE clearance. Fireworks, height work, excavation, and Audit: An audit compares real site work to standardized practice and safety regulations to verify that all procedures are followed. If misconduct is seen, tell the administration and the worker, Make-Report (Violence): If an HSE investigator finds a serious problem during an audit, the department must take remedial action within a certain time. If the problem persists beyond the specified period, the HSE inspector may file a violent complaint against the organization, Receive Respond: The responsible department must react to the HSE after receiving the violent

complaint. If the answer is accurate, closeout, Closeout: If the answer is right, close the summary and react to the affected department. If not, bring the event to the Project Manager's attention as a significant violation, and Data Record: Retaining DPR and WPR records and data (weekly progress report). Data is analyzed to avert future dangers.

Identified Factors:

It's crucial to discover elements that contribute to effective safety programs and safe results. Many elements need to be identified so a few crucial safety factors reflecting the source of many difficulties can be disclosed. The whole construction lifecycle must consider safety in order to maximize project safety. This study reviews motivation and works on building safety considerations. The identified risk factors are as follows:

- Initiative Nature
- Site arrangement
- use of modern technologies in building projects
- the illumination of the work area at night.
- emergency preparation and execution
- Create a strategy for emergency response.
- Give employees instruction on how to act in an emergency
- Barriers and signs
- Use caution signs
- Use cautious and obey all warning signals.
- Barricades may be used to enclose the work area.
- employee perception
- knowledge and awareness of workplace safety
- received training on workplace safety
- relationships with management and other employees
- Education and Experience of Workers
- commitment from management
- Existence of a uniform safety policy
- management's approach to the workforce
- Safety Knowledge about the company's executive team
- The foreman and site manager, who are responsible persons, always let me know about safety work.

- Risk on site is recognized, along with its likelihood and severity, and is covered in training.
- Safety Examination
- periodic safety inspections by the government
- Periodic management safety inspection
- Safety Conferences
- Talk about the Toolbox before each activity. Start
- Top executives should attend safety meetings.
- Financial Investment
- Set up enough money for safety
- Using an Insurance Firm to Insure
- Safety Education and Training
- proper direction and training for employees on safety Poster Control held Safety Training
- Educate and prepare employees in first aid
- Welfare Centers
- Offer enough space for resting and eating
- First Aid Station
- Offer a washing facility and restrooms that are clean.
- food and water supply infrastructure
- disposal of hazardous and waste items
- Create a risk management strategy.
- Create a waste management strategy.
- Quick Clear the area of the trash.
- Perception of Tools and Equipment
- Give the necessary tools so that the work may be completed securely.
- Using PPE (Personal protective equipment)
- Initiative Nature
- Site arrangement

Conclusion

For a construction project to achieve "Safe man Hour" (no fatalities or significant injuries), safety-impacting elements must undergo constant development. However, if appropriate measures and early safeguards are taken, it may be improved to some amount. To improve the performance of building projects in the future, this study analyses and identifies potential solutions that might be taken. To prevent this in the future, this study analyses a variety of elements that might affect construction safety on every given project.

References

Gohil, P., Malek, M., Bachwani, D., Patel, D., Upadhyay, D., and Hathiwala, A. (2022). Application of 5D Building Information Modeling for Construction Management. ECS Transaction, 107(1), 2637–2649.

Bakshi A., Kumar, K., Rani, Ekta. (2009). Organizational Justice Perceptions as Predictor of Job Satisfaction and Organization Commitment. Journal of International Business and Management, 4(9), 145–154.

Malek, M. S., Gohil, P., Pandya, S., Shivam, A., and Limbachiya, K. (2022). A novel smart aging approach for monitoring the lifestyle of elderlies and identifying anomalies. Lecture Notes in Electrical Engineering, 875, 165–182.

Che Hassan, C.R., Basha, O.J., Wan Hanafi, W.H. (2007). Perception of Building Construction Workers towards Safety, Health and Environment, Journal of Engineering Science and Technology, 2(3), 271–279.

Malek, M. S., Dhiraj, B., Upadhyay, D., and Patel, D. (2022). A review of precision agriculture methodologies, challenges and applications. Lecture Notes in Electrical Engineering, 875, 329–346.

Cooper, D. (2010), Safety Leadership: Application in Construction Site. Supplement A, Psychologia, 32(1), A18–A23.

Ng, S.T., Cheng, K.P., and Skitmore R.M. (2005). Framework for Evaluating the Safety Performance of Construction Contractors, Building and Environment, 40(10), 1347–1355.

Sawacha, E., Naoum S., and Fong, D. (1999). Factors Affecting Safety Performance on Construction Sites, International Journal of Project Management, 17(5), 309–315.

Tam, C. M., Zing, S. X., and Deng, Z. M. (2004). Identifying Elements of Poor Construction Safety Management in China, Safety Science, 42(7), 569–586.

Malek, MohammedShakil S., and Bhatt, Viral (2022). Examine the comparison of CSFs for public and private sector's stakeholders: a SEM approach towards PPP in Indian road sector, International Journal of Construction Management, DOI: 10.1080/15623599.2022.2049490

Project Management Techniques for Planning and Scheduling: A General Overview

Dhiraj Bachwani,[*,a] *Mohammedshakil Malek,*[b] *and Rahul Bharadiya*[c]

[a]Research Scholar, Gujarat Technological University, Ahmedabad, India
[b]Principal, F. D. (Mubin) Institute of Engineering & Technology, Gandhinagar, India
[c]PG Student, Construction Project Management, Parul University, Vadodara, India
E-mail: [*]199999912501@gtu.edu.in

Abstract

Effective and realistic project management techniques for planning and scheduling are essential in the construction industry. This paper provides an overview of project management techniques for managing infrastructure projects. The critical Path Method (CPM) & Programme Evaluation and Review Technique (PERT) are widely used but these techniques have certain limitations and are not effective for large-scale projects. To minimize the constraints in the project this paper introduced an advanced approach to the Critical Chain Project Management (CCPM) project management technique. Critical Chain Project Management is the effective and feasible construction scheduling method used to address constraints such as project uncertainties, complexity, and resource scarcity. Critical Chain Project Management develops a robust schedule using a buffer to minimize the time and cost overruns in the project. This study additionally shows that for the timely completion of a project, CCPM provides a more flexible schedule than traditional techniques (CPM and PERT).

Keywords: Project Management, Planning & Scheduling, Critical Path Method, Programme Evaluation and Review Technique, Critical Chain Project Management, Theory of Constraints, Risk Management, and Multi-Project Environment

Introduction

The successful completion of projects is crucial to the national growth and economic improvement of emerging countries. According to the Project Management Institute (2013): "The application of knowledge, skills, tools, and techniques to project activities to meet the project requirements." Every construction project has divided into five critical phases: Initial Phase, Planning Phase, Execution Phase, Monitoring & Controlling Phase, and Closing Phase. Scheduling and planning for construction are two of the most significant elements of any project. According to (Guofeng et al. 2015) construction projects are differentiated by their constraints, complexity, and uniqueness. Also Guofeng et al. (2014) describe that Constraints such as construction delays, resource scarcity, and time uncertainties are frequent in the construction industry. Research work by (Chester and Hendrickson 2005) said that inadequate project management techniques often experience schedule delays in construction.

The majority of scheduling techniques aim to reduce project length and optimise the organisation of project operations (Sarkar et al., 2018). However, the most significant worry for a contractor throughout the construction phase is to execute the project as planned despite limits such as the contract money, materials, etc. To make project scheduling more dependable, a contractor's primary scheduling purpose is to reduce the uncertainty of project activities and project duration. Taking a pragmatic view of (Habibi et al. 2018) Success in completing a project and a reduction in associated expenses can result from better scheduling as part of the project management process.

Traditional Project Management Techniques

The Critical Path Method (CPM) and Programme Evaluation and Review Technique (PERT) are used for project planning and scheduling. Program evaluation and review technique (PERT) enhances the critical path method (CPM) by integrating the commonly utilized aspect of uncertainty in predicting activity time. The CPM and PERT technique based on the assumption that unlimited resources are available for the execution of the project (Kastor and Sirakoulis 2008). In the context of real-life projects, these techniques are not feasible for managing uncertainties and constraints. According to (Guofeng et al., 2015) PERT is based on strong statical assumptions. However, PERT project duration predictions are often inaccurate since the assumptions underlying the method are not always viable.

What are the issues with Traditional Techniques?

There are several issues associated with traditional techniques. Out of all those issues the major five issues were identified that were highly associated with construction project management.

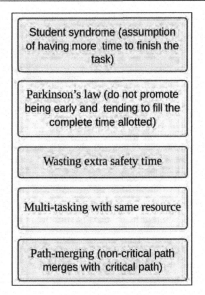

Fig. 1. Major Problems from Literature Review

Critical Chain Project Management

In 1997, Eliyahu M. Goldratt suggested an application of TOC to project management; this approach, known as Critical Chain Project Management, was developed to solve the problem of delay, which has been identified as a major shortcoming of standard project management methodologies. (Critical Chain Project Management) is a novel approach to scheduling projects efficiently and effectively with available resources. Activity duration is a primary consideration in CCPM planning and scheduling. (Azar et al. 2016) describe that a theory of constraints (TOC) tool, Critical Chain Project Management (CCPM) is employed in the phases of planning and managing projects. Theory of Constraints forms the basis of Critical Chain Project Management. In CCPM the project buffer ensures the time required to complete the project as specified by the critical path, while the feeding buffers better protect the time required to complete activities on non-critical paths. Additionally (Sarkar et al. 2018) said the management of buffers improves the accuracy of measurements and the quality of decisions made for project control. When CCPM is used on a project, the record of performance in terms of schedule, cost, and scope is significantly enhanced.

Critical Chain Project Management provides effective resource management in multi-project management. As analyzed in depth by Li et al. (2017) said, in multi-project environment resource management is the most challenging aspect of achieving the project's objectives. Resource management includes resource planning, allocation, monitoring, controlling and coordination. A study conducted by (Azar et al. 2016) conclude that construction companies using the CCPM Technique usually have more than 95 % of project completion within the time limit and minimize the project duration by 25% to 40%.

The Theory of Constraints

Eliyahu Goldratt's book "The Goal" popularised the theory of constraints (TOC), upon which the critical chain is founded. An essential principle is that the barrier of any system is defined by a specific constraint. If we want to make progress in bettering the system, we need to concentrate on reducing that limitation.

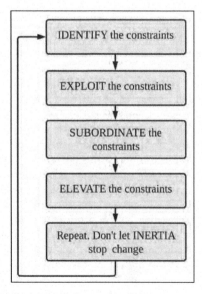

Fig. 2. Five focusing techniques for TOC

Identify the constraints: In construction, constraints are classified across construction stages. Identification of the constraint (material, manpower, money, equipment) must be done during/before the execution of the project.

Exploit: In order to complete the project within the stipulated duration the constraint must be utilized to the full extent, if possible, it should be shared/rented depending upon the usage. Various techniques like aggressive time estimation and critical chains can be utilized as a tool to maximize the use of constraints.

Subordinate: In extent to utilize the constraints many other resources become critical which should not be neglected. There are chances during such stages many non-critical activities shall be idle which should not be an issue.

Elevate: It is a stage where the resource shall be exploited to a maximum extent if the desired results are yet to be achieved. The least preferred option in the above situation is to extend/add on the availability of resources in order to achieve the desired objectives. This can be done by technical means, sharing of resources or by HRM.

Inertia: This stage highlights uncertainty coming across the above-mentioned steps which might have a negative impact on the project. The aim should be to eliminate the new constraint in the initial stages.

Buffer Monitoring and Management

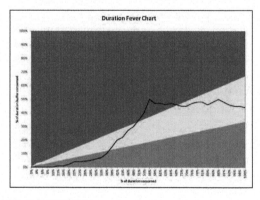

Fig. 3. Fever Chart
(Source: LYNX Scheduler)

When using the CCPM technique, the project is not tracked according to the date it is expected to be finished; rather, it is tracked based on the rate at which the activities are depleting the buffers. The progress report should be used to create color-coded fever charts that show the risk level in carrying out the project, with the buffers needing to be managed on a weekly or monthly basis. These are the meanings behind the various colored signals.

The green area represents safety; further action is not required.

When things go into the yellow zone, it's a signal that the main reason for the delay needs to be determined and a plan of action developed.

Zone red means things are getting problematic, so get going on those now and take care of any problems you find.

Steps for Critical Chain Project Management

The following figure shows the various steps involved in CCPM.

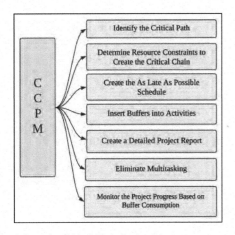

Fig. 4. Steps for Critical Chain Project Management

Conclusion

This study analyses the various project management techniques such Critical Path Method, Programme Evaluation and Review Technique, and Critical Chain Project Management. Appropriate project management technique is essential for the successful completion of projects but the majority of construction projects fail to achieve the project's requirements and objectives. Construction projects faced constraints during the execution phase which leads to project time and cost overrun. After analyzing various techniques this paper concludes, Critical Chain Project Management provides a relatively better schedule-based alternative keeping in view the resource limitation which helps project managers and contractors to achieve the target of completion of the project.

Future Direction for Research Work

As a future direction for researchers, taking a case study approach to infrastructure projects using an advanced technique (Critical Chain Project Management) can provide more effective and feasible planning and scheduling. Additionally, using software for Critical Chain Project Management such as ProChain and LYNX Scheduler gives better output for managing infrastructure projects.

References

Ma, G., Gu, L., and Li, N. (2015). Scenario-based proactive robust optimization for critical-chain project scheduling. Journal of Construction Engineering and Management, 141(10), 04015030.

Sarkar, D., Jha, K. N., and Patel, S. (2021). Critical chain project management for a highway construction project with a focus on theory of constraints. International Journal of Construction Management, 21(2), 194-207.

Ma, G., Wang, A., Li, N., Gu, L., and Ai, Q. (2014). Improved critical chain project management framework for scheduling construction projects. Journal of Construction Engineering and Management, 140(12), 04014055.

Chester, M., and Hendrickson, C. (2005). Cost impacts, scheduling impacts, and the claims process during construction. Journal of construction engineering and management, 131(1), 102–107.

Ghaffari, M., and Emsley, M. W. (2015). Current status and future potential of the research on Critical Chain Project Management. Surveys in Operations Research and Management Science, 20(2), 43–54.

Habibi, F., Barzinpour, F., and Sadjadi, S. (2018). Resource-constrained project scheduling problem: Review of past and recent developments. Journal of project management, 3(2), 55–88.

Izmailov, A., Korneva, D., and Kozhemiakin, A. (2016). Effective project management with

San Cristóbal, J. R., Carral, L., Diaz, E., Fraguela, J. A., and Iglesias, G. (2018). Complexity and project management: A general overview. Complexity, 2018.

Li, X. B., Nie, M., Yang, G. H., and Wang, X. (2017). The study of multi-project resource management method suitable for research institutes from application perspective. Procedia Engineering, 174, 155–160.

Kastor, A., and Sirakoulis, K. (2009). The effectiveness of resource levelling tools for resource constraint project scheduling problem. International Journal of Project Management, 27(5), 493–500.

Construction Supply Chain Management: A Literature Review

Deepak Yadav,[*,a] *Dhiraj Bachwani,*[b] *and MohammedShakil Malek*[c]

[a]PG Scholar, Construction Project Management-Pursuing, Parul University, Vadodara, India
[b]Assistant Professor, Department of Civil Engineering, Parul University, Vadodara, India
[c]Principal, F. D. (Mubin) Institute of Engineering & Technology, Gandhinagar, India
E-mail: [*]deepakryadav2022@gmail.com

Abstract

Due to the difficulties presented by increased global competition and the current economic crisis, managing supply chains is getting more and more complicated. Comparatively speaking, the construction business is characterized by high conflict, poor productivity, cost and time overruns, and significant levels of fragmentation. This paper highlights the different frameworks that are crucial in enabling an external SC integration of suppliers, owners, designers, and contractors to improve efficiency in decision-making throughout the building phases. Decisions of SC planning, IT planning, and logistics planning are offered to conduct the construction planning and design phase in response to trends. It also gives an insight of the future direction of work that can be carried out in the field of construction supply chain management in the construction industry.

Keywords: Supply Chain Management (SCM), Construction Industry, Lean Supply Chain Management (LSCM), Digital Supply Chain (DSC)

Introduction

SCM is a management procedure that allows organizations to manage the global network of stockholders, including suppliers, retailers, and distributors, through which raw materials are obtained, finished goods are produced, and consumers are served. Throughout every stage of the building process, procurement activities take place.

Construction supply chain management benefits businesses by enhancing their ability to compete, boosting their earnings, and giving them more command over the many project components and constraints.

The rivalry in the construction sector is significantly influenced by the supply chain. Construction supply chain management benefits businesses by enhancing their ability to compete, boosting their earnings, and giving them better command over the numerous project components and uncertainties. In the building sector, material management is crucial. Projects are interrupted and production levels drop without a good material management system. The management of materials is essential to increasing efficiency and attaining schedule, cost, and quality performance goals since materials often account for 50% to 60% of the expenses in building project budgets.

Over 38% of respondents, up from +29% in Q4 2021, indicated an increase in workloads in the first quarter of 2022, according to RICS Global Construction Monitor data for the whole industry. Given that 55% more respondents indicated an increase, workloads are mostly being fueled by a continuous emphasis on infrastructure projects. A strong commitment to transportation infrastructure was also mentioned by respondents, as almost half (44%) of them noted an increase in workloads in this area.

Methodology

There were two distinct phases of the research paper.

1. Analyzing the Primary Sources
2. Examination of Existing Works

Some of the most fundamental questions of supply chain management were investigated by reading through primary literature. Subsequently, research on supply chain management across industries has been compiled. Success rates of supply chain management in other departments and their implementation have been analyzed, as has the potential of supply chain management in the construction industry. Quality management papers have been outlined, along with papers on industries, commercial structures, residential developments, and infrastructural developments. This data also comes from prior studies conducted on supply chain management.

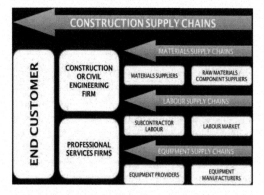

Fig. 1. Supply Chain Management

Source: The myriad of construction supply chains (Cox et al., 2006)

Examples of construction supply chain management

- *GoFor:* GoFor, a member of CEMEX Ventures' investment portfolio, is a pioneer in same-day, last-mile project delivery of products in North America. They offer their customers prompt and cost-effective delivery of products, from little loads to large, bulky items, either on demand or in advance.
- *LINKX:* A cloud-based logistics platform called LINKX is used for Mexican product delivery. By tracking their loads and deliveries in real-time and incorporating both sides of the transaction, their system provides the optimization of their operations.
- *Soil Connect:* By matching users based on distance, Soil Connect allows users to trade across shorter distances, saving thousands of dollars and significantly lowering the carbon emissions of the construction sector by lowering CO_2 emissions and avoiding trash from going to landfills.
- *Voyage Control:* Voyage Control has several advantages, including improved operational efficiency, increased security, lessened traffic, increased driver compliance, and environmental advantages.
- *CEMEX Go:* It's also important to note that CEMEX has its digital platform and a workflow-streamlining solution called CEMEX Go. The platform makes it simple and frictionless for customers to make purchases, monitor delivery, and handle requests.

Supplier networks in manufacturing and building industries compared:

The construction industry's supply chain differs significantly from the manufacturing industry's because:

- The construction product is usually intended for a single customer.
- Each project uses a unique product.
- Each project has its own unique production environment, set of tools, and workflow.
- There is a significant turnover rate among construction workers both while a project is being built and between builds.
- However, not all of the necessary components and supplies can be kept on-site.

Top Supply Chain Companies In India in 2022

There is a significant growth of different supply chain startups in India and they are helping a lot in the proper transportation facilities. They are:

- Rivigo
- Leap India
- Shadowfax
- 4tigo
- Delhivery
- Porter
- Edgistify
- FreightBro
- TruckEasy
- Blackbuck

Literature Review

Different research papers review papers and Research articles have been studied from renowned journals like Internal Journal of Project Management, Elsevier, Canadian Journal of Civil Engineers (CJCE), American Society of Civil Engineers (ASCE), and other esteemed journals.

While conducting this literature review, a number of useful frameworks, models, and techniques were uncovered, such as: a BIM-WMS integrated decision support tool; an RFID-aided tracking system to improve the productivity of scaffold suppliers in the Australasian supply chain; a SCO (smart construction object)-enabled logistics and SCM system; a qualitative data analysis of supplier quality-management practices; and so on. Management of sustainable supply chains, with blockchain technology into a case study using an agent-based simulation framework incorporating a threat model.

Fig. 2. Journal-wise literature review graph

Different frameworks, models, and techniques that were found out are as follows:

- Integrating 4D BIM & GIS
- RFID-Aided Tracking System
- RII Method
- BIM-WMS
- SCO-Enabled Logistics& SCM System
- Two-step SEM model
- Serious Gaming Approach
- Expert Choice Software
- Decision-Making Trial and Trial and Evaluation Laboratory (DEMATEL)
- Analysis Network Process (ANP)
- Technique for Order Preference by Similarity to Ideal Solution (TOPSIS)

Five Significant Barriers to SCM

The five significant barriers to SCM are as follows:

1. Insufficiently capable leadership
2. Parties in SCM are incompetent
3. Lack of comprehension of the idea of a supply chain
4. Passive Suppliers and Sub-contractors
5. Corporate opposition to SCM

Factors that are related to community and industries

Following are the main factors that are related to community and industries:

- Brief and temporary project length
- Supply chain includes as many participants as possible.
- Constant and unreliable project changes
- Constant and unreliable project changes
- More disagreements and conflicts occur in this industry than in others
- Lack of standardization of labor, equipment, and machinery
- Legal considerations, pricing changes, and workplace safety

Future Directions

Technology may play a significant role in the coordination of information requirements among stakeholders, and this trend can continue and increase throughout ETO-type sectors. Opportunity to expand the body of knowledge to incorporate more quantitative research. The chance to think about the ideal management structure and governance to get desired results. Investigation into the optimal ratio of centralized vs. dispersed control may provide useful information and also the BIM and Construction Chain Management Integration.

Conclusions

It can be concluded that a lot of scopes is there for the supply chain in the field of the construction industry. Several frameworks have been developed to overcome the different barriers of the industry but there is a requirement for a proper framework or model which can help to solve or overcome a number of barriers and help in the proper functioning of the supply chain of a construction sector.

References

Malek, MohammedShakil S., and Bhatt, Viral (2022). Examine the comparison of CSFs for public and private sector's stakeholders: a SEM approach towards PPP in Indian road sector, International Journal of Construction Management, DOI: 10.1080/15623599.2022.2049490

Malek, MohammedShakil S., and Gundaliya, P. J. (2020). Negative factors in implementing public–private partnership in Indian road projects, International Journal of Construction Management, DOI: 10.1080/15623599.2020.1857672

Malek, MohammedShakil S., and Gundaliya, Pradip (2021). Value for money factors in Indian public–private partnership road projects: An exploratory approach, Journal of Project Management, 6(1), 23–32. DOI: 10.5267/j.jpm.2020.10.002

Malek, MohammedShakil S., Saiyed, Farhana M., and Bachwani, Dhiraj (2021). Identification, evaluation and allotment of critical risk factors (CRFs) in real estate projects: India as a case study, Journal of Project Management, 6(2), 83–92. DOI: 10.5267/j.jpm.2021.1.002

Malek, MohammedShakil S., and Zala, Laxmansinh (2021). The attractiveness of public-private partnership for road projects in India, Journal of Project Management, 7(2), 23–32. DOI: 10.5267/j.jpm.2021.10.001

Gohil, P., Malek, M., Bachwani, D., Patel, D., Upadhyay, D., and Hathiwala, A. (2022). Application of 5D Building Information Modeling for Construction Management. ECS Transaction, 107(1), 2637–2649.

Malek, M. S., Gohil, P., Pandya, S., Shivam, A., and Limbachiya, K. (2022). A novel smart aging approach for monitoring the lifestyle of elderlies and identifying anomalies. Lecture Notes in Electrical Engineering, 875, 165–182.

Malek, M. S., Dhiraj, B., Upadhyay, D., and Patel, D. (2022). A review of precision agriculture methodologies, challenges and applications. Lecture Notes in Electrical Engineering, 875, 329–346.

Bachwani, D., and Malek, M. S. (2018). Parameters Indicating the significance of investment options in Real estate sector. International Journal of Scientific Research and Review, 7(12), 301–311.

Parikh, R., Bachwani, D., and Malek, M. (2020). An analysis of earn value management and sustainability in project management in construction industry. Studies in Indian Place Names, 40(9), 195–200.

Safety and Quality Management (TQM) – Implementation in the Construction

Mohit Tilokani,[*,a] *Jaydeep Pipaliya,*[b] *and MohammedShakil Malek*[c]

[a]MTech (Construction Project Management-Perusing) Parul University, Vadodara, India
[b](Assistant Professor - Department of Civil Engineering) Parul University, Vadodara, India
[c]Principal, F. D. (Mubin) Institute of Engineering & Technology, Gandhinagar, India
E-mail: [*]mohittilokani88@gmail.com

Abstract

Total refers to the whole thing, Quality refers to the degree of excellence supplied by a product or service, and Management refers to the way of regulating and directing. TQM is frequently used to enhance quality in manufacturing and other sectors. The fundamental aim of TQM is stakeholder participation. It increases client pleasure, the organization's economy, and its supremacy. New ideas, technologies, and approaches for employee management inside firms must be adopted. It is also useful in the building business. The study aims to investigate the most current research that focused on increasing quality via the use of TQM in construction and its applicability in different phases of building projects. There are various Quality and Management concerns throughout the construction. TQM is a cutting-edge (modern) quality and quality implementation method on building sites. TQM incorporates Juan, Deming, and Crosby's approaches, as well as ISO requirements.

Keywords: Building Construction, Quality Management, Total Quality Management (TQM)

Introduction

Total Quality Management (TQM) is a strategy for organizing, planning, and understanding all tasks that rely on all people at all levels. In TQM, the emphasis is on preventing problems rather than finding them.

India's construction industry is the country's second-biggest economic driver, behind agriculture. It supports around 51 million people in India and contributes 7.16 percent of the country's GDP. The construction sector has been able to take on several massive projects because of infrastructure development efforts during the last few decades. Due to industrialization, urbanization, economic development, and people's rising expectations for better living conditions, the industry has faced chronic problems like time and cost overruns, low productivity, poor occupational safety, and poor working condition and storage of skilled workers and investment, all of which are poised for growth. Therefore, building quality is crucial.

The lack of TQM implementation in Indian building projects is its primary flaw. Some of the largest construction firms are already using and benefiting from TQM. To better accomplish their objectives and goals, several large-scale enterprises have turned to TQM. Most construction companies, especially those of a smaller or medium size, don't put enough emphasis on TQM. Particularly in emerging nations like India, minor and middle-scale building enterprises perform a vital part in the financial prosperity and mental development of the country's expanding population. Total Quality Management, which emphasizes the need for continuous quality improvement, aims to constantly enhance existing strengths.

Safety During Construction

The decisions taken at the beginning of a construction project, in the planning and design stages, may have a lasting effect on worker safety. Some designers' or architectural designs are intrinsically difficult and risky to execute, even while equivalent alternatives may greatly reduce the chance of accidents.

For instance, keeping construction zones adequately isolated from traffic during roadwork may dramatically minimize the likelihood of accidents. While these design factors are certainly important, awareness, creativity, and teamwork are even more crucial to ensuring everyone's well-being on the job site. Workers should constantly be mindful of the possibility of mishaps and abstain from taking needleless risks.

Management for Quality and Safety

Different organizations may be used to monitor quality and security throughout the building process. It is common to practise for a company to divide responsibilities for ensuring quality and safety between two separate teams. In big organizations, quality assurance and risk management teams often assign specific individuals to oversee individual projects.

For less extensive initiatives, the project manager or an assistant could be responsible for these tasks and others. In both cases, the overarching project manager must deal with issues of staff, budget, schedule, and quality control, among other management concerns.

Many different organizations will be represented in the venture by examiners and quality assurance people. Parties involved for example the proprietor, and the engineer. The architect and the various construction companies may employ their own quality and safety assessors. Individuals with this expertise may be self-employed or engaged by specialized quality control firms.

In addition to on-site assessments, the sampling of products is frequently checked for conformity by specialized laboratories. Also engaged will be inspectors to ensure adherence to legal regulations. Common examples are inspectors for environmental agencies, occupational health and safety organizations, and the construction department of local governments

Assurance of Quality (Q.A.)

Quality assurance refers to the process by which the quality of a product or service is ensured by a series of planned and methodical activities taken inside the quality system. It has to be done continuously as the project progresses. Quality planning activities were widely employed in Quality Assurance before the development of the ISO 9000 series.

The presence of a Quality Assurance Division or equivalent is preferred but not needed.

Internal quality assurance can be given to the performing organization's management and the project management team, or it can be given to the customer and those who aren't actively working on the project.

Inputs to Q.A.

- Managing Quality in the Workplace
- The outcomes of quality assurance testing
- Conceptualizations of Operations

Outputs from Q. A.

A key component of quality improvement is taking steps to improve the efficiency and efficacy of the project so that the stakeholders can receive an additional advantage.

Procedures for managing general change disputes will be applied to the implementation of quality improvements, which may include filing a change request or enacting corrective measures.

Quality Control (Q.C.)

As part of quality control, we make sure that each project's output is meeting relevant quality requirements and we identify and eliminate the causes of any undesirable outcomes. It has to be done continuously as the project progresses.

The outcomes of a project are broken down into two categories: management outcomes, such as deliveries and cost and schedule performance, and product outcomes, such as product quality and customer satisfaction.

It's not necessary to have a dedicated Quality Control division or equivalent in your company, although it helps.

Input to Q. C.

1. Impact of Labor
2. Plans for quality management
3. Conceptualizations of Operations
4. Checklist

Output from Q. C.

1. Enhancement of Quality
2. Conclusions of acceptance
3. Rework
4. All items checked off
5. Adaptations to the Process

Factors That Affect Quality

As part of quality control, we make sure that each project's output is meeting relevant quality requirements and we identify and eliminate the causes of any undesirable outcomes. It has to be done continuously as the project progresses. The outcomes of a project are broken down into two categories: management outcomes, such as deliveries and cost and schedule performance, and product outcomes, such as product quality and customer satisfaction. It's not necessary to have a dedicated Quality Control division or equivalent in your company, although it helps.

Owners consider their wants with financial considerations and, in certain cases, the possibility of failure. The design experts have a responsibility to protect public health and safety within the scope of the completed projects. Means, methods, techniques, sequences, processes, and safety precautions and programs are all the responsibility of the builder.

The needs of the specific building project are the primary determinants of construction quality. Stages one through three of a typical construction project include the planning and design stage, the construction stage, and the maintenance and operation stage.

Review of Literature

Integrity in construction is the backbone of the industry. In the construction sector, it may be challenging to keep the job at a high-quality standard. The building sector relies heavily on effective implementation. Research conducted in the United States (Arditti and Gunaydin, 1997) found that a management commitment to excellence and to its importance to continue developing quality in construction can benefit the building industry. Professionals in the field understand how crucial it is to have clear instructions. To wit: (Haupt and Whiteman, 2001). Go through how everyone in the company works toward a TQM goal.

The commitment and involvement of upper management in the TQM process that engages workers is crucial to the success of construction sites. As reported by (Pheng and Teo, 2004). Recommendations for the development of a strategy for implementing TQM in the building industry. This was documented in 2008 (Tomoya Shibayama and Tarek Elgharawy). TQM is an administration philosophy that has been extensively executed in the construction industry due to its track record of success. Various features unique to Japanese architecture are discussed in this study. The construction industry in Egypt benefits from an innovative approach to TQM implementation and is a potential sector for usage in the Egyptian market.

Based on the work of Gul Polat et al. (2011), this study seeks to investigate the potential gains of implementing TQM and the challenges that prevent its broad use in the building sector. Ali and Ezialu Onyeizu, Khalid bin and Abu Hassan bin Abu Bakr (2011). As the latest technique in quality management, Total Quality Management (TQM) follows in the footsteps of quality assurance, quality control in the construction industry, and ISO 9001. The study's goal is to assess how well TQM concepts are being implemented within the construction industry in Oman, both by the companies' staff and by their external contractors.

Quality improvement has been an evolutionary process, according to Frank Voehl, Hal Wiggin, and H. James Harrington (2012). The TQM approach to quality assurance (QC). As the "war on waste" continues, construction companies must implement a quality management system, and the ten tenets of TQ remain central to this effort. Training and facilitative leadership are especially important, and emphasis on consultation client demands once again is long needed. It has been suggested by David Arditi and others in 1997 that the construction industry has a lot of opportunity for development in terms of quality. Quality must be improved at every stage of the building process, including the continuing work. Using quality management systems, one may innovate in design more than in other areas, as stated in the article "design" by Ilias Said and Nuruddeen Usman (2013). Effective management approaches, the author argues, not only foster invention but also lead to the progress of new, game-changing technology that may be used to further a company's goals. Those authors are: Hossein Rahmati, Davood Gharakhani, Arshad Farahmandian1, Mohammad Reza Farrokhi3 (2013). TQM, or total quality management, is widely regarded as an essential factor in any company's ability to succeed over the long term.

There is little doubt that implementing TQM has been essential to the success of many businesses in recent years. TQM is centered on a company-wide commitment to process improvement that both meets customers' demands and exceeds their expectations. The purpose of the essay by Ms. Aiswarys. K. Lalaji and Ms. Shivagami (2014) is to identify TQM practices in construction businesses, evaluate the efficacy of TQM in construction companies, and discuss problems associated with implementing TQM in construction organizations. Nevertheless, most companies end up learning about it. Preparation and Execution The implementation of TQM is not always smooth. Anantha Subramaniam believes that most small businesses are not adopting TQM because of their size (2014). Both management's understanding and support are deficient. Since there was not enough knowledge about TQM, it was abused by big and medium-sized organizations who tried to apply it. It is impossible to progress without training. Profitable in terms of both money and time, construction projects are possible with TQM knowledge. In this paper, Saurin kakkad and Pratik Ahuja (2014) discuss the broad use of TQM teams in corporate settings throughout the globe. It's a tried-and-true method for implementing a standard throughout the global organization's various vertical and horizontal levels. Implementing TQM helps businesses become more productive and competitive over time, provides customers with the goods they want, and maximizes the efficiency of their processes in relation to their inputs.

This article by Umair Mazher et al. (2015) assesses the value of total quality management (TQM) in the construction sector. The purpose of this study is to examine the four pillars upon which quality management in the building industry rests. Good management, quality assurance, quality control, and quality inspection are the metrics in question. According to P.P. mane and J.R. patil's (2015) study, TQM elements including quality, time, and money are essential to the completion of construction projects that serve their main goal. Quality management in the building industry is often referred to as quality planning, quality assurance, and quality control. Rahul S. Patil and Priyanka Hirave (2016) India's progress is in large part due to the work of the construction sector. The construction business employs the second-highest number of people in India, after only agriculture. In order to enhance construction management across the board, TQM should be used in the Indian construction industry.

Methodology Review

Use the online library database science direct to read up on the TQM implementation literature from reputable publications and conferences. Adopting a holistic approach to quality management and tapping into a variety of other useful tools. The internet search is conducted using research on the difficulties of adopting TQM in the building sector as well as other scholarly works. Managers and engineers at construction firms may fill out a survey on the

"Need for TQM in the industry," "Application of TQM in the industry," and "Implementation of a prepared questionnaire survey" either online (through e-mail or Google Forms) or in person. The generation of a result through the RII or weighted average approaches.

Conclusion

The paper will provide a summary of the research on Total Quality Management is a technique for coordinating and understanding each step in a process that involves everyone at every level. TQM is preventive rather than focusing on client satisfaction. The manufacturing industry often employs TQM to great effect. Construction is unique from other sectors in several ways that complicate the implementation of TQM. Increasing stakeholder satisfaction, product quality, market share, and customer satisfaction may all be attained via the use of TQM in the construction industry.

Construction site safety is heavily influenced by choices made in the project's early stages of planning and design. Education, attentiveness, and collaboration are all crucial to maintaining a safe work environment on a construction site. Employees should be alert to the possibility of accidents at all times and avoid taking any unnecessary risks on the job.

Acknowledgment

This research would not have been possible without the help of Asst. Prof. Jaydeep Pipaliya from the Department of PIET at Parul University in Vadodara. In addition, I'd like to express my appreciation to Dr. Lalit Thakur, Head of the Civil Department at PIT, Vadodara; and Assistant Professor Dhiraj Bachwani, Postgraduate Coordinator. In addition, we owe a debt of gratitude to Dr. Vipul Vekariya, the dean of Parul University's Institute of Technology. Thank you very much, Sir, for your insightful comments and recommendations that helped me finish my work.

Finally, I'd want to thank my fellow students for their invaluable assistance.

References

Arditi, David and Gunaydin, H. Murat (1997). Total quality management in the construction process. International Journal of Project Management. 15(4), 235–243.

Abu Bakar, Abu Hassan Bin, Ali, Khalid Bin and Onyeizu, Eziaku (2011). Total Quality Management Practices in Large Construction Companies: A Case of Oman. World Applied Sciences Journal 15(2): 285–296.

Harrington, H. James, Voehl, Frank, Wiggin, Hal (2012). Applying TQM to the construction industry. 24(4), 352–362.

Arditi, David and Gunaydin, H. Murat (1997). Total quality management in the construction process. International Journal of Project Management. 15(4), 235–243.

Gharakhani, Davood, Rahmati, Hossein, Farrokhi, Mohammad Reza, and Farahmandian, Arshad (2013). Total Quality Management and Organizational Performance. American Journal of Industrial Engineering. 1(3), 46–50.

Iqubal, Asif, Banerjee, Rajeev, Khan, Zeeshan Raza and Dixit, Raj Bandhu (2017). Construction Disputes in Construction Work Sites and Their Probable Solutions. International Journal of Civil Engineering and Technology, 8(3), 74–81.

Gohil, P., Malek, M., Bachwani, D., Patel, D., Upadhyay, D., and Hathiwala, A. (2022). Application of 5D Building Information Modeling for Construction Management. ECS Transaction, 107(1), 2637–2649.

Malek, M. S., Gohil, P., Pandya, S., Shivam, A., and Limbachiya, K. (2022). A novel smart aging approach for monitoring the lifestyle of elderlies and identifying anomalies. Lecture Notes in Electrical Engineering, 875, 165–182.

Malek, M. S., Dhiraj, B., Upadhyay, D., and Patel, D. (2022). A review of precision agriculture methodologies, challenges and applications. Lecture Notes in Electrical Engineering, 875, 329–346.

Malek, MohammedShakil S., and Bhatt, Viral (2022). Examine the comparison of CSFs for public and private sector's stakeholders: a SEM approach towards PPP in Indian road sector, International Journal of Construction Management, DOI: 10.1080/15623599.2022.2049490

Design and Modelling a Structure with a Comparison of Cost Estimation by Traditional Method and BIM (Revit)

Namira M Saiyad,[*, a] *Dhiraj Bachwani,*[b] *and MohammedShakil Malek*[c]

[a]MTech (Construction Project Management-Perusing) Parul University, Vadodara, India
[b](Assistant Professor - Department of Civil Engineering) Parul University, Vadodara, India
[c]Principal, F. D. (Mubin) Institute of Engineering & Technology, Gandhinagar, India
E-mail: *saiyednamira30@gmail.com

Abstract

A prevailing technology in construction is called Building Information Modelling (BIM). To boost productivity and efficiency in construction, real-time models are used. This technology easily accomplishes projects on schedule and within the stipulated budget. Reworks can be reduced by improving project comprehension and visualization. Thus, BIM is not just a tool for 3D modeling but a methodology that can be applied from pre-construction to post-construction stages. This research is on the advantages of using BIM technology for architects, engineers, and contractors for scheduling and cost estimation. The key objective of this article is to calculate the quantities manually and on software, lastly, a comparison of manual and software-based estimating is provided. We found that there is a 6% overall cost difference between the manual and BIM-based estimation approaches. Thus, it may be said that BIM-assisted estimations perform better than Traditional approaches.

Keywords: Autodesk REVIT, BIM, 3D, Benefits, Challenges, Cost estimation, Visualization

Introduction

Building Information modeling (BIM) is advanced and widely acquired in collaboration construction projects with multiple different firms providing facility and software applications. In recent years, BIM has enhanced one of the best and faster fastest-growing multidisciplinary data sharing in Architecture, Engineering, and Construction firms. Its success is largely attributable to a unique design methodology that enables tracking of the project's complete life cycle, with additional data that makes it possible to see every area in depth, including the timetable, financial management, computations, and simulations. (Oktem, 2018)

In comparison to 3D design CAD, Every element in the Building information model has a specified function and access to project-related data. It also represents space in three dimensions. The BIM model created by an architect may be easily converted into an analytical model for use in structural analysis by a builder. Stresses and deflections may be calculated in a matter of minutes by a structural engineer, who can then go on to a complete review of code requirements and a cost estimate (Bute 2018).

Since modifications made to the design, or 3D digital model, are automatically reflected in all associated construction documents, such as drawings and schedules, BIM-based cost estimation is particularly effective and reliable. Its utilization range varies from 50–80 % the usage of BIM in the cost-estimating process has a larger potential to result in significant time savings (Haider, 2020).

Traditionally during the tendering phase is when cost estimation for construction projects begins. Manually estimating cost is subject to humans and has the propensity to spread errors. The accuracy of cost estimation is improved by the usage of BIM software. Systems for computer-based estimating (Revit/BIM software) are now widely used in the building sector. In wealthy nations, construction management uses BIM extensively. The use of computer software for cost estimates is still in its infancy in underdeveloped nations. (Sarkar 2015.)

This paper will demonstrate how Autodesk Revit technology will help architects, engineers, and contractors with scheduling, cost management, and estimation. It starts with a broad introduction to BIM technology and how it differs from the Traditional CAD (Computer Aided Design) method, and then it moves on to an analysis of the Autodesk Revit tools. The application of Scheduling and Cost Estimating in Autodesk Revit is then explained, and a G+2 design is provided to demonstrate how Autodesk Revit may benefit Architects, engineers, and contractors.

BIM Software

There is a wide variety of BIM software available. Autodesk Revit Advanced, GraphiSOFT ArchiCAD, Bentley Architecture, Nemechek All plan Architecture, Gehry Technologies- Digital Project Designer, Nemetschek Vector works Architect, MSA IDEA Architectural Design (IntelliCAD), CADSoft Envisioneer, Softtech Spirit, and Rhino BIM are all examples of software used in the architectural and engineering industries (Beta).

In addition, Autodesk Revit MEP (mechanical, electrical, engineers) is compatible with the structural design tools Autodesk Revit Structure, Bentley Structural Modeler, STAAD and Prosteel,

Tekla Structures, CypeCAD, Bentley RAM, and Nemetschek Scia (electrical and plumbing).

Different types of BIM dimensions are as follows:

- 3D – For Geometry
- 4D – For time data
- 5D – For cost data
- 6D – For sustainability
- 7D – Life cycle management of a building

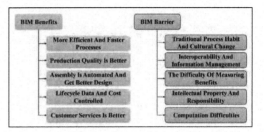

Fig. 1. Benefits and barriers of BIM

Source: compiled by Author from BIM Implementation in Infrastructure Projects: Benefits and Challenges

The key aim and objective of this research are To Design and model a residential (G+2) building in Autodesk Revit with the cost estimation, Bill of Quantities with BIM applications, and a comparison of Manual and software-based cost estimation of the building.

Literature Review

According to Sarkar and Modi (2015) survey and responded 60% of the architect, Engineers, and consultants; the software used to create the 3D models, Revit 2015 would lessen issues with coordination between the structural engineers, building builders, and architects' service companies (MEP). Using virtual reality as well the customer would be able to interact with the building model. Before the structure is built the likelihood of the interactions between the various utilities and services is minimized this would decrease the length of time and expense of the project and would raise the likelihood that Successful project completion within the allotted period and budget.

(Haider et al., 2020) identified the main objective is to take the plan, section, and elevation and calculate quantities with and Manual and by software and got to know that is very easy to calculate quantities in software and it is not time-consuming, if we are using the manual method of estimating cost then there are chances of human errors and by visualize software is so easy to found any error and clashing.

While it is true that many people (Kumar et al., 2017) believe that BIM exclusively relates to buildings, the project's authors correctly discovered and showed that BIM was equally relevant and, more crucially, helpful to infrastructure projects. The project team used both CAD and BIM in tandem to conduct an in-depth analysis of the differences between the two methods of design and change.

Two case studies were analyzed with an eye on the practical effects of building information modeling (BIM) on the design and construction processes and the roles played by professionals. The former example demonstrates how a BIM strategy was used to resolve tensions that arose during the design and procurement phases.

Research Methodology

Residential building (G+2)

Plot area 20m x 22m = 440 m²

Built up area 11m x 14m = 154 m²

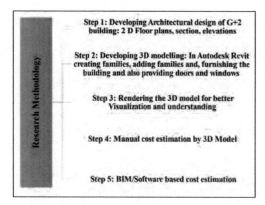

Fig. 2. RM steps

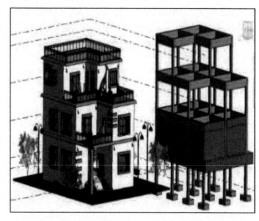

Fig. 3. G+2 Residential building

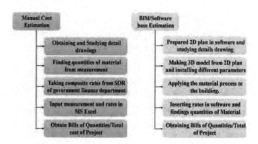

Fig. 4. G+2 Comparison of cost estimation of Manual and software/BIM

Result Comparison

Finding an appropriate value for the project before construction is the major goal of cost estimation. Traditional calculations of quantities are less accurate than those made using BIM or Revit software. According to Table 1, the project's total estimated cost by Traditional cost estimation is 3289031.863 RS, and using BIM/Revit Software, it is 3222813.145 RS.

Table 1 displays the quantitative comparison of the various products. The Project Cost Difference works out to be 6%. Because of the quality of the BIM/Revit software estimation, it has been discovered that the amounts estimated by these programs are more accurate.

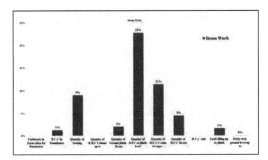

Fig. 5. Percentage Difference of Both Estimation Method

Conclusion

Table 1. Comparison of manually calculated cost verses BIM software-based costs

Item Description	Versus	BIM/Software Estimated Cost (Rs)
1. Earthwork in Excavation for foundation	436510.62	436650
2. P.C.C In Foundation	6887.409	6976.8
3. Footing	116640	116640
4. R.R.C Column up to plinth	96759.36	96788
5. Ground plinth Beam	51904.83	52986.2
6. D.P.C at plinth level	9849.6	12750
7. R.C.C Column of superstructure up to terrace	387037.44	347616.96
8. R.C.C Beam up to the terrace	360960	378000
9. R.C.C slab	245786.1	245805
10. Earth filling up to plinth	170107.2	167332
11. Brickwork ground level up to the terrace	1406589.30	1475613.9

This study shows that manual labor is demanding, time-consuming, and inaccurate, whereas Revit software estimates quickly, simply, effectively, automatically, and without human error.

In addition to optimizing pre-existing models from other applications like AutoCAD, this study aims to provide more effective G+2 Residential buildings, from furnishings to light fixtures.

For a small-scale project the BIM software has effected a net 6% of cost savings based upon which, if utilized in various infrastructure sectors, it can result in cost-efficient and budgeted costs.

Future Scope of Work

In this project for a small scale, the BIM software has effective cost and time saving so, we can adapt the BIM software in infrastructures and utilize into a result cost efficient and in the budget also it is recommended the research that can be expanded to different BIM application software like Naviswork, Primavera.

References

Malek, MohammedShakil S., and Bhatt, Viral (2022). Examine the comparison of CSFs for public and private sector's stakeholders: a SEM approach towards PPP in Indian road sector, International Journal of Construction Management, DOI: 10.1080/15623599.2022.2049490

Malek, MohammedShakil S., and Gundaliya, P. J. (2020). Negative factors in implementing public–private partnership in Indian road projects, International Journal of Construction Management, DOI: 10.1080/15623599.2020.1857672

Malek, MohammedShakil S., and Gundaliya, Pradip (2021). Value for money factors in Indian public-private partnership road projects: An exploratory approach, Journal of Project Management, 6(1), 23–32. DOI: 10.5267/j.jpm.2020.10.002

Malek, MohammedShakil S., Saiyed, Farhana M., and Bachwani, Dhiraj (2021). Identification, evaluation and allotment of critical risk factors (CRFs) in real estate projects: India as a case study, Journal of Project Management, 6(2):83–92. DOI: 10.5267/j.jpm.2021.1.002

Malek, MohammedShakil S., and Zala, Laxmansinh (2021). The attractiveness of public-private partnership for road projects in India, Journal of Project Management, 7(2):23–32. DOI: 10.5267/j.jpm.2021.10.001

Gohil, P., Malek, M., Bachwani, D., Patel, D., Upadhyay, D., and Hathiwala, A. (2022). Application of 5D Building Information Modeling for Construction Management. ECS Transaction, 107(1), 2637–2649.

Malek, M. S., Gohil, P., Pandya, S., Shivam, A., and Limbachiya, K. (2022). A novel smart aging approach for monitoring the lifestyle of elderlies and identifying anomalies. Lecture Notes in Electrical Engineering, 875:165–182.

Malek, M. S., Dhiraj, B., Upadhyay, D., and Patel, D. (2022). A review of precision agriculture methodologies, challenges and applications. Lecture Notes in Electrical Engineering, 875:329–346.

Bachwani, D., and Malek, M. S. (2018). Parameters Indicating the significance of investment options in Real estate sector. International Journal of Scientific Research and Review, 7(12):301–311.

Parikh, R., Bachwani, D., and Malek, M. (2020). An analysis of earn value management and sustainability in project management in construction industry. Studies in Indian Place Names, 40(9), 195–200.

Parameters Impacting Duration and Price of Ahmedabad Metro Project

Shibli khan,[*,a] *Farhan Khan Pathan,*[b] *Dhiraj Bachwani,*[c] *and MohammedShakil Malek*[d]

[a,b,c]Parul University of Vadodara, India
[d]Principal, F. D. (Mubin) Institute of Engineering & Technology, Gandhinagar, India
E-mail: [*]dimpeekhan22@gmail.com

Abstract

Ahmedabad Metro is a rapid transit system that serves Ahmedabad, Gujarat, India. The system has been in the works since 2015 and is scheduled to be completed by 2020. The first phase of the project, which will consist of two lines, will cost Rs 10,773 crore (US$1.6 billion). The project spans 38 kilometers (24 miles) and includes 27 stations. The Gujarat Metro Rail Corporation Limited (GMRCL), a 50:50 joint venture between the governments of Gujarat and India, is carrying out the project. The project is being funded through a combination of equity and debt, with the Government of India and the Government of Gujarat each contributing 2,600 crores (US$390 million) in equity and the remainder being raised through debt. The project is expected to benefit the city of Ahmedabad by reducing traffic congestion and pollution and boosting the city's economy. The project is also expected to create jobs during its construction and operation phases.

Keywords: Parameters Impact Duration, Cost of Metro Project, Indian Economy

Introduction

The Indian government is making every effort to advance the infrastructure industry. The Ministry of Urban Development has already spent approximately 2,863 crores (US$ 433 million) under the Atal Mission for Rejuvenation and Urban Transformation (AMRUT) initiative, indicating that the Government of India intends to make significant investments in the infrastructure sector. This program will build fundamental urban infrastructure over the 2017–20 fiscal year.

Infrastructure projects experienced significant delays, causing them to take longer than anticipated and cost more money.

Several factors influenced the cost and duration of the Ahmedabad Metro Project. The cost of land, the availability of skilled labor, the price of raw materials, the interest rate environment, and the rate of construction were among the parameters considered.

The cost of land was a significant factor in the project's cost. The Ahmedabad Metro Project was built on land purchased from the Ahmedabad Municipal Corporation (AMC) for Rs 1,600 crore. The land was initially valued at Rs 3,600 crore by the AMC, but the final cost was lower due to the AMC's discounted price. Another factor that influenced the project's cost and duration was the availability of skilled labor. Many trades people including masons, electricians, and plumbers were needed for the Ahmedabad Metro Project. There were a lot of competent employees brought in from Gujarat and other regions of India for the job.

One of the most significant contributors to the project's final price tag was the purchase price of raw materials. The project made extensive use of iron and steel bars, cement, and other building materials. The project was successful in obtaining a contract from Jindal Steel for the supply of iron and steel bars.

Infrastructure projects are suffering from the adverse effects of multiple issues and intractability issues. This includes customers, contractors, specialists, and some groups of controllers. Project managers in the construction industry face the challenge of finding the balance between schedule, budget, and quality. Project effectiveness can be quickly assessed by comparing actual time and costs against expectations. If his time slot on the project is postponed, it will not be possible to complete it on time-run time is very important in the construction industry because "time is money." In general, there is no other way to identify project milestones. A fundamental part of project management is uncertainty. However, uncertainties can be reduced by taking appropriate actions at design time.

During the planning phase, a predictive project plan is created and critical routes are identified when multiple tasks are dependent or interdependent. Lack of certainty stems from a variety of factors, including activity duration, available resources during execution, and task dependencies on the completion of other activities. During the execution phase, uncontrollable circumstances such as weather, resource constraints, and management decisions can alter the expected schedule, causing delays and cost overruns, especially when work is on the critical path. There is a nature.

Critical or near-critical delays lead to project delays. Non-critical activities can be delayed to

some extent, but offshoring critical work quickly leads to project delays and costs.

Ahmedabad and Gandhinagar metro projects are currently under construction. In October 2014, received project approval. The establishment of such a special vehicle company took place on February, 2010. Ahmedabad city metro line is shown from Visat to APMC.runs mostly along the Ashram route. The second axis from Thaltej to Vastral, connects east and west of Ahmedabad, passing through Kalupur, Ashram Road, Thaltej, and major intersections of the industrial area.

Future growth of this sector is expected to accelerate between Ahmedabad and Gandhinagar, according to the draft Greater Ahmedabad Zone Development Plan and Integrated Mobility Plan. Demarcation of the study area takes into account anticipated future expansion of the study area.

Literature Review

1] Adnan Enshassi. "Delays and cost overruns in the construction projects in the Gaza Strip." Journal of Financial Management of Property and Construction. Published by Emerald July 2013.

Construction project delays and cost increases in the Gaza Strip were analyzed to determine their root causes. Strikes and border closures, material-related reasons, material shortages in markets, and delays in transporting goods to building sites were his top four causes of schedule delays. Price fluctuations for construction supplies, contractor supply delays for essential components and equipment, and general price increases were the primary contributors to budget overruns. Furthermore, municipal governments are urged to introduce laws to address issues caused by construction material monopolies.

2] A.S. Ali, S.N. Kamainzzaman "Cost performance for building construction projects in Klang Valley" Journal of Building Performance ISSN: 2180-2106 Volume 1 Issue 1 2010 January

To stay under budget for a project's construction, expenses must be managed carefully throughout its execution. Poor initial project cost estimates and underestimating of construction costs by volume surveyors were two main contributors to construction project cost overruns.

3] Fahad Saud Allahaim and Dr. Li Liu "Understanding Major Causes Cost Overrun For Infratsrtucture Projects; A Typology Approach" AUBEA 2012 Conference, the University

The reasons of infrastructure project cost overruns were investigated here. Based on the review, kinds of causes and cost overrun relationship patterns were defined in a theoretical framework, and evolved from the cost overrun cause typology. That reveals the intricate web of interconnections.

4] Panagiotis Ch. Anastasopoulos, John E. Haddock and Srinivas Peeta "Cost Overrun in Public-Private Partnerships: Toward Sustainable Highway Maintenance and Rehabilitation" Journal of Construction Engineering and Management 140(6):04014018 June 2014

Compare bid costs to win contracts to end-of-stock costs for highway maintenance and remediation projects, identify potential project cost overruns, and employ appropriate PPPs for various types of maintenance and remediation work. High costs or high implementation risks increase the likelihood of cost overruns.

Major Findings of Literature Review

Infrastructure is a capital-intensive industry, and the constantly shifting dynamics continue to be difficult. The discriminating consumer expectations modify the character of demand, which has a significant impact on the Indian infrastructure market. There are several delays and cost overruns in the infrastructure project, but adequate management practices are lacking. Additionally, we are unable to determine the ideal remedy. Such a disruption will affect their ability to make wise investing decisions. Indian businesses must thus handle infrastructure projects in a more methodical and organized manner. In order to thrive in this cutthroat and transformative moment for the Indian construction industry, it is crucial to understand the elements that influence time and cost overruns.

Research Methodology

Research is often described as the quest for knowledge. Clifford Woody (Kothari, 1988) described studies as scientific and methodological searches for relevant knowledge on a particular topic. Briefly, research is the process of seeking information on a systematic methodological approach to a topic. Research also includes methodological approaches to generalization and theory building.

Research must use the scientific method to provide solutions to unsolved problems. There are multiple options for conducting research, including polls, surveys, conferences, and interviews.

According to Fig. 4.1, there are two basic research approaches: quantitative and qualitative. Proper quantitative analysis and the production of quantitative data that can be combined in a clear and well-defined way are key elements of any quantitative research approach. This research method can also be divided into inference, experimentation, and simulation techniques. Qualitative research methods subjectively assess attitudes, beliefs, and behaviors. As a result, research methods produce results that are non-quantitative or cannot be precisely quantified. Focus group interviews, projection techniques, and in-depth interviews are commonly used.

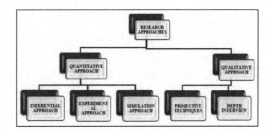

Fig. 1. Research Approaches

Source: C. R. Kothari, *Research Methodology, Methods, and Techniques.

Criticality Index Method

A methodology termed the criticality index approach is used to prioritize different criteria according to the feedback gathered. In this study, a five-point Likert scale is used for all questions. A unidimensional scaling technique often employed to measure ordinal variables is the Likert scale.

Each risk's criticality index was assessed based on how critical each group of respondents rated each danger.

The index utilized for calculating Criticality Index (CI) is:

$$CI = \frac{5n_5 + 4n_4 + 3n_3 + 2n_2 + n_1}{5(n_5 + n_4 + n_3 + n_2 + n_1)}$$

In which,

n_1 = no. of persons who responded "*most critical*";

n_2 = no. of persons who responded "*very critical*";

n_3 = no. of persons who responded "*critical*";

n_4 = no. of persons who responded "*somewhat critical*";

n_5 = no. of persons who responded "*not critical.*"

While evaluating the criticality of the factors on time overrun and cost overrun, all those factors having CI<0.50 (i.e. 50%) have been considered less significant. means they have less effect on the project time overrun and cost overrun.

Conclusion

In conclusion, the main parameters impacting the duration and price of the Ahmedabad Metro project are the construction schedule, the land acquisition process, the availability of construction materials, the availability of skilled labor, and the weather conditions. All of these factors need to be carefully monitored and managed in order to ensure the successful completion of the project within the desired timeframe and budget.

Future Scope of Work

These infrastructure initiatives are relevant to this study. A short list of factors from literature reviews was corroborated by face-to-face interviews. Based on the variables selected for the shortlist, the paper contains an analysis of "Ahmedabad Metro Rail Project"

Phase I'. Gujarat conducted a study using Ahmedabad Metro Rail Project Phase I as case study. Construction sector researchers and other stakeholders will use the results of this study and his proposals. It is used to learn more about the variables that affect time and cost overruns in infrastructure projects in India.

References

Bachwani, D., and Malek, M. S. (2018). Parameters Indicating the significance of investment options in Real estate sector. International Journal of Scientific Research and Review, 7(12), 301–311.

Malek, MohammedShakil S., Saiyed, Farhana M., and Bachwani, Dhiraj (2021). Identification, evaluation and allotment of critical risk factors (CRFs) in real estate projects: India as a case study, Journal of Project Management, 6(2), 83–92. DOI: 10.5267/j.jpm.2021.1.002

Haddock, John E., and Joan P. OâBrien (2013). Hot-Mix Asphalt Pavement Frictional Resistance as a Function of Aggregate Physical Properties. Road Materials and Pavement Design (RMPD) Journal. 14 (2), 35–56.

Tanaka, Alison, Jon D. Fricker, and John E. Haddock (2013). Determining Viable Sizes for Indiana Communities Based on Essential Establishment and Services. Journal of Urban Planning and Development. 139(1), 33–39.

Tanaka, Alison, Panagiotis Ch. Anastasopoulos, Neal Carboneau, Jon D. Fricker, John A. Habermann, and John E. Haddock (2012). Policy Considerations for Construction of Wind Farms and Biofuel Plant Facilities: A Guide for Local Agencies. State and Local Government Review. 44(2), 140–149.

Anastasopoulos, Panagiotis Ch., Fred. L. Mannering, and John E. Haddock (2012). Random Parameters Seemingly Unrelated Equations Approach to the Post-Rehabilitation Performance of Pavements. Journal of Infrastructure Systems. 18(3), 176–182.

Malek, MohammedShakil S., and Bhatt, Viral (2022). Examine the comparison of CSFs for public and private sector's stakeholders: a SEM approach towards PPP in Indian road sector, International Journal of Construction Management, DOI: 10.1080/15623599.2022.2049490

Malek, MohammedShakil S., and Gundaliya, P. J. (2020). Negative factors in implementing public–private partnership in Indian road projects, International Journal of Construction Management, DOI: 10.1080/15623599.2020.1857672

Gohil, P., Malek, M., Bachwani, D., Patel, D., Upadhyay, D., and Hathiwala, A. (2022). Application of 5D Building Information Modeling for Construction Management. ECS Transaction, 107(1), 2637–2649.

Malek, M. S., Dhiraj, B., Upadhyay, D., and Patel, D. (2022). A review of precision agriculture methodologies, challenges and applications. Lecture Notes in Electrical Engineering, 875, 329–346.

Review Paper on LEAN Construction Techniques

Jaydeep Pipaliya[a,*] and MohammedShakil Malek[b]

[a]Assistant Professor, Department of Civil Engineering, Parul University, Vadodara, India
[b]Principal, F. D. (Mubin) Institute of Engineering & Technology,
Gujarat Technological University, India
E-mail: *jaydeep.pipaliya21306@paruluniversity.ac.in

Abstract

Lean production has not only shown itself superior to established methods of mass production in the auto industry, but it has also shifted the balance between productivity and quality, Toyota's creation framework was first presented by Japan after The Second Great War when Japan required delivering little clumps of vehicles in numerous assortments in opposition to the Passage rule of large scale manufacturing. Toyota reasoned that the rule of large-scale manufacturing isn't proficient any longer, particularly, after the breakdown in deals that Toyota experienced and prompted delivering a huge piece of their labor force. Thus, they thought of novel thoughts and presented the Toyota Creation framework, known as LEAN. The motivation behind this review is to build efficiency in development projects by utilizing LEAN apparatuses and likewise further develop time, cost, quality, and security in development by the utilizing of LEAN instruments and after that decide execution at the building site later and before the utilization of LEAN instruments.

Keywords: LEAN Construction, Just-in-time, Impact on productivity, LPS

Introduction

Financial growth that satisfies the demands of both the present and future generations has been defined as "manageable." Supportable development is likewise viewed as a way advances for the development business to accomplish manageability being developed while taking ecological, financial, and social issues into thought. In appearing this work, the development business is asked to move from conventional wet development techniques towards natural well disposed of, energy production, and less waste age strategies for development.

The remarkable requirements on the framework projects in Malaysia, which aim to help the nation qualify as a developed country by 2020, have resulted in a substantial quantity of development squandering, including 28.34% squanders generated by the development exercises.

Because of this anomaly, the company must alter its standard procedures for waste reduction and disposal. Lean Construction is an alternative approach that may be used in the construction industry due to its remarkable potential to achieve the aims of increasing the worth and efficiency of a development project.

Fundamental Knowledge of Lean

The idea of lean assembling was made popular by Toyota, thinking back to the 1980s. Their constant improvement culture and endeavors to take out squandering made them one of the most productive organizations around.

Numerous corporate pioneers and business investigators attempted to study and duplicate these Toyota creation frameworks, including organizers behind the Lean Endeavor Establishment, James Womack, and Dan Jones.

During the nineties, Womack and Jones distributed "Lean Reasoning: Expel Squander and Make Abundance in Your Enterprise'.

This book is viewed as a definitive manual for running a lean venture and set out five key rules that supported the interaction. This attention on boosting the worth of the client while taking out squandering route. They are presently known as the standards of lean assembling and have impacted the universe of sequential construction systems and creation plants.

The Assessment of Lean Development Execution

Since Koskela proposed the idea of lean creation in construction, lean development has encountered a long improvement time. Hypothetically, the Change Stream Worth (TFV) model can make sense of the principal considered lean development. The possibility of T(transformation) implies the info and result processes.

Its accentuation is to accomplish esteem-added exercises. F(flow) implies data and material stream, and its accentuation is to lessen non-esteem-added exercises. V(value) implies the method involved with accomplishing the clients' prerequisites, and its accentuation is to expand the task's worth.

The TFV model can assist individuals with understanding lean development without any problem. At the point when we assess the execution of lean development in ventures, some incline construction advances are expected to assemble an assessment record framework. Taking into account the execution level and development of lean construction advancements in China, we chose the accompanying instruments for assessment.

Background and Review

Kosela's pioneering work on the establishment of the TFV hypothesis of creation to overcome the failure of conventional development board conveyance methodologies to meet schedule, cost, and quality targets is often credited as the impetus for the shift and fusion of lean standards in development. T in this technique stands for "Transformation of materials into a completed office," F for "progression of the material through the development cycle," and V for "respect age" and "creation," both of which are achieved primarily via the simultaneous elimination of loss and waste.

Lean development is an approach to organizing production frameworks that minimize wasteful use of resources (such as time, money, and human effort) to provide the greatest possible amount of value.

To create something "lean," you need to plan and make it in a way that distinguishes it from "mass" and "art" production in terms of goals and process, and then you need to streamline the execution of your production framework against a flawlessness standard that allows you to meet unique client needs.

Applications

Just in Time

In just-in-time manufacturing, production begins only when a customer requests it, thanks to an on-demand system. This means that businesses don't have to stock up on unnecessary supplies, lowering the risk that certain components or things would be overloaded or damaged in storage.

Lean experts could use JIT if their company can benefit from interest shaving and can reduce the risk of just transporting merchandise when necessary. Although JIT may be an effective approach for managing inventory, it can also make it more challenging to meet customer demand in the case of a disruption in the production network.

For instance, the independent publishing area frequently utilizes this model, just printing books as they are requested. Advanced dispersion for media items has additionally assisted with limiting the expenses related to overabundance of materials.

Last Planner System

Improving the planning and management of building projects is key to raising industry standards.

Within the framework of Lean Development, planning and monitoring are seen as continuous loops that must be maintained for the duration of the project.

While planning establishes the parameters within which the project will operate and the processes that will be used to achieve its objectives, controlling the project's progress guarantees that every event will occur as planned. When the predetermined successions are no longer beneficial, it is time to reevaluate.

Confronting the reality that life doesn't always go as planned is a key component of criticism. The Last Organizer Framework is one of the most notable examples of a fully realized Lean methodology; it has proven to be an extremely useful tool for the management of the development process and the constant monitoring of the arranging effectiveness, which together aid in the creation of foresight, the smoothing of work process varieties, and the reduction or elimination of vulnerabilities that plague development in favor of cess.

It consists of managing both the production line and the workforce. The look-ahead method is primarily responsible for controlling the workflow, whereas weekly work planning is mostly responsible for managing the production unit.

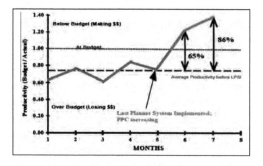

Fig. 1. Productivity Improvement using LPS

Source: Applying lean thinking in construction and performance improvement)

Lean Thinking Standards

For maximum benefit from the lean accomplishment, it is essential to adhere to the following five guidelines for lean thinking:

(1) Establish Value: Value exercises that add value to the final product, and highlight client input on how much that input is valued.

(2) Recognize the Worth Stream: After all, activities that do not add value to the final product, you will see the worth stream. This means you need to immediately put a halt to the creative process and make adjustments if things aren't looking well.

Miss production, overproduction (repetition of support of duction of a similar item, etc.), the capacity of materials and pointless cycles, transport of materials, development of work labor forces and items, the final creation of items which do not satisfy the wished norm of the client, and all sorts of superfluous holding up time are all processes that should be avoided.

(3) Flow: Focus on the whole retail infrastructure to ensure a constant supply of goods and services. It's important to keep your focus on the cycle, not the end outcome. However, unless customer value is established and the value stream is understood, the flow will never be optimal;

(4) Pull: rather than push, should be used in the process of production and development in support of the duction. This means always being flexible and open to consumer feedback and alterations, and producing exactly what the client wants when the client needs it. The goal is to cut down on wasteful production while using the management tool "just in time";

(5) Perfection: Pays close attention to getting things just right and improving things all the time. Deliver a product that meets the client's requirements and expectations within the agreed-upon time frame and in perfect shape, free of mistakes and concessions. This is best achieved by close communication between the client/client, chiefs, and employees. In Fig. 1, we see a compilation of many types of lean equipment in use today.

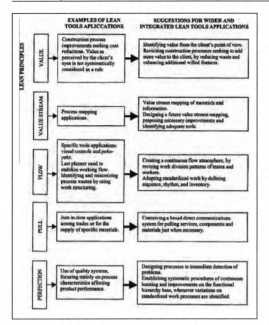

Fig. 2. Example of LEAN tools in construction)

Source: Applying lean thinking in construction and performance improvement)

Literature Review

Many templates for implementing Lean Creation in large and medium-sized projects can be found in the literature. Regardless of how big, medium or small an organization is, the core concepts of Lean Assembling are within reach.

Lean Creation principles do not seem to be appropriate for use in projects employing less than ten employees. However, for most modest endeavors with 10 to 49 employees, adopting Lean Assembling might be one crucial step toward enhancing efficiency and becoming more competitive in the market.

Small and medium-sized businesses, in particular, are known for their adaptability and innovation, not just in terms of the products they produce but also in terms of the methods they use to mass-produce such products. As they become aware of mounting threats, small businesses are taking the initiative to expand their operations, which is a good first step toward implementing lean practices.

Many authors agree that there are significant difficulties in implementing lean manufacturing and other measures for boosting efficiency. Hayes found that preparing ahead of time is essential for the implementation of Lean Creation initiatives in SMEs in order to avoid such difficulties and setbacks.

In the last decade, researchers have looked at Lean Assembling in great detail. White, Conner, and Achanga are among the few authors who have researched Lean Assembling's effectiveness in small and medium-sized enterprises (SMEs).

Small and medium-sized enterprises (SMEs) are unsure about the total cost of implementation and the potential magnitude of the results and benefits.

Many businesses worry that adopting lean manufacturing procedures would be too costly and time-consuming. Smaller businesses are less likely to get costly service packages since they have less resources and, in many cases, less access to finance than their larger counterparts.

It is said that small and medium-sized enterprises (SMEs) that use Lean Assembling may contribute to the development of intensity by accelerating development and production, increasing flexibility, and decreasing costs.

Conclusion

The paper will provide a summary of the research on LEAN construction and also Impact of Just in Time tool in construction and also it helps to reduce the wastage of material and improve the productivity of work in construction. So many times, to improve customer satisfaction we can take the help of different LEAN tools which also it gives a huge impact on customer satisfaction. We can take a survey on it for next project.

References

Koskela, L. (1992). Application of the New Production Philosophy to Construction, CIFE Technical Report #72. Department of Civil Engineering, Stanford University, Stanford, USA.

Sacks, R., Dave, B. A., Koskela, L., and Owen, R. (2009). Analysis Framework for the Interaction Between Lean Construction and Building Information Modelling. In: Cuperus, Y., and Hirota, E.H., 17th Annual Conference of the International Group for Lean Construction. Taipei, Taiwan, 15–17 Jul 2009. pp. 221–234.

Orihuela, P., Orihuela, J., and Ulloa, K. (2011). Tools for Design Management in Building Projects. In: Rooke, J., and Dave, B., 19th Annual Conference of the International Group for Lean Construction. Lima, Peru, 13–15 Jul 2011.

Alarcon, L. F. (2012). Last Planner SystemTM, GEPUC, Pontificia Universidad Católica de Chile.

Koskela, L. (2000). An Exploration Towards a Production Theory and its Application to Construction, VVT Technical Research Centre of Finland.

Mann D. (2009). The Missing Link: Lean Leadership. Frontiers of Health Services Management. p. 15–26.

Ohno T. (1988). The Toyota Production System - Beyond large scale production, Portland: Productivity Press.

Womack, J., Jones, D., Roos, D. (1991). The Machine that Changed the World: The Story of Lean Production, 1st Harper Perennial Ed., New York, 1991.

Wright, G. (2000). Lean construction boosts productivity, Building Design and Construction 41(12), 29–32.

Eurostat (2009). European Business – facts and figures 2009. Eurostat statistical books.

Factors Affecting Efficient Highway Infrastructure Projects

Dhruvi Shah,[a] Rajesh Gujar,[*,b] Jaykumar Soni,[c] and MohammedShakil S. Malek[d]

[a,b,*]Pandit Deendayal Energy University, Gandhinagar, India
[c]L. J. University, Ahmedabad, India
[d]F. D. (Mubin) Institute of Engineering and Technology, Gandhinagar, India
E-mail: [a]dhruvi.smtcl21@sot.pdpu.ac.in; [*]rajesh.gujar@sot.pdpu.ac.in;
[c]jay.soni_ljiet@ljinstitutes.edu.in; [d]shakil250715@yahoo.co.in

Abstract

The construction business is known for its fragmentation. Highways are significant assets due to their considerable share of a country's financial, social, and governmental progress. Highway building projects have always had cost overruns owing to delayed project completion, resulting in a loss of public funds. It is vital to assess the influence of key risk factors on building projects so that timely alternatives or solutions may be implemented to avoid them. Several studies have detected the crucial variables driving road projects to define the areas that need improvement. Various strategies and methods have been utilized in studies on project quality aspects to understand the correlations between certain variables. The purpose of this study is to identify and analyze the numerous risk variables connected with building projects. The approach used for the study is a review of the previous year's research articles for the international construction sector.

Keywords: Highway Project, Project Performance, Success Directive Factors

Introduction

The infrastructure is the economic sector engaged in the construction, maintenance, and destruction of civil engineering projects worldwide (Seninde et al., 2021)time, scope and quality. The aim of this study was to assess the factors influencing performance of road construction projects in Uganda. The study adopted a descriptive research design and data were collected using questionnaires from 147 purposively selected respondents from construction companies, consultancy firms, and government. Relevant literature was reviewed to establish actual factors influencing performance of road construction projects in Uganda. Data were coded and entered into statistical packages for social scientists (SPSS. In recent decades, India has placed a significant focus on infrastructure development

In India, construction and infrastructure projects are critical components of the country's productive capability and efficiency. With the recent decrease in economic activity, the rate of highway building has reduced to 27 km per day in the current fiscal year, down from 30 km per day the previous year. As a result, the government has reduced its highway construction objectives for the current fiscal year to 6000 km (Deep et al., 2022)causing a loss of public funds. Since highways are the backbone of a nation, the purpose of this study is to measure the criticality of the factors that influence the performance of highway projects. A survey instrument was prepared and distributed to 185 project managers. To achieve the aim of the study, exploratory factor analysis was used and the standard factor loading was the criteria to measure the criticality. From the analysis, it was identified that the factors were grouped under four categories: (a.

As a result, the study intends to assess the importance of factors contributing to the success of the performance results in Indian highway construction projects.

Factors Influencing the Success of Road Infrastructure Projects

Project success criteria are the concepts or standards used to assess project success. Project success criteria may be defined as anything against which projects can be measured. If success is the favorable outcome of the project, the project's approved objectives are the conditions for achieving the goal. Success criteria are defined as expectations based on participation, service scope, project scale, use of the technology, and other variables.

Previously, the "Iron Triangle" (Fig. 1) of time, cost, and quality have been only commonly employed as a significant criteria. Although the literature says that consideration of only these three factors may lead to a biased opinion, they are still being used for determining project success. Additional factors and methods for determining success evolved.

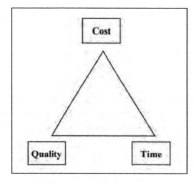

Fig. 1. Iron Triangle

Poor concepts of measurement of success have led stake holders to rely only on the Iron Triangle as the criterion. This perspective on success criteria may result in a skewed mapping of project success, for improved or deteriorated. Many times, a project is seen as failing while meeting all of the recognized success criteria. As a result, the Iron Triangle is not the tool to assess the success of the operation stage of any project, if it is appreciated by sponsors, or whether it improves the organization's effectiveness or efficiency. As a result, continuing to use an insufficient collection of success factors will lead to revised failures. Because highway projects consist of many diverse stakeholders, such as customers, managers, contractors, workers, and users, the appropriate success criteria must reflect several points of view. Contractors' and clients' differing perspectives on success criteria impede an efficient working relationship (Johansson, 2013).

Project-related elements are those that characterize the features of the project, such as the kind of project, its nature, the complications of the project, and the magnitude of the project. Project management factors are characterized by project management acts. The success of the project depends on these actions. The characteristics by which the project success is assessed are a transparent system, regulatory systems, review abilities, problem resolution, teamwork performance, decision-making performance, surveillance, task organization, timeline, related prior management experience, and so on. Human-related factors include characteristics of all important stakeholders, including the project manager, user, planner, contractor, consultants, subcontractor, suppliers, and fabricators. Client experience, understanding of construction project organization, project funding, customer confidence in the construction team, and so on are examples of client variables. The factors of the participants may be classified into two categories: client and project team. Furthermore, one study argues that team spirit is critical for project performance and emphasizes the need for team building among the various members. External factors are external impacts on the project that include the financial, social, and governmental surroundings and level of technological progress.

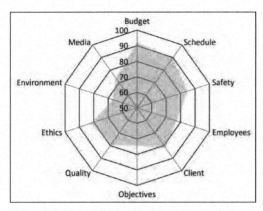

Fig. 2. Importance of various factors (Johansson, 2013)

The above radar was created after conducting the surveys with the project managers. The plotted factors are significant to decide the success of the project.

Plan alterations, stoppages because of disagreements between stakeholders, insolvency, noncompliance with regulatory requirements, and inspection delays are all factors that impact highway construction performance (Otim and Alinaitwe, 2013) Palestine suffer from many problems and complex issues. Consequently, the objective of this paper is to identify the factors affecting the performance of local construction projects; and to elicit perceptions of their relative importance. A comprehensive literature review was deployed to generate a set of factors believed to affect project performance. A total of 120 questionnaires were distributed to 3 key groups of project participants; namely owners, consultants and contractors. The survey findings indicate that all 3 groups agree that the most important factors affecting project performance are: delays because of borders/roads closure leading to materials shortage; unavailability of resources; low level of project leadership skills; escalation of material prices; unavailability of highly experienced and qualified personnel; and poor quality of available equipment and raw materials. Based on these findings, the paper recommends that: 1.

Poor management and planning, a lack of experience, inadequate controls during construction, a shortage of materials, a delay in commencing work, a delay in making choices, and a failure to address road safety are all issues that can impair the performance of road construction projects (Seninde et al., 2021)time, scope and quality. The aim of this study was to assess the factors influencing performance of road construction projects in Uganda. The study adopted a descriptive research design and data were collected using questionnaires from 147 purposively selected respondents from construction companies, consultancy firms, and government. Relevant literature was reviewed to establish actual factors influencing performance of road construction projects in Uganda. Data were coded and entered into statistical packages for social scientists (SPSS.

Stakeholders' inconsistent participation, service-oriented attitudes, numerous plans and drawing modifications, poor monitoring and feedback, and a lack of project leadership competencies are also seen. Procurement procedures, project team performance, conflict among project participants, substandard work, and external factors all contribute to poor project performance. The project managers' incompetence and lack of understanding, incorrect project conceptualization, and severe tendering competitiveness are all issues that might impair the success of highway development projects (Seninde et al., 2021)time, scope and quality. The aim of this study was to assess the factors influencing performance of road construction projects in Uganda. The study adopted a descriptive research design and data were collected using questionnaires from 147 purposively selected respondents from construction companies, consultancy firms, and government. Relevant literature was reviewed to establish actual factors influencing performance of road construction projects in Uganda. Data were coded and entered into statistical packages for social scientists (SPSS.

Methods Used for Analysis of Factors

There has been a substantial quantity of international and Indian literature dealing with the breadth and complexity of the concerns of project success/failure, and different studies have utilized different frameworks, but the majority of them used ranking analysis based on mean scores or the RII or its derivatives for finding crucial success/failure determinants.

Road infrastructure projects, like other projects, have a project life cycle during which project operations suffer from unsuccessful/successful completion, delays, and escalation, which are attributable to many variables functioning at the project level. As a result, many aspects have been evaluated at various phases of PPP projects in the Indian road sector for further research on project success/failure

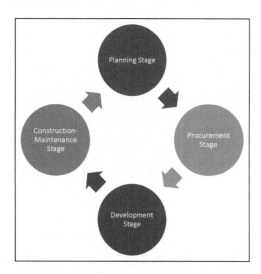

Fig. 3. Phases of Road Infrastructure PPP Projects

The Planning Stage is the initial stage of road infrastructure PPP projects that comprises studies such as project need, feasibility studies, and risk assessment. This is a preparatory stage in which a strategy for implementing the PPP project is developed. As far as the Procurement Stage is concerned, the bidding procedure and related activities are the main activities involved at this stage. The contract conditions and transparency of the bidding procedure are important factors in the project's competitiveness. Factors like Acquisition of land, obtaining clearance from various government departments, Defining the scope etc, are parts of the Development Stage. In

the last stage of the COM Stage (Construction-Operations and Maintenance Stage), the major factors are related to time overrun and cost overrun. Factors that play an important role in the COM stage are tariff rates, cost-overruns, time overruns, technical and financial closures

Conclusion

It has been observed that in the previous times, the major factor contributing towards the decision making for any highway infrastructure projects was the time-cost overruns. However, recent advancements have shown the existence of a variety of factors that can significantly affect the efficiency of projects. Many important factors such as.... have been mentioned in the paper. The authors have tried to incorporate majority of prominent factors which can affect the efficacy of highway infrastructure projects. Considering these factors can notably increase the chances of high efficiency during the execution of such projects. Efforts can be made to prioritize the factors based on statistical tools which are currently being undertaken by the authors as a part of the progressive work scope.

References

Deep, S., Banerjee, S., Dixit, S., and Vatin, N. I. (2022). Critical Factors Influencing the Performance of Highway Projects: Empirical Evaluation of Indian Projects. Buildings, 12(6), 1–14. https://doi.org/10.3390/buildings12060849

Johansson, N. (2013). Success Factors in Large Infrastructure Projects : The contractor's perspective. 51.

Nallathiga, R., Shaikh, H. D., Shaikh, T. F., and Sheik, F. A. (2017).

Otim, G., and Alinaitwe, H. M. (2013). Factors affecting the performance of pavement road construction projects in Uganda. Makerere University, 15(3), 269–280.

Rahman, R. A., Radzi, A. R., Saad, M. S. H., and Doh, S. I. (2020). Factors affecting the success of highway construction projects: The case of Malaysia. IOP Conference Series: Materials Science and Engineering, 712(1). https://doi.org/10.1088/1757-899X/712/1/012030

Seninde, S., Muhwezi, L., and Acai, J. (2021). Assessment of the Factors Influencing Performance of Road Construction Projects in Uganda: A Case Study of Ministry of Works and Transport. International Journal of Construction Engineering and Management, 2021(4): 101–115. https://doi.org/10.5923/j.ijcem.20211004.02

Malek, MohammedShakil S., and Bhatt, Viral (2022). Examine the comparison of CSFs for public and private sector's stakeholders: a SEM approach towards PPP in Indian road sector, International Journal of Construction Management, DOI: 10.1080/15623599.2022.2049490

Malek, MohammedShakil S. and Gundaliya, P. J. (2020). Negative factors in implementing public–private partnership in Indian road projects, International Journal of Construction Management, DOI: 10.1080/15623599.2020.1857672

Malek, MohammedShakil S., and Gundaliya, Pradip (2021). Value for money factors in Indian public-private partnership road projects: An exploratory approach, Journal of Project Management, 6(1), 23–32. DOI: 10.5267/j.jpm.2020.10.002

Malek, MohammedShakil S., and Zala, Laxmansinh (2021). The attractiveness of public-private partnership for road projects in India, Journal of Project Management, 7(2), 23–32. DOI: 10.5267/j.jpm.2021.10.001

Safety Performance on Construction Sites of Gujarat

Ms. Anuja Rajguru,*,a MohammedShakil Malek,b and Lalit S. Thakurc

[a]Ph.D.Scholar, Parul University, Vadodara, Gujarat, India
[b]Principal, F. D. (Mubin) Institute of Engineering and Technology, Gandhinagar, India
[c]Associate Professor, Parul University, Vadodara, Gujarat, India, India
E-mail: *anujarajguru1@gmail.com

Abstract

The research examines how effective safety management is in the building sector. This research was conducted to provide insight into the topic of construction site safety. To achieve this, a field survey was conducted with a representative sample of local construction industry experts working in an area with a high density of ongoing building projects. The data were collected using structured questionnaires, and the findings were tallied to display the breakdown of responses. The results demonstrate that construction professionals understand that the scope of their operations affects their capacity to adhere to health and safety laws. According to the poll, there is a significant problem with accidents and injuries in the construction sector in Gujarat. Therefore, the results provide insight into the policy for implementing workplace safety. In light of these findings, it is clear that the construction sector in Gujarat needs stricter regulation and greater levels of inspection to ensure worker health and safety.

Keywords: Cost, Safety management, Construction work

Introduction

Any industry that wants to succeed needs to operate in a way that is safe, dependable, and sustainable. Construction is a very dangerous industry on a global scale compared to other industries. As with other professions, construction professionals have greater difficulties when carrying out their jobs. The construction industry in India employs the most people—44 million—and is the second-largest contributor to GDP, behind agriculture, at around 9%. Semi-skilled and unskilled labor is used extensively in India's construction sector. Compared to other industrial sectors of the nation, the workers of an Indian Construction company are less protected.

Occupational hazard statistics are also accessible in developing nations; however, they are based on statistics from affluent nations.

Occupational health and safety have never been prioritized above everything else in the Indian construction sector.

Safety Performance in Indian Construction Industry

The building industry has been making efforts to improve its health and safety record for decades. However, the focus of these initiatives has switched from assessing safety performance to taking preventive steps to enhance performance. In India, there are roughly 48,000 worker fatalities due to workplace accidents each year, and there are about 37 million incidents that result in 4 days of missed work (Hamalainen et al., 2006). A report by the International Labour Organization (ILO) indicates that the construction sector in India has the highest accident rate worldwide. According to the study, 165 out of 1,000 workers suffered on-the-job injuries, and around 7,600 workers every day lose their lives as a result of workplace fatalities. (Singh et al., 2017).

A 2019 article from Counterview.net states that Gujarat is the state with the second-highest investment in the construction sector (13%), after only Maharashtra (25%). The number of fatal accidents in Gujarat rose to 137 in 2018, the highest number in ten years. A total of 67 people died in 2017, 55 people died in 2016, 62 people died in 2015, and 69 people died in 2014. According to data obtained through Right to Information (RTI) requests made to the Gujarat Police and media clippings, 49% of workplace fatalities in Gujarat are caused by falls from heights, while 21% are caused by being buried beneath debris.

Safety Management

The philosophy of safety makes an effort to identify and coordinate common causes and corrective actions for both general and specific safety issues. It covers the fundamental need or requirement for safety, its causes and effects, strategies for spotting risky conditions and behaviors, and the causes and sequence of accident occurrence. Along with this, it discusses safety devices, tactics, and measures, as well as the roles and responsibilities of various authorities in terms of safety and the quick response to accidents.

In other Countries, safety is defined as, the protection and prevention of individuals and people from any form of accident and risk that comes as a result of injury, harm, unhealthy environment for working at the place of work. Also, law of safety is defined as a way in which the working environs can strengthen to secure the safety and health of people who are liable to be affected by the working environment. (Fredrick Ahenkora Boamah et al., 2019).

A condition in which the potential for harm to people or property is low enough to be acceptable (IS 18001:2007).

National Safety Council (NSC) defines safety as "the control of recognized hazards to attain an acceptable level of risk."

In other words, safety is the condition of not being threatened by any harmful or harmfully occurring events. It's when risks aren't present. Accident, Risk, Hazard, Near Miss, and Danger are some of the most fundamental phrases in safety (Jani, 2017).

The Research Methodology Adopted for the Study

This study opted to use quantitative methods of inquiry after reviewing the literature on research strategies and techniques, as these can be used to test a theory made up of variables, measured with numbers, and examined using statistical procedures to see if the theory's predictive overview is accurate in investigating social or human problems.

Sampling Size and Technique: For this study, a sample size of Hundred (100) construction industry workers was intended; however, only eighty-six (86) of the target population could participate. The study uses probability sampling, commonly referred to as a random sampling technique that ensured that each respondent in the target group had an equal probability of participating in distributing the questionnaire. This was accomplished by making sure the number of questionnaires given to each responder was equal to the size predicted.

Research Population

The target audience of interest for this research includes construction professionals, such as project managers, consultants, contractors, engineers, owners/developers, and all other personnel who are directly engaged in construction project sites in Gujarat, India. Information was acquired from experts working on construction projects.

Questionnaire Design/Survey

Close-ended questions and questions with predetermined response categories were both included in the questionnaire that was created and used in the study. Questionnaire asked for information about the construction company/respondents such as names, positions held, locations, dates in business, number of employees, safety policies in place, and compliance levels.

Method of Analysis of Data: Descriptive statistical methods were employed to analyze the data to attain these aims.

Data Analysis

Profile Analysis of the Companies/Interviewees: During the course of this study, a total of 100 questionnaires were sent out, however only 86% (86) were returned. 13 owners/developers, 12 managers/project coordinators, 8 builders/suppliers, 16 designers/architects, 3 safety officers/heads of departments, and 11 other numbers (24%) responded to the survey.

Table 1. Rate of Reply from Selected Population

Respondent Profile	Questionnaire Response rate/	Percentage (%) of response
Owner/ Developer	13	15
Project managers	12	14
Contractors	5	6
Engineer	42	49
Responsible persons for Safety	3	3
Other	11	13

Table 2. Respondent's experience details

Years of experience	Response	Percentage (%)
0 to 3 years	30	35
3 to 5 years	18	21
5 to 10 years	19	22
More than 10 years	19	22

Compliance with Safety Policies and Procedures: The results are listed in the table below: was asked what they thought about how well health and safety policy was being implemented in their workplace, of the respondents said the place of construction. The "YES," "NO" and "Maybe" boxes will determine their response.

Table 3 clearly states that 31% responded says that 51 to 75% and 29% responded says 26 to 50% safety is implemented on construction sites of Gujarat.

Table 4 reveals that 72% responded says there is health and safety reporting in their company and 22% say not available. As per 63% responded there is an occurrence of accidents during the construction work carried out and 22% says safety measures follows which leads to no accidents. 79% responded says their employee and employer take safety seriously and follows the rules and regulation, whereas 12% says don't take it seriously. 63% responded that says safety committee is recommended in their company and 24% says they don't have any safety committee whereas 13 don't know about the safety committee.

Table 5 reveals that 67% responded that there is safety policies available in their company whereas 29% says safety policies are not available. Sixty-six percent of those surveyed said they believed that serious measures had been taken against anybody who did not comply with or violate the current health and safety norms on the building site. However, 27% of respondents indicated no severe action was taken against contractors or subcontractors who did not follow safety regulations when required to do so.

Table 3. Safety implementation construction sites in Gujarat

Question	Response To					%Response To				
	0%	10 to 25 %	26 to 50%	51 to 75%	76 to 100%	0%	10 to 25 %	26 to 50%	51 to 75%	6 to 100%
Do you think how much percentage of safety is adequately applied on the construction sites of Gujarat?	4	17	25	27	13	5	20	29	31	15

Table 4. Safety implementation policy

Question	Response "Yes"	Response "No"	Response "Maybe"	(%) Response "Yes"	(%) Response "No"	(%) Response "Maybe"
Is there health and safety reporting in your company?	62	19	5	72	22	6
Did Accidents happen in your executed projects in the last five years?	54	19	13	63	22	15
Employees and employers take Health and Safety seriously?	68	10	8	79	12	9
Do you have a safety committee in your company?	54	21	11	63	24	13

Table 5. Prohibitive action against actions that compromise safety

Question	Response "Yes"	Response "No"	Response "Maybe"	(%) Response "Yes"	(%) Response "No"	(%) Response "Maybe"
Safety policy is available for your company	58	25	3	67	29	4
Is there a recording and action taken for accident and violations during your executed projects?	57	23	6	66	27	7

Conclusion

It was found that while most construction firms have some sort of health and safety policy framework in place, few use it. While the components of a healthy and safe building site have been established, compliance levels remain below the norm. According to the research, health and safety initiatives on a building site typically wind up increasing the project's final price tag. Eighty-two percent of all respondents believed that worker health and safety are major factors in determining the quality of the construction project's final product. Some other factors like worker-related factors explore the impact of individual characteristics, second group focuses primarily on the organizational factors from the administration point of view. Then, the job factors that describe the working conditions where the worker needs to perform the work may affect the safety performance on the construction site. All projects should priorities worker safety and health.

The following things need to be taken into account:

Workers' health and safety, and how it affects the project's outcome, should be the subject of both on-the-job and off-the-job training.

It is important to recognize the occurrence of accidents and injuries so that adequate accident

Contractors that disregard safety regulations should face severe penalties, and in cases of repeated offenses, their certification could be revoked.

References

Fredrick, Boamah (2019). Measures and Strategies for Managing Safety on Construction Site. International Journal of Advance Research. 7(8), 96–102.

Gohil, P., Malek, M., Bachwani, D., Patel, D., Upadhyay, D., and Hathiwala, A. (2022). Application of 5D Building Information Modeling for Construction Management. ECS Transaction, 107(1), 2637–2649.

Hämäläinen, P, Takala Jukka and Saarela (2006). Global Estimates of Occupational Accidents. Safety Science. 44(2), 137–156.

Malek, MohammedShakil S., and Bhatt, Viral (2022). Examine the comparison of CSFs for public and private sector's stakeholders: a SEM approach towards PPP in Indian road sector, International Journal of Construction Management, DOI: 10.1080/15623599.2022.2049490

Malek, MohammedShakil S., and Gundaliya, P. J. (2020). Negative factors in implementing public–private partnership in Indian road projects, International Journal of Construction Management, DOI: 10.1080/15623599.2020.1857672

Jani Nisarg and Patel Jitendra. (2017). Current Practice of Safety in Construction Industry: Case Study of Ahmedabad and Gandhinagar City. IJSRD. 5(9), 1002–1004.

Patel Divya, Prof. Bhavsar, and Dr. Pitroda Jayeshkumar. (2017). A Critical Review on Safety Management in Construction Projects. IJCRCE. 3(4):148–154.

Priya Mohana, Dr. Kothai P.S., and Ms. Kohilambal. (2016). Analysis of Safety Management System in the Construction Industries. IJRTER. 2(3), 198–203.

Priye Gaurav and Gupta Jaykant (2020). A Study on Safety Management in Construction Projects in India. International Research Journal of Engineering and Technology. 7(7), 1208–1216.

Singh Karan. (2014). Safety in Indian Construction. International Journal of Engineering Research & Technology. 3(11), 1564–1566.

Utilization of Space Between Metro Pillars

Keshvi Shah,[a,] Rutvi Vaghani,[a] Manthan Pandya,[a] Karan Maru,[a] Dhruvil Ghelot,[a] Mr. Uzair Shaikh,[a] and Ankur Shah[b]*

[a]Adani Institute of Infrastructure, Ahmedabad, India
[b]Lecturer, R. C. Technical Institute, Ahmedabad, India
E-mail: *keshvishah.cie19@gmail.com

Abstract

Population is increasing immensely in recent times. But the space for utilization is limited. To overcome this obstacle, we must come up with new, non-conventional ways to accommodate the population and their needs. To overcome this problem, we can utilize the spaces between the metro pillars and accommodate various amenities and facilities not available in certain areas. This research makes these inaccessible amenities accessible in neighborhoods where the metro tracks run. These amenities will occupy the empty spaces between the pillars of the metro rail making good use of space. Various amenities like a gymnasium, libraries, small medical clinics, grocery stores, homeless shelters, etc. can be built in these spaces. Through this research, we can achieve better and more hospitable neighborhoods.

Keywords: Infrastructure Development, Urban Planning, Space Utilisation, Metro Rail Pillars

Introduction

To have 24*7 access to basic amenities is still a far-fetched dream for many localities In metropolitan cities. Land is a non-renewable resource and in today's world with an increasing population, we are bound to use every inch of it. Especially in high-density areas where people are struggling to get basic livelihood things nearby, so we must stop the wastage of land.

In third-world countries, metro rail is considered a necessity, so the government of every country is focusing on it more and more because they are the fastest, easiest, and cheapest form of transportation on land. But the construction of metro rail takes a lot of space so there is a portion of unused land between the pillars that can be used for benefit of people.

So here we have come up with the solution to use the unused space by creating small cabins that can be used for public use, which will not only help people from the neighborhood but will also create employment opportunities and will also encourage people to start their small businesses. These small cabins can be used by the government to create medical shops, Coffee shops, or other beverage shop owners can have their Drive-Thru's, We can also use those cabins to decorate our cities and show our cultural history with that we can attract more tourists and on the long term that can help small businesses. We can also use that space as a Wi-Fi junction or a charging point. Grocery Stores can also be used as space a parking spaces in monsoon season can be used as a control room for government officials and can also be used as shelter for poor people and stray animals. And most important it can be used as a CCTV camera station and as a women's help safety desk.

All this can be achieved will small construction costs and with help of some government permissions. We can use this place as an important addition to our neighborhood and that will not only help local people but will help the GDP of the entire city.

Literature Review

Becky P. Y. Loo, Cynthia Chen, Eric T.H Chan in their research paper "Rail- Based Transit-Oriented Development: Lessons From New York City And Hong Kong" shows that The idea of using transit-oriented development (TOD) in reducing automobile dependency and improving the sustainability of transportation activities has gained wider support. Residents living in a TOD neighborhood used transit more frequently than people having similar socio-economic characteristics. The result suggests that higher car ownership may be associated with more pick-ups and park-and-ride activities. Car ownership may be associated with more pick-ups, drop-offs, and park-and-ride activities to the transit stations for the longer transit trip legs. In the future, more fruitful cross-country comparisons can be made to better understand factors affecting metro ridership.

Batara Surya in their research paper "Social Change, Spatial Articulation In The Dynamics Of Boomtown Construction And Development (Case Study Of Metro Tanjong Bunga Boomtown, Makassar)" shows that Globalization contributes to changes in local architecture and local pattern in urban areas. Urban and out-of-town transport have a role to play in the uncertainties of developing countries, and they are growing slowly. With the advent of capitalism, the partnership between the government and capitalists in the development of the Metro Tanjung Bunga region creates a rapid transformation of the visual space.

D. P. Sari, and M. R. Alhamdani in their research paper "The Urban Design Guidelines Of Sungai Pinyuh Steet Corridor" shows that The idea of an urban design that aims to create a spatial and open space to create a balanced ecosystem emerges, in addition, provides guidance on how to design a regional land use in a three-dimensional way. The result is 2 principles, which are an integrated corridor and a green urban corridor.

Wahyuni Zahrah, Achmad Delianur Nasution, and Novi Rahmadhani in their research paper "Perception And Utilization Of Urban Corridor As Public Space In Medan, Indonesia" examines the concept and application of street corridors as public spaces. The study's goal was to assess the planning and design process, visual quality, user perception, and corridor usage. People believed that using public space on the street and in parking lots was illegal, so they used private urban areas individually. The city corridor has been viewed as a "vehicle for cars and tables," rather than a public space where people converse. There was no choice but to meet their needs, particularly in terms of economic interests and the "car-dependent" trap.

Xue Jianf, Xiaoya Song, Hongyu Zhao, and Haoran Zhang in their research paper "Rural Tourism Network Evaluation Based On Research Control Ability Analysis: A Case Study Of Ning'an, China" shows that Many rural areas are growing rapidly in cities, which have led to economic growth but also to local building inequality. This study provides a way to accelerate the integration of regional tourism resources by providing a calculation of resources for the reorganization of resources and a promotional approach to tourism resources. Using the findings of the study as a guide, we can effectively explore tourist network districts and provide recommendations for the development of remote network areas.

Batara Suryam Agus Salim, Hernita Hernita, Seri Suriani, Firman Menne and Emil Salim Rasyidi in their research paper"Land Use Chanfe, Urban Agglomeration, And Urban Sprawl: A Sustainable Development Perspective Of Makassar City, Indonesia" shows Due to the increase in socio-economic activity, the expansion of Makassar City to suburban areas contributes to land use changes and features of the transport system. Due to urban sprawl, informal settlements with limited volume in relation to available space tend to grow. This study suggests that urban sprawl, land use change, and urban integration should be considered in the development of development policies.

Paulus Bawole in their research paper "Appreciating the Growth Of Informal Utilization Of City Space For Sustainable Urban Development In Yogyakarta City" shows More than 30% of all urban dwellers live in slums or isolated areas. This paper will address patterns of informal urban use by low-income communities. The method used is to conduct spatial surveys and to look for items that are specifically determined by other cities. Based on the existing construction, a planning and design approach may be developed that can be later proposed as a sustainable development strategy for low-income urban areas.

Methodology

In today's interconnected world, online survey tools or web-based survey tools have become popular for information-gathering techniques. Online survey platforms are used by researchers to obtain data. The benefit of web technology has been useful in Creating, developing, and getting input from targeted audiences in an easier method. An easy-to-use web interface is provided by Google Forms for creating web-based survey assessment questions. The Google Form offers several ways to collect data from various responses. One can include multiple choice choices, check boxes, scales, grids, text, and other elements. The precise number of questions that must be gathered can be set up by the analyst. The template option offers pre-built templates for giving the questionnaire a distinctive appearance. Google form also variously provides response output.

To collect data across Ahmedabad city for this project we use google forms an online surveying tool. Our questionnaire contains what can be done in free space of metro pillars in terms of infrastructure development and urban development. The title of the form was "Your opinion for space utilization between gap of metro pillars in our Ahmedabad City - survey form". After the title there was description for the form in which we had described - "This is survey form for Utilization of space between the metro pillars in Ahmedabad city to overcome day to day difficulties of nearby people and

to improve livelihood. For more progress in infrastructure development in Ahmedabad City we can utilize the space between metro rail pillars. We can develop infrastructure between this series of pillars according to the needs of people and government for that particular region. By doing this we can also implant good architecture appearance. please share your review and fill out the form.'

Results: After description some basic required information was asked in form such as Name, Email address & Gender. Name and Email were asked because if in any case someone had selected another option and they raise any quires or someone provides any interesting suggestions apart from this so, we can easily contact them. Asking gender questions in this survey Provides insight into patterns across different gender categories. These questions help determine how gender influences individual choices and thereby influences the survey. After these basic questions, there was a question regarding the nearby metro station. This question all area's name was mentioned in which metro stop /station was constructed. Person can choose their nearby metro station. Through this, we got particular results according to the area and get to know its requirements. At the end of this form, we put questions related to facilities that can be possible to provide between the metro line's pillars gap. All recommendations of facilities and utilities are made considering dimensions, available space between pillars, all possible obstacles, vehicle occupancy, road width and dimensions, noise level, aesthetical view, and surrounding environmental conditions.

Utilities and facilities we had suggested between metro pillar gap. Drive thru café, My bike, Primary health center, Dairy Store, Provision Store, Book Store, Library, Water room, and rest room, Mini Police control room, Gym, Rough carpentry house for workers and below poverty people, Medical Store, Wi-Fi Centre, Game zone, VR center, Water, storage point, Aesthetic view- Ahmedabad as a heritage City theme, Night lights Parking spot, Government common documentation Centre, Parking spot. Through this google form, we had collected data from different area along with facilities and utilities which can be implemented for infrastructure development and urban planning by means of better growth of the city can be done.

The overall results show that the amenities needed the most in these areas are parking spots, government document centers, police control rooms, and mini gardens.

Areas Amraiwadi, APMC, Apparel Park, Ashram Road, Gurukul Road, Gyaspur Depot, Kalupur Railway Station, Paldi, Rabari Colony, Rajiv Nagar, Ranip, Shreyas, Stadium, Thaltej Gam, Thaltej, Usmanpura and Vadaj, Vastral Gam, Vijay Nagar show requirements of parking spaces.

Government documentation centers are required in areas like Amraiwadi, APMC, Apparel Park, Ashram Road, Commerce Six Roads, Gurukul Road, Gyaspur Depot, Jivraj, Kankaria, Nirant Cross Road, Old High Court, Rajiv Nagar, Ranip, Shreyas, Thaltej, Vadaj, and Vastral. Police control rooms are required in areas like Amraiwadi, Apparel Park, Ashram Road, Commerce Six Roads, Gandhigram, Gurukul Road, Old High Court, Rajiv Nagar, Sabarmati, and Vadaj. Some other highly demanded amenities are primary health care centers, medical stores, provision stores, book stores, dairy stores, and libraries.

Shortcomings:

- In a country like India with more than 1.3 billion people it will always be difficult to find something that benefits everyone as there is no "one size fits all" approach. With the methodology, we have approached and the solution's we have come up with it will definitely solve most people's problems and will also have an improvement in majority people's life but as our solution include using space between metro pillars for different purposes where commercial or recreational or either just as parking space, this will have an impact on traffic level to a certain extent as shops will have visitors and people may park there

vehicles haphazardly which will cause problems to general people on road and this will slow down the nearby commute.

- And one more problem we may face in future is that if we decide to commercialize the space the management of space like purchasing, leasing or renting will require the government authority to phase a tendering process which can be hassle and it is a long tedious process too.

- Other short coming of our survey can be derived from that this is a very general survey so the choices of people are heavily dependent on their mood and they can easily be misguided by other's and can also be influenced by others' choices.

- In our research paper there are some choices where we are using the space's between metro pillar as a commercial space and there might be some cases like the one in small beverage shops and drive-thru's which may cause an increase in pollution and that might cause problems.

Gap in Data: In a country like India it is always difficult to do such surveys as people are reluctant to participate in surveys as a matter of privacy and a matter of choice. Here we conducted the survey taking Ahmedabad as base. We have conducted the survey and have received not enough number of responses even though the survey was done online but because of people's reluctance and their ignorance we didn't manage ideal amount of required responses. But as this is a very general survey the amount of sample size presented can paint the general image of people's choice's and there requirements. This can justify our gap in data. But with help of government and required authorities we can easily match the required amount of sample sizes.

Future Scope: Researchers can analysis and get future demand from this data and develop a detailed implementation plan, prepare a cost estimation sheet for the project, compute return on investment, how much percent of GDP should be allotted to this kind project? calculate life cycle cost analysis and implementation and construction planning.

Conclusion

This paper discusses space utilization as a public space in Ahmedabad. In a developing country like India where the population is increasing exponentially and the Human density per Km is also increasing rapidly, To overcome this hurdle space utilization becomes an important factor. And one of the advance but simple to implement ways is the utilization of space between metro pillars. With help of google forms and local surveys people response to what was required in their area. This can help the neighborhood drastically because having access to basic amenities is still a far-fetched dream for many areas. The easy solution to this is the usage of space between metro pillars by creating small cabins which can be used by the government to create stalls for necessary products and can also be used by private owners to own a small beverage shops and Drive-Thru's.

This can not only help the cities make equipped but can also help the neighborhood financially as it is going to increase the GDP, it will also open door of employment. The small cabins can also be used to decorate the cities and can also be used for enhancing the security of area by installing CCTV camera stations and as a women's help safety desk. This study shows how with help of government permissions and effective creation of small cabins between metro pillars can help the city in financially and also improve the livelihood of the area.

References

Becky P.Y. Loo, Cynthia Chen, Eric T.H Chan. Rail-Based Transit-Oriented Development: Lessons From New York City And Hong Kong. ELSEVIER July 2010.

Surya, Batara. Social Change, Spatial Articulation In The Dynamics Of Boomtown Construction And Development (Case Study Of Metro Tanjong Bunga Boomtown, Makassar). Modern Applied Science, 2(4), 2014.

Sari, D. P., Alhamdani, M. R. The Urban Design Guidelines of Sungai Pinyuh Steet Corridor. IOP conference series.: Earth Environ. Sci. 409 012028.

Zahrah, Wahyuni, Nasution, Achmad Delianur, and Rahmadhani, Novi (2014). Perception And Utilization Of Urban Corridor As Public Space In Medan, Indonesia. ELSEVIER Volume 153, 16 October 2014.

Jianf, Xue, Song, Xiaoya, Zhao, Hongyu, and Zhang, Haoran (2021). Rural Tourism Network Evaluation Based On Research Control Ability Analysis: A Case Study Of Ning'an, China. MDPI April 2021.

Batara Suryam Agus Salim, Hernita Hernita, Seri Suriani, Firman Menne and Emil Salim Rasyidi. Land Use Chanfe, Urban Agglomeration, And Urban Sprawl: A Sustainable Development Perspective Of Makassar City, Indonesia. MDPI May 2021

Paulus Bawole. Appreciating The Growth Of Informal Utilization Of City Space For Sustainable Urban Development In Yogyakarta City. IOP Conf. Ser.: Earth Environ. Sci. 402 012013

Gohil, P., Malek, M., Bachwani, D., Patel, D., Upadhyay, D., and Hathiwala, A. (2022). Application of 5D Building Information Modeling for Construction Management. ECS Transaction, 107(1), 2637–2649.

Malek, M. S., Gohil, P., Pandya, S., Shivam, A., and Limbachiya, K. (2022). A novel smart aging approach for monitoring the lifestyle of elderlies and identifying anomalies. Lecture Notes in Electrical Engineering, 875, 165–182.

Malek, M. S., Dhiraj, B., Upadhyay, D., and Patel, D. (2022). A review of precision agriculture methodologies, challenges and applications. Lecture Notes in Electrical Engineering, 875, 329–346.

Human Resource Management: An Essential Resource for Organizing Construction Workforces

Hiral Borikar Dhal,[a,*] Rutu Patel,[b] and MohammedShakil Malek[c]

[a]Assistant Professor, Karnavati University, Gandhinagar, India
[b]Planning Engineer, Shree Balaji Developers, Ahmedabad, India
[c]F. D. (Mubin) Institute of Engineering and Technology, Gandhinagar, India
E-mail: *hiraldhal@karnavatiuniversity.edu.in

Abstract

The goal of this review article is to determine the significance of Human Resource Management in labor management in the construction sector. Human Resource Management and its practises can be used effectively in the setting to improve labor management. Scrutinizing HRM and labor management information from multiple sources was done, and pertinent information was filtered based on that. The study revealed various elements influencing labor management and human resource management. Labor management strategies should be effective enough in context to the type of projects to increase labor productivity. The topic's scope of research is based on information and findings from many sources. Labor management in a company comprises manpower planning. Human Resource Management strategies are used in the process of manpower planning. The research summarises studies on the topic of labor productivity. It also outlines the techniques of Human Resource Management among labour migrants in a systematic and structured manner.

Keywords: Human Resource Management, Labor Management, Labor Productivity, Man Power Planning

Introduction

Labor-management relations refer to the dialogue between labor unions and their employer members. Workers in a certain field or at a given firm or across many firms might band together to form a labor union to advocate for their collective needs and interests.

Human resource management (HRM) and labor management (LM) may each affect the building trades. The efficiency and effectiveness of building projects can be improved via careful management of both labor and human resources. The goal is to determine the relationship between HRM and personnel administration. Possible correlations can be highlighted by reviewing the relevant literature.

Human Resource Management (HRM)

HRM is a method for overseeing a company's staff. HRM in the construction industry focuses on finding and retaining workers with the right mix of experience and expertise to carry out tasks in a timely and effective manner. High levels of employee engagement, job happiness, and low turnover are all possible thanks to human resource management.

Figure 1 below describes the HRM components that together provide a high-level overview of the field of human resource management.

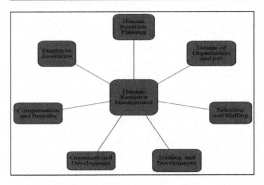

Fig. 1. Components of HRM

Construction Management

The building business relies on a high level of craftsmanship and materials to finish projects successfully and affordably. The goal of construction management is efficient coordination and arrangement of all necessary resources and personnel. Utilizing the most up-to-date methods in project management entails making the most of available resources and equipment to finish the project on time and within budget. Management in the construction industry must be strong, focused on making a profit, and carried out by trained professionals. The success of every construction project relies on the management team's ability to coordinate the efforts of multiple divisions, cultivate a skilled workforce, maximize productivity, and ensure that all materials and labor meet strict quality standards.

Labor Management

Work performed by humans manually or physically, as opposed to mechanically, is called "labor." Management of workers is a crucial function for those in charge of building projects. Appropriate utilization of manpower is a cornerstone of good construction project management. Foresight, development, and management are the three phases of manpower planning that ensure an organization's workforce is optimized for productivity. A sufficient number of workers will be on hand to accomplish the project on time if their management is efficient. That way, both time and money can be controlled.

Methodology and Objectives

The literature review and online questionnaires are the primary sources of data collection. The useful data is being sorted through that. Human resource management (HRM) and workforce administration (WFM) face a variety of challenges. These criteria were validated by in-person interviews with over twenty specialists.

The purpose of this study is to assemble data on HRM and labor management. The impact of worker productivity on the building sector is also included here.

Literature Review

Bhagwat, A. (1989): The author argues that inequalities in treatment cause countries to stagnate economically. According to the author, the construction sector has both universal regulations that apply to all workers and specialized regulations that only apply to women. A questionnaire was designed and used to conduct the survey. The author concludes that laws are being written, but not executed properly. Issues arise because employees cannot advocate for themselves. When comparing different parts of a business, such as paid leaves (Holidays), there is none for employees. The workers aren't getting the basics like clean water, urinals, lunch breaks, adequate shade, and adequate rest periods. Creches should be available so that working mothers may focus on their careers without feeling guilty about leaving their children at home.

Those three women are A. W. (2009) Several variables, the author argues, shape the modern job market. There was a survey conducted for the factors, and the data was analyzed accordingly. Changes in manufacturing and the transfer of personnel to other sectors are included. This may lead to a drop in the need for workers with lower levels of education and training. There is a newfound heightened focus on education.

It's Changyoon Kim, T. P. (2013) The author argues that advances in mobile computing present an opportunity to enhance traditional approaches to project management on building sites. In this article, the author focuses on three crucial concepts of project management: site monitoring, task management, and real-time information sharing. Preparation of the questionnaire and analysis was in progress. According to the data, important factors influencing website success are inadequate communication between site administrators, inadequate site documentation, and frequent design updates. Because of this time and expense can't be saved and hence job can't be completed as per the planning. Work that requires skill can be completed in less time without a decrease in quality, and this has been shown to have an impact on labor productivity. Because of the advantages that cell phones have brought to on-site management, these methods must be put to use in the workplace. The authors suggested allowing staff to utilize these tools too, to keep things flexible. The implementation of a work task management system and a site monitoring module resulted in savings in time, money, and operational flexibility.

D.J.C. Edison (2014) Human resource management (HRM), personnel administration, training, employment legislation, etc. are all discussed in detail. Human resource management, as outlined by the author, entails formulating, implementing, and overseeing plans to make use of human capital. Relationships among managers, supervisors, and employees are at the heart of personnel management. Specifically, its actions are associated with the evolution of Human resource management, staffing strategies, hiring practices, career advancement, compensation and benefits, office culture, line of communication, problem-solving, employee relations, and workplace conditions. A company's manpower often consists of a wide range of skill sets represented by distinct employee subsets. Predicting the need for human resources, in general, would be challenging, and it would offer little help to management in ensuring a steady stream of workers with the appropriate set of abilities for each position.

This article was written by Jair de Oliveira, E. E. (2015) The author argues that both the process approach and the role approach can improve managerial efficiency and effectiveness. Changes in the labor market and the level of competition in the business world can have an impact on the nature of the administrative job. Therefore, management must make adjustments and take necessary actions in response to shifts in external conditions. Managers would be responsible for everyday tasks such as checking in with employees, fixing broken equipment, processing data, training new hires, and so on.

Those authors are Naresh Kumar Prajapati, P. J. (2015) Human resources, as the author explains, are the construction industry's "living properties," subject to a wide variety of special considerations and dangers. Because they are real people and are not always permanent as in other businesses, managing them is difficult and dangerous. That's why it's important to take a closer look at HRM in the building trades. Workforce personnel may be highly skilled, intermediately skilled, or unskilled. Employees of any given enterprise are responsible for carrying out the bulk of the enterprise's daily operations. The author argues that HRM is impacted by numerous elements, including formalized methods of instruction, specified job responsibilities, compensation structures, etc. The author's identified factors were examined with the RII technique.

Labor factors, external factors, owner/consultant factors, execution plan factors, designer factors, working time factors, equipment factors, financial variables, quality components, leadership and coordination factors, organizational factors, safety and health factors, and health and safety factors are all identified as independent groups for the variation in labor productivity in the construction industry. As such, the questionnaire was developed, and the elements were assigned their proper priority. There was a lack of development materials, payment delays, change orders from the creators, improper gear, and a lack of sanitation on the construction site and in the temporary shed. There were also neighboring labor injuries, alcoholism, overtime hours, and a shortage of development supplies. Problems with the quality of the development resources; Confusion among the staff. Scheduling strategies, the use of a motivation system, a productivity study, a project procurement system, better contract conditions, trade schools, training in product development programs, and the transfer of novel techniques are all necessary for productivity improvement.

One Rohit Trivedi, M. P. (2016) According to the author, increasing the efficiency of construction workers is a top priority for many construction-related businesses and endeavors. The parameters were known and the questionnaire method was utilized. According to the author's suggestions, (1) workers should not have to wait around for payment because it is essential to their survival. Overtime is counterproductive since it causes workers to put in more time for less pay. Rather, personnel should be assigned and controlled according to cycles of labor. Thirdly, construction machinery should be used to boost output. (4) If you want to boost production, you need to implement effective material management strategies. Fifthly, employee absences can harm efficiency and scheduling. Sixth, a financial incentive scheme should be implemented to incentivize employees to produce better work. It is important to keep an eye on the inventory of tools and other equipment to make sure there is never a shortage. It is recommended to avoid designs that are incomplete or intricate drawings that are unclear.

I. Saleem; A. B. (2016) One of the most telling indicators of a society's economic health is its migration pattern, which also acts as a vehicle for cultural transformation and the spread of innovative ways of living. The characteristics of labor mobility are intrinsically linked to the pattern of economic development. Migrant workers in this industry are particularly vulnerable to exploitation and can endure deplorable working and living circumstances. Neither their well-being nor their whereabouts are priorities for them. Using the Chi-square test, Analysis of Variance, and Factor Analysis, the author concludes that it is essential to make the working environment fully appropriate for the needs of all workers by confirming leadership styles that inspire decentralization and allocation of duties, implementing worker evaluation and appraisal systems on an equitable basis, and providing adequate employee benefits, rewards, and credit structures. Human resources must devise policies that anticipate the needs of their workforce, eliminate bias and discrimination, implement a fair and transparent compensation structure, and offer a range of incentives and perquisites, including health insurance, to retain and motivate workers.

S. Acharya (2018) According to the author, in today's culture, the pay rate plays a crucial role in ensuring individuals can make a living and share wealth. Sustainable wages are the most effective strategy for increasing fairness and equality in an economy and society. The importance of minimum salaries in maintaining social stability and leveling income disparities is well acknowledged. The author employed a survey approach. Wage rates are an effective policy instrument in lowering revenue disparity and alleviating poverty in developing nations. In this age of the information economy, minimum wages are successfully set and implemented to increase employees' living conditions and, more importantly, to allow them to properly invest in the future of their families and children.

This is Sreeresh. R, S. K. (2018) One of the most serious and widespread issues, according to the author, is the decline in construction productivity. Despite the author's claims that technical progress, construction supplies, and tools are all readily available, projects are not being completed on time. The information was gathered through the use of questionnaires and in-person interviews. The author analyzed the factors based on the available data. Several variables can impact construction site productivity.

Stephan Kampelmann, Initials: F. R. (2018) The author argues that the impact of education on productivity is greater than its impact on salary levels. Ordinary Least Squares and the Generalized Method of Moments (OLS and GMM) were employed in the writing of this piece. He claims that increasing the proportion of college-educated workers on a project increased its profitability. The author argues that this has a greater impact on younger workers and women.

As described in the figures below different factors of labor productivity were mentioned by the author.

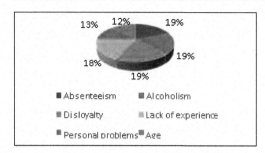

Fig. 2. Manpower factors affecting labor productivity

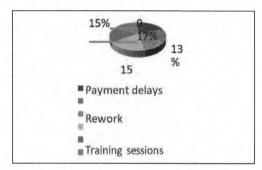

Fig. 3. External factors affecting labor productivity

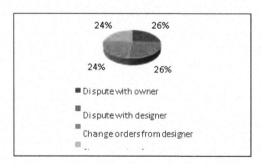

Fig. 4. Communication factors affecting labor productivity

Fig. 5. Resources that affect labor productivity

Conclusion

Knowing how to maximize productivity on a building site can help save money and time. To increase worker happiness, HRM methods and organizational structure need to transform. Furthermore, the separation of the sexes was monitored, with an emphasis on the difficulties encountered by women in senior positions. Background happens in the construction industry's human resource management; labor migrants in terms of recruiting, selection, training, and performance management labor welfare measures.

Labor productivity can be maximized by careful planning and administration of available workers. Human resource management (HRM) allows for the achievement of targeted work efficiency through the implementation of proper training and other HR practices. Consequently, this suggests a strong connection between HRM and the administration of workers.

Future Scope

Other inducing factors affecting construction productivity at all stages of the procurement process can be studied in greater depth. These factors include but are not limited to: proper training given to the workers; advanced site design; a structured flow of work; on-time payment to the workers; etc. For the author's part, future research on the small-scale sector may represent a comparison of HR performance across time, allowing for the identification of the factors that most significantly affect employee contentment. To make systematic use of the technologies available on-site, thorough construction scheduling and planning should be performed. More study is needed, focusing mostly on the more dynamic segments of the youth labor market, to determine whether or not there is a correlation between current labor market information and forecasts or scenarios for the future of the labor market.

References

Malek, MohammedShakil S., and Bhatt, Viral (2022). Examine the comparison of CSFs for public and private sector's stakeholders: a SEM approach towards PPP in Indian road sector, International Journal of Construction Management, DOI: 10.1080/15623599.2022.2049490

Malek, MohammedShakil S., and Gundaliya, P. J. (2020). Negative factors in implementing public–private partnership in Indian road projects, International Journal of Construction Management, DOI: 10.1080/15623599.2020.1857672

Malek, MohammedShakil S., and Gundaliya, Pradip (2021). Value for money factors in Indian public-private partnership road projects: An exploratory approach, Journal of Project Management, 6(1), 23–32. DOI: 10.5267/j.jpm.2020.10.002

Malek, MohammedShakil S., Saiyed, Farhana M., and Bachwani, Dhiraj (2021). Identification, evaluation and allotment of critical risk factors (CRFs) in real estate projects: India as a case study, Journal of Project Management, 6(2), 83–92. DOI: 10.5267/j.jpm.2021.1.002

Malek, MohammedShakil S., and Zala, Laxmansinh (2021). The attractiveness of public-private partnership for road projects in India, Journal of Project Management, 7(2), 23–32. DOI: 10.5267/j.jpm.2021.10.001

Gohil, P., Malek, M., Bachwani, D., Patel, D., Upadhyay, D., and Hathiwala, A. (2022). Application of 5D Building Information Modeling for Construction Management. ECS Transaction, 107(1), 2637–2649.

Malek, M. S., Gohil, P., Pandya, S., Shivam, A., and Limbachiya, K. (2022). A novel smart aging approach for monitoring the lifestyle of elderlies and identifying anomalies. Lecture Notes in Electrical Engineering, 875, 165–182.

Malek, M. S., Dhiraj, B., Upadhyay, D., and Patel, D. (2022). A review of precision agriculture methodologies, challenges and applications. Lecture Notes in Electrical Engineering, 875, 329–346.

Bachwani, D., and Malek, M, S. (2018). Parameters Indicating the significance of investment options in Real estate sector. International Journal of Scientific Research and Review, 7(12), 301–311.

Parikh, R., Bachwani, D., and Malek, M. (2020). An analysis of earn value management and sustainability in project management in construction industry. Studies in Indian Place Names, 40(9), 195–200.

Gap in Expected and Actual Services Offered by Restaurants in Virudhunagar District, TN, India

Praveen Paul Jeyapaul,[a] and S. T. Jaya Christa[,b]*

[a]Mepco School of Management Studies, MepcoSchlenk Engineering College, Sivakasi, India
[b]Department of Electrical and Electronics Engineering, MepcoSchlenk Engineering College, Sivakasi, India
E-mail: [*]stchrista@mepcoeng.ac.in

Abstract

This paper aims to study the gap in the customer perception of services offered in terms of the SERVQUAL factors in restaurants. In an attempt to achieve this, a survey was conducted using a structured questionnaire among 254 customers in Virudhunagar District of Tamilnadu State in India, who regularly visit restaurants. Paired sample t-test is employed to examine if there is a gap between the expected and actual services perceived by the customers. The results of the research indicate that there is a gap between customer expectation and the actual performance of the restaurants in the tangibility, reliability, assurance, and empathy facets of SERVQUAL except for the responsiveness factor. Hence, it is essential that decision-makers need to take up the issues pertaining to the service quality in the restaurants to better serve the customers since most of the SERVQUAL factors are lacking in the restaurants.

Keywords: SERVQUAL, Restaurants, Paired sample t-test, reliability, assurance, empathy

Introduction

Restaurant business is one of the biggest businesses in India. The food services market in India is worth Rs 3,09,110 crore, according to the National Restaurant Association of India's (NRAI) India Food Services Report(IMARC 2022). This market is estimated to grow to more than Rs. 5,00,000 crore by 2023, at a compounded annual rate of 10%. As far as the fine and casual dining customer segments are concerned, the segment is expected to record a market growth of more than 20% and 17%, respectively. This growth rate in the restaurant business is fuelled by young customers and upwardly mobile middle-class individuals who crave good quality food, novel ambiance, courteous service and new eating-out adventures.

There are numerous restaurants in Virudhunagar, both big and small. Many of these hotels operate in their own premises and are in operation for more than two or three decades. They offer excellent ambience, seating capacity and offer good quality as well as varieties of tasty food. There are also numerous small hotels which operate in smaller spaces, typically 1000 to 1500 square feet. Further, there are restaurants that operate in spaces less than 500 square feet too. All these restaurants cater to a wide variety of customer groups. Whatever the size of the restaurant, customers will expect a decent ambience and good quality food apart from courteous service from restaurant employees.

Some of the important factors that customers expect from restaurants are quality and variety of food, courteous service from staff, appropriate pricing, and a good atmosphere (Bichler et al., 2021). Along with the above factors, others such as a comfortable and welcome feeling when visiting a restaurant, neat appearance of staff, friendliness, and courtesy of staff, attention given by staff, cleanliness of the restaurant, attractive decor, variety in the menu, high quality of food offered, high degree of hygiene of food and reasonable prices are also considered to indicate the quality of restaurants (Chen et al., 2008) (Tsang and Qu, 2000).

Understanding the gap between what is expected by a customer from a restaurant service and what is perceived as service received is vital in providing satisfaction to customers. According to Parasuraman et al. (1985), the quality of services offered is a comparison between customer expectations and the actual service delivered by the organization (Parasuraman et al., 1985). Gronroos in his research implied that there could be a gap between the expectation of customers on services offered and their perception of the actual service received (Gronroos, 1984). Followed by this, many authors have proposed models to understand this service gap including Parasuraman et al., (1985). The importance of understanding the service gap is so alive and happening and cannot go out of fashion as is evident from recent research (Frost and Kumar, 2000; Tsang and Qu, 2000; Nakhleh, 2019; Esmalian et al., 2021; Jamkhaneh et al., 2022). This is because service organizations cannot afford to lose focus on the quality of services they offer to customers and also need to always be in touch with the sentiments or perceptions of the customers' expectations and continuously strive to satisfy their expectations.

Normally there are five gaps proposed in the service gap model, which are (a) the gap between customer's expectation of the services offered and the organizations' perception about what they are offering to the customers (b) the gap between management perception of what service quality customers expect and defining those expectations of service quality in terms of clear specifications to be delivered (c) the gap between service quality specification given by the organization on the modus operandi of delivery of services and the actual delivery of service(d) the gap between actual service delivered by the organization and expectation it had created among consumers through external communication in mass media and other communications (e) the gap between the customer expectation of the services offered and actual experience the customers perceive about the services being delivered (Parasuraman et al., 1985).

This research focuses on the fifth gap propounded in the services gaps model with respect to restaurant businesses in the suburban areas of Virudhunagar district in Tamilnadu state of India. The importance of the research is that there could always be a gap in customer expectation and the actual customer experience in services when they come to a restaurant since customers develop their expectations based on external communication through mass media and word of mouth. Thus understanding the services created by the restaurant and the actual perception by the customers is essential in reducing the dissonance that the customers may experience.

The gap between the expected service quality and the perceived service quality (SERVQUAL) can arise in the dimensions of services namely the tangibility factors in the service offered, reliability of services, customer responsiveness, assurance, and empathy is shown to customers. Thus in this research, the gaps in the above-mentioned SERVQUAL dimensions are studied, so that the services in the restaurant business can be improved.

Methodology

Primary data was collected from 254 respondents utilizing a structured questionnaire incorporating the factors of SERVQUAL and measuring the expected as well as actual services offered by restaurants for the same item in order to address the objectives considered for the research. The items in the questionnaire were measured using 5 point scale by getting agreement or disagreement on each item. The sampling unit is the customers of restaurants in Virudhunagar district of Tamilnadu, state in India. The questionnaire was administered through online mode. Paired sample t-test is employed to examine if there is a gap between the expected and actual services perceived by the customers and also to test if there is a significant difference in mean values between an expected level of services by the customers from the restaurants and the actual services the customers receive from restaurants. If there is a gap in the mean values of the paired sample statistics, then the statistical significance in this gap is clarified from the paired sample t-test results.

Analysis and Discussion

In an ideal service delivery situation, it will be only appropriate that the expectation of customers on the service from restaurants and the actual performance of the restaurants in the customers' viewpoint are congruent and matched. However, in reality, there could be gaps in the above-mentioned expectation and the actual performance of restaurants. To

analyze this gap, initially, the mean ($\bar{x}$) values of the opinion of customers and the standard deviation (σ) on both the expected and actual performance of restaurants are examined. The gap in the SERVQUAL variables considered for the study (Tangibility, Reliability, Customer Responsiveness, Assurance, Empathy) is tabulated in Table 1. If the mean values of expectation and the actual reality are quite close then it can be concluded that there is congruence in the customers' expectations and reality. If the differences in the means are wide, then the opposite is true.

Generally, a gap between the expected service and the actual service is acceptable. However, if the gap is too wide, then the organization will need to look into the gap quite seriously. To analyze the significance of the gap, a paired-samples t-test is carried out. The null hypothesis (H_o) in each of the cases is that there is no significant difference between the expected level of services by the customers from the restaurants and the actual service the customers receive from restaurants.

Table 1. Paired Sample Statistics

Factors	Metrics	$\bar{x}$	N	σ	Standard error mean
Pair 1	E*-Tangibility	4.093	254	0.559	0.047
	A#-Tangibility	3.877	254	0.615	0.052
Pair 2	E-Reliability	4.155	254	0.667	0.056
	A-Reliability	3.840	254	0.630	0.052
Pair 3	E-Responsiveness	3.803	254	0.612	0.051
	A-Responsiveness	3.713	254	0.681	0.057
Pair 4	E-Assurance	4.114	254	0.572	0.048
	A-Assurance	3.903	254	0.658	0.055
Pair 5	E-Empathy	3.998	254	0.671	0.056
	A-Empathy	3.773	254	0.713	0.059

* E is expected services by the customers
# A is customer opinion as actual service enjoyed

Table 2. Paired sample t-test

Metrics / Factors	Paired differences					t	Sig.
	x̄(Difference)	σ	Std error mean	95% confidence interval of the difference			
				Lower	Upper		
Pair 1 E-Tangibility A-Tangibility	0.215	0.845	0.704	0.075	0.354	3.054	0.003
Pair 2 E-Reliability A-Reliability	0.314	0.806	0.067	0.181	0.447	4.684	0.000
Pair 3 E-Responsiveness A-Responsiveness	0.090	0.824	0.068	0.045	0.226	1.314	0.191
Pair 4 E-Assurance A-Assurance	0.211	0.801	0.066	0.079	0.343	3.161	0.002
Pair 5 E-Empathy A-Empathy	0.225	0.785	0.065	0.096	0.355	3.447	0.001

The results of the paired samples statistics show that in all the SERVQUAL variables, there is a gap or difference between the expected and actual service from restaurants as perceived by customers. In all cases, the expectations of the customers are higher than the actual service they receive. Since the opinion of the customers is expressed on a five-point scale, values closer to five indicated higher expectations and higher satisfaction. While observing the opinion of customers on the factor tangibility (Table I), there is a difference of 0.215 (Table II) in the mean values. This shows that the expectation of the customers" is more than what they actually experience. Similarly, in all the factors, the expected value is more than the actual service received by the customers of Virudhunagar. This indicates that the restaurants need to upgrade their services to match the customers" expectations.

However, observing the factor "responsiveness," it can be seen that both the customer expectation and the actual responsiveness exhibited by the restaurants are quite close albeit the expectation is slightly higher. The difference in the value is only 0.09, which can be assumed that there is a match.

In order to check if the difference or congruence between the expected and actual experience is statistically significant, the values shown in paired samples test (Table 2) are investigated. As stated previously, the null hypothesis is that there is no significant difference between an expected level of services by the customers from the restaurants and the actual service the customers receive from restaurants. In all the cases except Pair 3 which is responsiveness, the p-value (Sig.) is less than 0.05.

Considering the p-value, the null hypothesis is rejected and it can be concluded that in these cases (except responsiveness), there is a significant difference between the expectation of customers on services offered by the restaurants and the actual service they receive from these restaurants. In case of "responsiveness," since the p-value is greater than 0.05, the null hypothesis is that there is no significant difference between the customers' expectation on "customer responsiveness" and the actual responsiveness the customers receive from the restaurants.

Conclusion

This research set out to study the gap in the customer perception of service offered in terms of the SERVQUAL factors in restaurants in Virudhunagar district of Tamilnadu state in India. Survey was conducted with a structured questionnaire among 254 customers who regularly visit restaurants. The results of the research indicate that there is a gap in customer expectation and the actual performance of the restaurants in the tangibility, reliability, assurance, and empathy facets of SERVQUAL. However, the research shows that there is a match between what customers expect and what actually experience in the responsiveness factor. The implications are that, as far as the tangibility factors are concerned, customers expect better facilities in the restaurants such as comfortable and clean tables and chairs, better space between tables, clean walls, privacy for family dining, better and cleaner toilet facilities, clean dress on employees and spacious vehicle parking space. The factor reliability has the biggest gap among all the factors.

This implies that restaurants in Virudhunagar should deliver the expected services correctly and consistently. Here, customers expect consistent taste in the same dish offered at different points in time, the restaurant staff to deliver required services to customers, and timely delivery of food. The customers also feel that employees should be polite, gentle, and kind with the right attitude while they serve the food. As far as the empathy factor is concerned, customers expect restaurant employees to remember their food preferences and serve accordingly at least in restaurants they visit often. In the responsiveness factor, most customers agree that the restaurants serve food quite quickly and the employees respond immediately to any queries customers may have. Thus it can be concluded that decision-makers need to take up the issues pertaining to the service quality in the restaurants to better serve the customers since most of the factors are lacking in these restaurants.

References

Bichler, B. F., Pikkemaat, B., and Peters, M. (2021). Exploring the role of service quality, atmosphere and food for revisits in restaurants by using an e-mystery guest approach. J. Hosp. Tour. Ins., 4 (3), 351–369.

Chen, C. P., Deng, W. J., Chung, Y. C., and Tsai, C. H. (2008). A Study of General Reducing Criteria of Customer-Oriented Perceived Gapfor Hotel Service Quality. The Asian J. Qual., 9(1), 113–133.

Esmalian, A., Dong, S., Coleman, N., and Mostafavi, A. (2021). Determinants of Risk Disparity Due to Infrastructure Service Losses in Disasters: A Household Service Gap Model. Risk Anal., 41 (12), 2336–2355.

Frost, F. A., and Kumar, M. (2000). INTSERVQUAL – an internal adaptation of the GAP model in a large service organization. J. Serv. Mark., 14 (5), 358–377.

Gronroos, C. (1984). A Service Quality Model and its Marketing Implications.Eur J Mark, 18 (4), 36–44.

IMARC (2022). India Food Service Market: Industry Trends, Share, Size, Growth, Opportunity and Forecast 2022–2027. IMARC Serv. Pvt. Ltd.

Jamkhaneh, H. B., Shahin, R., and Shahin, A. (2022). Assessing sustainable tourism development through service supply chain process maturity and service quality model. Int. J. Product. Perform, 71(4), 696–721.

Nakhleh, H. (2019). Closing the Expected and Perceived Service Gap in the Jordanian Hotel Industry: A Gap-driven Analysis. In Azid, T., Alnodel, AA., and Qureshi, MA., Research in Corporate and Shari'ah Governance in the Muslim World: Theory and Practice (pp. 331–340). Emerald Publishing Limited, Bingley.

Parasuraman, A., Zeithaml, V. A., and Berry, L. L. (1985).A Conceptual Model of Service Quality and Its Implications for Future Research.J. Mark., 49(4), 41–50.

Tsang, N., and Qu, H. (2000). Service quality in China's hotel industry: a perspective from tourists and hotel managers. Int. J. Contemp. Hosp. Manag., 12(5), 316–326.

Sustainable Growth of Banks Through SERVQUAL Model

Praveen Paul Jeyapaul[a] and S. T. Jaya Christa[,b]*

[a]Mepco School of Management Studies, MepcoSchlenk Engineering College, Sivakasi, India
[b]Department of Electrical and Electronics Engineering, MepcoSchlenk Engineering College, Sivakasi, India
E-mail: [*]stchrista@mepcoeng.ac.in

Abstract

Ensuring service quality in banks is important to ensure the sustainable development of banks. In order to measure the service quality and its impact on customer satisfaction, the SERVQUAL model is used. Data is collected from 144 regular customers of commercial banks and the profile of the respondents based on their demographics is observed. Chi-square test is conducted to find if there is a significant difference between the different groups of customers in their satisfaction with banking services. It is found that there is no difference between the different genders, ages, occupations, and average monthly income of respondents. However, there is a difference between the customers belonging to different education levels. Further, regressions were carried out and it is found that all the Servqualvariables have a significant impact on satisfaction. It is suggested strategists need to concentrate on the Servqual factors to ensure the sustainability of banks.

Keywords: SERVQUAL, Banks, Tangibility, Reliability, Assurance, Empathy

Introduction

To ensure the sustainable development of banks it is important on the part of the strategists to make sure that the banks offer the best services to their customers. The best tool for introspection into the quality of banks' services provided to customers could be the SERVQUAL model. This model is quite popular and could be considered a tried and tested model in measuring the service quality of banks. It was first proposed by Parasuraman et al. (1985) and further bolstered by various authors such as Bitner et al. (1990), Rust and Zahorik (1993), Asubonteng et al. (1996), and Faria et al. (2022) among others. The factors of SERQUAL are tangibility factors in the service offered, reliability of services, customer responsiveness, assurance, and empathy is shown to customers. The study of SERVQUAL in banks is reinforced in the recent literature (Raza et al., 2020; Zhou et al., 2021; and Mir et al., 2022). However, there are also alternative approaches to measuring the service quality in banks, as proposed by Jeyapaul & Sharmila, (2021). Here the services that are specific to banks such as interest rate for savings and interest rate for loans offered by banks, range of banking products offered, availability of security of locker, security of deposited funds, various types of loans offered to suit customer needs and courteous employees were considered as factors that influence service quality offered by banks and in turn affecting customer satisfaction.

The implications and significance of service quality pertaining to banking services are well documented in extant literature. Concentrating on service quality in banks is linked to profitability of banks, better customer retention, enhanced customer loyalty, and increased customer satisfaction (Adil, 2013).

The objective of the present research is to profile the respondents based on their demographics and find out the factors they consider important in banking services so that banking strategists can tailor make their marketing strategies.

Methodology

Primary data was collected from 144 respondents using a well-organized questionnaire incorporating the factors of SERVQUAL to address the objectives taken up for the study. The items in the questionnaire were measured using 5-point scale by getting agreement or disagreement on each item. The sampling unit is the customers of banks distributed in Tamilnadu state of India. The questionnaire was administered through online mode. Chi-square test is used to observe if there is any difference between groups of respondents differentiated on demographic variables. To study the effect of the service quality variables on the opinion of satisfaction of the respondents, simple linear regression is employed. The results of regression are presented as $Y = a + b_n x_n$, where Y is the criterion variable, "b" is the regression coefficients for the matching predictor variables and "a" is the constant.

Data Description

To get a better understanding of the profile of respondents taken up for the research, a preliminary observation of the different groups of respondents is made. Here, the major demographic factors that may affect the contentment of bank customers with the services provided by the bank are considered. Thus the demographics such as the gender of respondents, their age, education level, occupation, average monthly income, mode of banking relationship, average number of visits to bank in a month, and the number of banks with which they hold an account are considered. This information is presented in the form of graphs for better understanding. It is seen from the graphs

that the sample has almost equal numbers in both genders albeit females being slightly more.

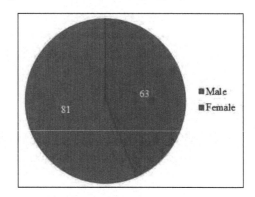

Fig. 1. Gender distribution of respondents

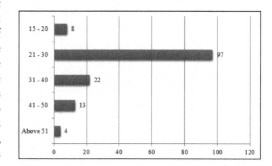

Fig. 2. Age distribution of respondents

Maximum number of respondents is in the age category of twenty to thirty years. These are the respondents who use the technical aspects related to banking services such as net banking, and mobile banking applications among others.

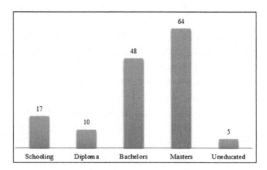

Fig. 3. Educationlevel of respondents

The distribution of the respondents based on their education level shows that most of them have graduate degrees showing that they will be in a better position to judge the service quality aspects of banks.

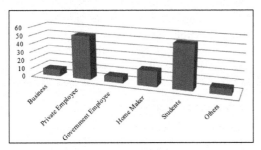

Fig. 4. Occupation of respondents

While observing the occupation of the respondents it is seen that most of them are private employees and students. These individuals are more active in using the online services of the bank.

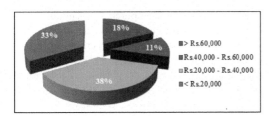

Fig. 5. Average monthly income of respondents

Most of the respondents considered for the study have an average monthly income of INR 20,000 to 40,000.

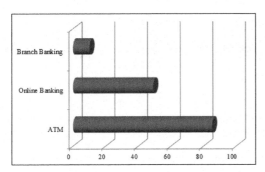

Fig. 6. Preferred mode of banking relationship by respondents

Banks these days most often do not encourage their customers to withdraw or deposit money physically in the bank branches. Instead, they urge their customers to use ATMs and money deposit kiosks. This is reflected in the opinion of the customers too. The customers also prefer to predominantly use ATMs.

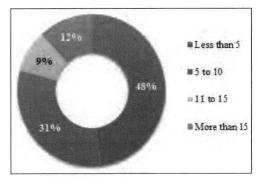

Fig. 7. Average number of visits to bank branches in a month by respondents

When asked about the number of times the bank customers visit the bank branches physically, nearly 80% of the respondents said that they visit the bank branches up to 10 times a month. However, almost half of the respondents have said that they visit the banks less than 5 times a month. In this category, many respondents visit bank branches once or twice a year. This also shows how much the banking needs of the customers can be satisfied without human intervention.

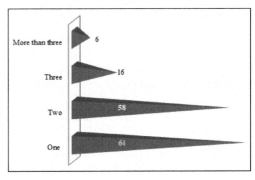

Fig. 8. Number of banks used by respondents

While some respondents seldom visit banks physically, there are a few respondents who have bank accounts with multiple banks. They are usually individuals who operate businesses. Many have accounts with two banks. This happens when they have their salary account in one bank and get loans from another bank.

To understand if there could be any difference in satisfaction between the different groups considered based on the demographics; a chi-square test is carried out. The null hypothesis that there is no significant difference in satisfaction between different groups of respondents is considered. The results of the chi-square test are discussed below.

Initially, the difference in contentment with the services provided by the banks between the male and female respondents is computed. It is found that there is no difference between the genders in their opinion of overall satisfaction. This implies that both genders have equal satisfaction with the services of banks. While looking into the opinion expressed by the respondents, it is observed that most of the respondents irrespective of gender express a medium (not so high not so low) level of satisfaction. Similarly, when the age of respondents is considered, here too there is no noteworthy difference between the different groups of respondents differentiated by their age. Further, when the education level of the respondents is observed, it is seen that respondents with higher education show better satisfaction than the respondents with lower education. This may be because better-educated individuals may have a better understanding of

the procedures and processes in the banks than lesser-educated individuals. The difference in the perception of satisfaction based on the occupation of the respondents is analyzed and it is found that almost all of the respondents" groups have the same satisfaction. Here too, the satisfaction of the respondents is medium. Further, when the average monthly income is considered, there is no group difference between the respondents in their satisfaction with the services provided by the banks.

Analysis and Discussion

To analyze the influence of the Servqualvariables considered for the study (Tangibility, Reliability, Customer Responsiveness, Assurance, Empathy) on the satisfaction of services provided by banks, as perceived by bank customers, a simple linear regression model is employed. Regression analysis is useful in understanding the effect of the independent variables(predictor variables) on the dependent variable (criterion variable).

Regressions are computed individually for each predictor variable and the criterion variable which is the customers' satisfaction with the services of banks. The results of these individual regressions carried out are compiled and presented in Table 1. The null hypothesis (H_o) in each of the cases is that there is no significant effect of the independent variables on the contentment of customers towards the services provided by banks (Y).

Table 1. Influence of SERVQUAL factors on satisfaction from services offered by banks

Model	R	R²	Constant	Coefficient	Adjusted R²	F Change	p
Tangibility	0.772	0.074	2.340	0.679	0.596	11.364	.001
Reliability	0.879	0.032	1.378	0.735	0.773	44.694	.000
Responsiveness	0.883	0.034	1.241	0.813	0.779	54.948	.000
Assurance	0.732	0.054	1.819	0.641	0.536	18.094	.001
Empathy	0.870	0.029	1.088	0.766	0.757	34.229	.000
Dependent – Overall contentment of customers on services provided by banks (Y)							

The results of the regression show that all the Servqual variables (antecedents)have a noteworthy and non-contingent effect on the perception of satisfaction with the services provided by the customers of banks. This is evident from the fact that the alpha (p) value for all the factors is less than 0.01(at p < 0.01) hence H_0 is rejected in all the cases and it is concluded that the Servqualfactors have a significant impact in influencing the satisfaction of bank customers. The R values and R^2 are high indicating a bigger impact of the variables on customer satisfaction. The regression equations are presented below.

$$Y = 2.340 + 0.679x_1 \tag{1}$$

$$Y = 1.378 + 0.735x_2 \tag{2}$$

$$Y = 1.241 + 0.813x_3 \tag{3}$$

$$Y = 1.819 + 0.641x_4 \tag{4}$$

$$Y = 1.088 + 0.766x_5 \tag{5}$$

Where:

x_1 is Tangibility, x_2 is Reliability, x_3 is Customer Responsiveness, x_4 is Assurance, x_5 is Empathy

The regression coefficients in all the regression equations indicate that when customers' opinion on the Servqual factors increases, their satisfaction also increases.

Conclusion

This research paper aims to study the effect of Servqualfactors in banks on the satisfaction of their customers so that sustained growth of banks is ensured. A survey was conducted among 144 regular customers of banks. It was found that overall; the maximum number of the respondents are content with the services offered by banks. Further, inspite of them regularly using the services of banks, they seldom visit the bank branches physically. Most of their banking needs are fulfilled by ATMs and other kiosks. From the results of the regressions, it can be concluded that all the Servqual factors have a consequential impact on the perception of satisfaction with the services provided by the banks. This indicates that the physical facilities (tangibles) such as the bank layout, the ATM facilities, air-conditioning in the ATM kiosk, etc. gain importance to satisfy customers. Also, customers expect reliable service time and again from banks. They also want the banks to respond to their requests and serve them promptly and expect the bank employees to be knowledgeable and helpful thereby conveying a sense of trust and confidence. Further, they also expect empathy from bank personnel in understanding the difficulties that they face while using the services of banks. Thus it can be concluded that strategists need to address the issues pertaining to the service quality in the banks to better serve the customers.

References

Adil, M. (2013). Modelling effect of perceived service quality dimensions on customer satisfaction in Indian bank settings. Serv. Oper. Manag. 15(3), 358–373.

Asubonteng, P., McCleary, K. J., and Swan, J. E. (1996). SERVQUAL revisited: A critical review of service quality. J. Serv. Mark. 10(6), 62–81.

Bitner, M. J., Booms, B. H., and Tetreault, M. S. (1990). The service encounter: Diagnosing favorable and unfavorable incidents. J. Mark. 54(1), 71–84.

Faria, S., Carvalho, J. MS., and Vale, V. T. (2022). Service quality and store design in retail competitiveness. Int. J. Retail. Distrib. Manag. 50 (13), 184–199.

Jeyapaul, P. P., and Sharmila, S. (2021). A study on service quality in private banks – An alternate approach. ICTACT J. Manag. Stu. 7 (1), 1349–1358.

Mir, R. A., Rameez, R., and Tahir, N. (2022).Measuring internet banking service quality: an empirical evidence. TQM J. 34 (7), 22–34.

Parasuraman, A., Zeithaml, V. A., and Berry, L. L. (1985). A conceptual model of service quality and its implications for future research. J. Mark. 49(4), 41–50.

Raza, S. A., Umer, A., Qureshi, M. A., and Dahri, A. S. (2020). Internet banking service quality, e-customer satisfaction and loyalty: the modified e-SERVQUAL model. TQM J. 32 (6), 1443–1466.

Rust, R. T., and Zahorik, A. J. (1993). Customer satisfaction, customer retention, and market share. J. Retail. 69(2), 193–215.

Zhou, Q., Lim, F. J., Yu, H., Xu, G., Ren, X., Liu, D., Wang, X., Mai, X., and Xu, H. (2021). A study on factors affecting service quality and loyalty intention in mobile banking. J. Retail. Consum. Serv. 60, 102424.

Analysis of Indian Stock Market Using Neo4j and Machine Learning

Akash Sivakumar,[a] and G. Hannah Grace[b]

[a,b]Vellore Institute of Technology Chennai, Vandalur Kelambakkam Road, Chennai, Tamilnadu 600 127

Abstract

The stock market is a rapidly changing industry, due to which it is difficult for investors to choose which company to invest in. This paper is an attempt to analyze the trend of companies in terms of both their performance in the stock market as well as the nature of current activities going on in them. The latter is obtained as news captions from online news websites, and their sentiments are analyzed. The stock market data is converted into visualizable graphs from which patterns are observed. Both these factors will be very helpful for investors to choose their companies.

Keywords: Neo4j, Stock market, Model, Prediction

Introduction

The stock market industry has been booming in the past decade. Many companies are becoming public day by day to allow people across the world to own parts of their company. The success of an investor in stocks depends on the company's stock he has bought, and other factors that affect that company. The task of predicting the stock value of a company at the current instant is difficult. Nevertheless, it has been an area of research for a long time. Different ML and DL-based approaches have been introduced to study the pattern of variation of companies to predict their stock value. This paper explores a rather generalized way of predicting not the stock values themselves, but the trend in which these values change. Neo4j and Cypher Querying are used for the next step. Apart from considering stock data, stock market companies' current news captions are also considered, to give a comprehensive insight into the task. The sentiments of those captions help determine what is currently happening with that company and if it is safe to invest in that company based on current estimates. For that, a well-known DL model is used.

Methodology

Neo4j and Cypher Querying

Neo4j is a database management system based on graphs. It allows data to be stored and represented in the form of graphs. Here, graphs are a network of interconnected nodes that gives an identifier to each node as well as the connection between two nodes. The nodes can be grouped based on labels, and so can the relationships. Every node can have properties relating to what it represents. The dataset used here has relationship between rows of data and Neo4j is a good choice for such kinds of datasets. One powerful feature of this DBMS is that it allows graphs to be created from any kind of data presented to it, including tables, spreadsheets, etc.

Neo4j graphs have their own querying language to create, modify and delete graphs, nodes in a graph, and relationships between nodes. This language is the Cypher Querying language. We can create nodes from spreadsheets too. Every row from a CSV file is loaded in separate rows, and for every row, individual nodes of a particular type can be created and assigned.

Machine Learning Model Basics and VADER Model

ML models are used to learn a pattern amongst the data given to it, in order to either cluster them into groups or else correctly learn the label for each data for future prediction. Many ML and DL models are used for complex tasks like the sentiment classification of text. One such model is the Vader or Valence Aware Dictionary for Sentiment Reasoning model. This model helps provide polarity values about the sentiments of a sentence whether it is positive or negative or neutral. The model uses a lexicon of sentiment-related words to predict the overall sentiment of the sentence taking into consideration adverbs like not and other grammatical considerations. Once the polarity values are found out, the overall sentiment is the one with maximum polarity.

Literature Survey

The authors in Hegazy et al. (2014) have proposed an ML model for stock market prediction. The algorithm integrates particle-swarm optimization (PSO) and Least square SVM (LS-SVM). PSO selects free parameters in order to optimize the LS-SVM which can then accurately predict stock values. The input data is historical stock data and technical indicators. This model was applied and tested on 13 financial benchmark datasets and compared with an artificial neural network based on the Levenberg-Marquardt (LM) algorithm. The authors Hiransha et al. (2018) have discussed a buying and selling time prediction of stock value of companies on the Tokyo Stock Exchange (TOPIX) and an analysis of internal representation. Pre-processing steps include conversion to space patterns, log approximation (to remove irregularities), and normalization to [0,1] range. The algorithm used is modular neural networks. The system achieved accurate predictions and simulation on real-time stock data showing an excellent profit.

The authors in Kim and Lee (2004) have developed a feature transformation method using a genetic algorithm (GA) with two conventional ANN models. The GA is incorporated to improve learning and generalizability of the ANN models. Three feature transformation methods for ANNs are compared, one with genetic algorithm-based transformation model, and existing feature transformation methods–linear transformation model and fuzzification-based transformation model. A comparison shows that the performance of the proposed model is better, in that it reduces the dimensionality of feature space and decreases the importance of irrelevant factors with respect to stock market prediction. The authors in Zhang and Wu (2009) have proposed an improved bacterial chemotaxis optimization based on bacterial foraging behavior that works based on velocity and trajectory of a bacterium that follows a specific probability density function, the new direction, and position. This algorithm is integrated with the backpropagation step of ANNs. Data is obtained from S&P500 companies, and technical indicators like current price, smoothing factor, time period etc are computed from raw data. The advantage of this model is that it reduces training time and computational complexity while improving prediction accuracy.

The authors in Gupta and Dhingra (2012) have used Hidden Markov Models to include properties relating to time-dependence, volatility and other interesting aspects of stock data. In particular, the authors have used Maximum a Posteriori HMM approach for forecasting stock values for the next day based on historical data. The datasets used belong to TATA steel, Apple Inc., IBM, and Dell Inc. Every stock data is represented as a 4-D vector of fractional difference of closing, highest and lowest value with respect to opening value. The MAP approach is used in testing phase with a fixed number of latency days. An assumption that the model for one particular stock is independent of the other stocks in the market is made, however in reality it is not the case. The authors in Pang et al. (2020). aimed to develop an innovative neural network approach for better stock market prediction. They have demonstrated stock vectors—an improvement on word vectors, that has high-dimensional multi-stock historical data as input. They have used a deep LSTM layer network with embedded layers and an autoencoder. The autoencoders were used with a continuous restricted Boltzmann machine (CRBM). Their results showed that the approach that used the embedded layer was better but by a mere 0.3% only.

The authors in Kimoto et al. (1990) have developed a buying and selling time prediction system for stocks. The dataset used is TOPIX, the well-known Tokyo stock dataset. They have used modular neural networks for this work, which is a normal DNN network but with some intermediaries for monitoring purposes. For backpropagation, they have used a method called supplementary learning, which automatically schedules pattern presentation and changes learning constants based on the backpropagation value. Their model showed good results with the simulations on stock trading showing excellent profits. The authors in Fu et al. (2022) aimed to use different ML/DL approaches to the Tehran Stock Exchange dataset. Historical data of upto 10 years was used. The authors considered four main categories within the dataset to work namely financials, petroleum, non-metallic minerals, and basic metals. Prediction was done for 1,2,5,10,15,20 and 30 days in advance. For preprocessing, they cleaned the data and used interquartile range metric to detect and remove outlier. The models tried out include decision tree, bagging, random forest, Adaboost, gradient boosting, XGBoost, DNN, RNN and LSTM networks. For evaluation, four different measures were used namely mean absolute percentage error (MAPE), mean absolute error (MAE), relative root mean square error (RRMSE) and mean squared error (MSE). Amongst all the models used, LSTM showed better results.

Data Description

The primary dataset used in this work is the stock market historical data. This was taken from bseindia website's historical entries known as Bhav_Copy[11]. Here, they have provided equity data of around 2000 companies every day the stock market is working. It provides the following details for a particular company: Name of each company and its SC code, SC group it belongs to, type of SC, opening and closing values, highest and lowest values, previous closing value, number of traders in that company for that day, number (or volume) of shares sold that day and net turnover. The data has been taken for all 21 working days in July 2022, excluding weekends and holidays (since data is not available for those days).

The main aim of the authors is to observe trends in companies based on two factors: market capital and volume of shares. The former is the net turnover a company has for a given day, and the latter is the number of shares sold for a given day. Thus, the data (.csv file) for each observed day has undergone the following process: duplicated, then in the first one, sorted according to decreasing order of net turnover, then removed all companies except the top 100 companies, and additionally added two columns, one denoting what the current leading factor is i.e market capital, denoted by "MCAP", and the current date value (note that these two values are same for all 100 companies in a given sheet), followed by the same procedure in the duplicate sheet but this time sorted according to the volume of shares, and the current leading factor is "VOLUME" for each record. Every of the 21 sheets has undergone this process, and in the end, we have 42 sheets. The two columns added for each sheet are helpful during graph formation via cipher querying in order to properly establish relationships among nodes.

The secondary data used here is the news captions for 10 companies. The 10 companies considered here include: HDFC Bank, Ashok Leyland, ITC Ltd., JSW Steel, Tata Consultancy Services, Reliance, Infosys, WIPRO, PayTm and SBI Bank. News captions for these 10 are considered. These captions have been taken from news headlines of chrome news pages based on current news. No pre-processing has been carried out for these captions, since the Vader model takes care of that by itself.

Workflow

Fig. 1. shows the workflow of the entire methodology

For this work, the authors have used Python to connect and directly query into Neo4j. This is done due to ease of processing since the Neo4j terminal took a lot of computational time when it directly interacted with the spreadsheets. It is possible to connect the Python instance to Neo4j using the neo4j module of python. A driver has to be set up to python for neo4j. Every graph gets a unique name and password which can be used for external connection purposes like this. Next, a session is launched using the driver. Every time a query has to be run, it can be passed as a string to the run() function of the current session object. After all queries are run, the session can be finally terminated. For each of the 21 dates available, the following process was carried out: a string is created in the format of date + "/07/2022". Then, a neo4j graph node of type "TRADEDATE" is created, with the name "_"+datestring. Then, the market capital file for the same date is retrieved and every row is created as a node. Finally, a relationship is defined from each company to the tradedate node, denoting that it is a market capital leader ("MCAP_LEADER") of that particular date.

Every node of type Company is matched to every node of type TRADEDATE with the condition the current lead factor is MCAP and date is current

date (two new columns earlier created), and then a relationship is defined from company to the tradedate node specifying the relationship name in between.

The process of opening the file and creating the nodes and mapping them with the current date, is carried out for the volume leader file of the same date. In the end, we would have 200 companies for each date, and 21 such sub-graphs.

Figure 2 shows how the whole system would look. As visible in that figure, we have 5 sub-graphs. For each one, the central blue node denotes the TRADEDATE node, surrounded by 200 nodes for each company.

Fig. 2. Graph representation for each working day. The graphs of only 5 days are displayed here, but the actual output is an extension of the above output

Note that only 5 tradedates are visible since neo4j has limited the number of visible nodes to 1000.

This means 200 + 1 = 201 nodes for each of the 21 dates, hence 201*21 = 4221 nodes. The overall visualization process is over here.

Next comes the processing part. For this, the authors once again use Python to connect to Neo4j, create a session with this particular graph, and get all the company names for each date, MCAP lead separate and VOLUME lead separate. In the end, there'll be 42 lists.

Based on the relationship defined, we filter out the companies related to a particular tradedate node, and get it's SC_NAME filed (name of the company) alone This is carried out for all 21 dates. Next, we access the company names using appropriate indexing and store them according to volume leaders and market capital leaders. Following this, everything is processing of python lists and data frames (which will be created) to get formatted lists. The program, in the end, will be able to print results in 6 different categories: Companies that lead in market capital, those that perform poorly in market capital, those that lead in volume of shares, those that perform poorly in volume of shares, those that lead both in market capital and volume of shares and those that have performed poorly on the whole (since they performed poorly on both the aspects). Based on user's wish, results for any of the above categories can be printed, like. Note that to be a leader in any of the two categories i.e market capital and volume of shares, a company has to be in the database of at least 15 days for that category. Likewise, for it to be a poor performer in either of the categories, it has to be in the database for no more than 5 days for that category.

The next part of processing is the vader model. For this, Google Collaboratory was used as the Jupyter notebook editor. The Vader model can be imported as a simple python library. Then, the 10 captions can be retrieved and stored in a list, and for each caption, the Vader model will present an output, from which the polarity can be chosen.

On performing the Neo4j part, we get results as described in the figure above. We then print results for each category, and check for the occurrence of each of the 10 companies under experimental consideration to be present in those results. Based on whatever categories they are found, a conclusion can be derived about that company for the month of July 2022. Also, the results from the VADER model can be obtained for each of the 10 companies. The results have the polarity scores for positive, negative and neutral sentiments, as well as the overall sentiment. The final results for all the 10 companies are shown in Table 1.

Experimental Analysis and Results

Table 1. Final results compiling the neo4j work as well as the sentiment analysis work

S.NO	NAME OF COMPANY	SENTIMENT NATURE	NEO4J WORK REMARKS
1	HDFC Bank	Neutral	Leader in market capital
2	Ashok Leyland	Positive	Poor performance overall
3	ITC Ltd.	Positive	Leader in market capital
4	JSW Steel	Negative	Poor performance in volume of shares
5	Tata Consultancy Services	Positive	Leader in market capital
6	Reliance	Positive	Leader in market capital
7	Infosys	Neutral	Poor performance overall
8	WIPRO	Neutral	Poor performance in market capital
9	PayTm	Negative	Poor performance overall
10	SBI Bank	Positive	Leader in market capital

Conclusion

This paper presents a method to observe trends in companies in terms of their stock values for a particular month, as well as the sentiments of the news captions related to them on the internet. The main application of this is to help investors and stock brokers choose the companies suitable for investing. Future works can be to include more days and more companies than considered here, to give an even more comprehensive view.

References

Hegazy, O., Soliman, O. S., and Salam, M. A. (2014). A machine learning model for stock market prediction. arXiv preprint arXiv:1402.7351.

Hiransha, M., Gopalakrishnan, E. A., Menon, V. K., and Soman, K. P. (2018). NSE stock market prediction using deep-learning models. Procedia computer science, 132, 1351–1362.

Kim, K. J., and Lee, W. B. (2004). Stock market prediction using artificial neural networks with optimal feature transformation. Neural computing & applications, 13(3), 255–260.

Zhang, Y., and Wu, L. (2009). Stock market prediction of S&P 500 via combination of improved BCO approach and BP neural network. Expert systems with applications, 36(5), 8849–8854.

Gupta, A., and Dhingra, B. (2012, March). Stock market prediction using hidden markov models. In 2012 Students Conference on Engineering and Systems (pp. 1–4). IEEE.

Pang, X., Zhou, Y., Wang, P., Lin, W., and Chang, V. (2020). An innovative neural network approach for stock market prediction. The Journal of Supercomputing, 76(3), 2098–2118.

Kimoto, T., Asakawa, K., Yoda, M., and Takeoka, M. (1990, June). Stock market prediction system with modular neural networks. In 1990 IJCNN international joint conference on neural networks (pp. 1–6). IEEE.

Fu, W., Sun, J., and Jiang, Y. (2022, July). Stock Selection via Expand-excite Conv Attention Autoencoder and Layer Sparse Attention Transformer: A Classification Approach Inspire Time Series Sequence Recognition. In 2022 International Joint Conference on Neural Networks (IJCNN) (pp. 1–7). IEEE. https://www.bseindia.com/markets/marketinfo/B havCopy.asp

Blockchain Technology as a Panacea in Tourism Industry

R. Hariharan,[a] R. Vedapradha,[b] D. David Winster Praveenraj,[c] and J. Ashok[d]

[a,c]Asst Prof. School of Business & Management, CHRSIT (Deemed to be University),
Yeshwanthpur Campus, Bangalore
[b]Asst Prof. Department of Commerce. Mount Carmel College, Autonomous, Bangalore
[d]Prof. School of Business & Management, CHRSIT (Deemed to be University),
Yeshwanthpur Campus, Bangalore
E-mail: hari712@gmail.com; Vedapradha123@gmail.com; David.winster@christuniversity.in;
Ashok.j@christuniversity.in

Abstract

The purpose of this study is to investigate how a customer's travel experience in the tourist sector affects customer satisfaction based on service value through the implementation of Blockchain Technology. 250 questionnaires were administered to tourists, tour operators, travel agents, hoteliers, and transportation companies using the Cluster Sampling method to obtain a sample size of 238. Blockchain technology, Customer Satisfaction, and Service Value are the critical variables observed to validate the hypothesis using Correlation, ANOVA & Regression Analysis. There is a strong correlation between the value of a service and blockchain technology. Secure payment is the most important component in developing and improving service value among customers and is the best predictor of service value.

Keywords: Blockchain, Tourism Experience, Service Value, Customer Satisfaction

Introduction

The tourism experience of a tourist is subjective to examining and undergoing a series of events (cognitive, affective & behavioral) pertaining to their activities from various perspectives. The hospitality and tourism industries are under increased pressure from competitors as a result of technological advancements and an increasingly sophisticated, wealthy, and demanding consumer base. Blockchain Technology is very young in the travel industry, potentially disrupting business operations across the globe (Bossink, 2002). A Blockchain is a list of transaction bundles called blocks that attaches and forms a distributed database. Information added to the Blockchain has to be validated and approved by all the network parties. Customers' opinions of the value of the services they receive are influenced differently by service quality. Customer Satisfaction is typically defined as the relationship between the customer and a product or service with the service provider (Gundersen, Heide, and Olsson, 1996). The difficulties that the various stakeholders today face in the tourism business include luggage management, identity services, safe payments, consumer loyalty, and pricing. RQ1: Is there any relationship between Blockchain Technology, Service Value, and Customer Satisfaction? H01:

Blockchain Technology, Customer Satisfaction, and Service Value don't share any relationship. Technology plays a critical role in the tourism and travel sectors, enhancing the client experience in tourism. Blockchain Technology is one such innovative disruption supporting the travel industry that enhances the customer experience to a greater level (Willie P, 2019). RQ2: Does blockchain technology has any impact on the tourism experience among customers? H02: Blockchain technology, traditional & combined modes of service delivery will have no significant effect on Tourism Experience of the customers.

The best way to increase the value of a service is to focus on long-term quality improvements while keeping an eye on how effective an integrated view of perceived service value will be when Blockchain Technology is used leading to happier customers. RQ3: Can Customer Satisfaction be explored through Block chain-driven Service Value in the Tourism industry? H03: Blockchain technology has no impact on the determination of service value.

Fig. 1. Proposed Model of Tourism Experience of Customers

Tour operators, travel agents, hoteliers, and transportation companies form the tourism industry's key players. The current study contributes significantly to the literature by examining how using Blockchain Technology, a customer's travel experience in the tourism industry affects customer satisfaction based on service value, which is grounded on the combination of Agency Theory and Transaction Cost Theory (Figure 1) integrated with Blockchain Technology in understanding the service value created, leading to customer satisfaction (Treiblmaier & Onder, 2019).

Methods & Materials

A self-administered 250 questionnaires were circulated to collect the responses from the respondents. It consists of 20 items based on the constructs of Technology (Barua et al., 2017), Customer Satisfaction (Li et al., 2006), and Service value (Petrick, 2002) which were refined to suit the current study on a five-point Likert scale. Demographic and Construct variables were considered for the deductive study. A sample size of 238 and the Cluster Sampling method, the main responses came from travelers, tour operators, travel agents, hoteliers, and transportation companies in and around the major cities, such as Bangalore, Pune, Hyderabad, and Chennai.

Service value was observed through payments, identification, customer loyalty, baggage management, and accommodation. Blockchain technology was captured on real-time, cost-effective, scalability, verification, and security questions. Traditional tourism observation comprised of general questions. SPSS V.21 was used to analyze the tests comprising Simple Percentage Analysis, Analysis of Variance (ANOVA), Correlation, and Multiple Regression Analysis.

Results

The Cronbach's alpha reflected with acceptable reliability (N = 238, α = 0.919 > 0.70, Kanal, et.al, 2018) confirming the items' high and excellent (α ≥ 0.9) reliability. A simple percentage analysis resulted that 54.2% males & 44.2 % females. 84.2 % belonged to 18–25 years, 11.7% between 26–35 years, 1.7% between 26–45 years, and 75.8% above 46 yrs. 85% were salaried & rest businessmen. 75% were married and rest single respondents.

A positive and highly significant correlation between Blockchain (BCT) & Customer Satisfaction (CS) (r = 0.893, p < 0.001), Service Value (SV) & Customer Satisfaction (r = 0.897, p < 0.001), Service value & Blockchain (r = 0.901, p < 0.001) can be observed in Table 1.

Table 1. Relationship between Blockchain Technology, Service Value & Customer Satisfaction

Variables	CS	BCT	SV
CS	1		
BCT	0.893**	1	
SV	0.897**	0.901**	1

**Significant @ 1 percent level.

One-Way ANOVA indicates a significant effect of tourism experience when integrated with blockchain technology (F = 13.417, p < 0.001), when compared with traditional methods of service (F = 8.792, p < 0.001), and combined effect reflect (F = 1.781, p < 0.01).

Levene's test reflected that the groups' variances were not equal toward tourism experience (F = 10.524, p > 0.01) can be observed in Table 2.

Table 2. The association between Traditional and Blockchain Technology in Tourism Experience

TE	Sum of Squares	Mean Square	F Statistic	p Value
Traditional	118.959	7.931	8.792	<0.01
Blockchain	290.471	12.103	13.417	<0.01
Trad*BCT	44.985	1.607	1.781	<0.01
Levene'Test			10.524	>0.05

A statistically significant regression equation was found (F = 33.255, p < 0.001), with R2 of 0.817 (R2 > 0.75, Mason & Perreault, 1991) can be observed in Table 3.

Table 3. Regression Analysis to predict Service Value using Blockchain Technology

Service Value	B	SE of B	β	t- Value
Constant	1.778	0.182	-	9.777**
Payments	1.416	0.147	0.588	9.614**
Identification	1.221	0.201	0.376	6.079**
Customer Loyalty	0.946	0.140	0.423	6.738**
Baggage Management	1.025	0.147	0.426	6.982**
Accommodation	1.092	0.137	0.495	7.992**

**Significant @ 1 percent level

Indices	Value
R	0.904
R²	0.817
F – Statistic	33.255
Durbin-Watson	1.921
Sig.	0.000**

**Significant @ 1 percent level.

The fitness of indices (R = 0.904) and Durbin-Watson (1.921) confirm the model's suitability in predicting the Service value by the predictors. Hence, the estimated Service value (Y) = 1.778 + 1.416 (Payments) + 1.221 (Identification) + 0.946 (Customer Loyalty) + 1.025 (Baggage Management) + 1.092 (Accommodation).

Discussion & Conclusion

Secure payment is the most significant predictor (β=0.588) and strongest predictor of increasing service value with blockchain technology. There is a strong association between service value and blockchain technology, which helps clients have an efficient travel experience. In terms of secure payment options, blockchain technology has the potential to dramatically enhance the services offered by the tourism sector.

The blockchain has the potential to change current passenger verification processes. Handling luggage becomes easier when artificial intelligence and sensor technology are used to locate lost bags and customers. Passports could be abolished by providing an advantage and raising customer happiness. Therefore, it can be stated that the use of Blockchain technology in the tourism industry does have an impact on how satisfied customers are with the services they receive.

References

Ambika, P., Vinoth Kumar, N., Hellan Priya, J. DOM-CHRO Number In Operations On Intuitionistic Fuzzy Graphs, Journal of algebraic statistics, 13(2), 1058–1062.

Bossink, B. A. (2002) The development of co–innovation strategies: stages and interaction patterns in interfirm innovation. R&D Management, 32(4), 311–320.

Barua, V. B., and Kalamdhad, A. S. (2017), Effect of various types of thermal pretreatment techniques on the hydrolysis, compositional analysis and characterization of water hyacinth. Bioresource Technology, 227, 147–154.

Kanal, I. Y., Keith, J. A., and Hutchison, G. R. (2018). A sobering assessment of small-molecule force field methods for low energy conformer predictions. International Journal of Quantum Chemistry, 118(5), e25512.

Li, B., Riley, M. W., Lin, B., and Qi, E. (2006), A comparison study of customer satisfaction between the UPS and FedEx: an empirical study among university customers. Industrial Management & Data Systems.

Mason, C. H., and Perreault, W. D. (1991), Collinearity, power, and interpretation of multiple regression analysis. Journal of Marketing Research, 28(3), 268–280.

Treiblmaier, H., and Önder, I. (2019), The impact of blockchain on the tourism industry: A theory-based research framework. In Business Transformation through blockchain (pp. 3–21). Palgrave Macmillan, Cham.

Önder, I., and Gunter, U. (2022). Blockchain: Is it the future for the tourism and hospitality industry?. Tourism Economics, 28(2), 291–299.

Balasubramanian, S., Sethi, J. S., Ajayan, S., and Paris, C. M. (2022). An enabling framework for blockchain in tourism. Information Technology & Tourism, 1–15.

Treiblmaier, H., and Önder, I. (2019). The impact of blockchain on the tourism industry: A theory-based research framework. In Business Transformation through blockchain (pp. 3–21). Palgrave Macmillan, Cham.

Barua, V. B., and Kalamdhad, A. S. (2017). Effect of various types of thermal pretreatment techniques on the hydrolysis, compositional analysis and characterization of water hyacinth. Bioresource Technology, 227, 147–154.

Zhou, R., Wang, X., Shi, Y., Zhang, R., Zhang, L., and Guo, H. (2019). Measuring e-service quality and its importance to customer satisfaction and loyalty: an empirical study in a telecom setting. Electronic Commerce Research, 19(3), 477–499.

Campbell, M. E., Keith, V. M., Gonlin, V., and Carter-Sowell, A. R. (2020). Is a picture worth a thousand words? An experiment comparing observer-based skin tone measures. Race and Social Problems, 12(3), 266–278.

Bucci, N., Luna, M., Viloria, A., Hernández García, J., Parody, A., Varela, N., and Borrero López, L. A. (2018, June). Factor analysis of the psychosocial risk assessment instrument. In International Conference on Data Mining and Big Data (pp. 149–158). Springer, Cham.

Jin, X. L., Zhang, M., Zhou, Z., and Yu, X. (2019). Application of a blockchain platform to manage and secure personal genomic data: a case study of LifeCODE. ai in China. Journal of medical Internet research, 21(9), e13587.

Exploring Virtual Draping and Digital Pattern Making in Fashion Design Institute

Meenakshi Verma*

*Assistant Professor, Karnavati University, Gandhinagar, India
E-mail: *meenakshiverma@karnavatiuniversity.edu.in

Abstract

Traditional approach to garment designing requires a lot of tangible materials like drafting papers, cloths, trims, machines, workspaces, etc. Traditional approaches are time-consuming. By using CAD software for garment designing, students can save a lot of time. Imagine that we no longer need to have a whole bunch of clothes cut into different shapes and sizes while working on a new design, with CAD, all we need is different stencils, and you can pick the best one for the design with just a mouse click. Isn't that fascinating? The new generation has always wished for digital advancement in the garments industry. Technology is playing a major role where we no longer collaborate physically, instead exchange ideas and work virtually. Even though the digital method is more efficient & accurate we have always been skeptical about relying on these technologies and preferred a more tangible approach in garment making.

Keywords: 3D Simulation, virtual Draping, Virtual Reality, Rapid Prototyping, Digital Repository/archiving, Optimization, Reduction in processing time, Accuracy, Sustainability

Introduction

Background: Design industry across the globe has shown significant growth over the last couple of decades. Design as a practice has touched almost every industry ranging from automobile to lifestyle to clothing. The nature of the process requires visualization of ideas, rapid digital prototyping, and digital assets which can be converted/used into physical products. The current areas of application of computers in the textile industry include weaving, knitting, printing, fashion illustrations, texture mapping, embroidery, pattern making, pattern grading, marker-planning, and cutting. La Mode and Der Mode (2009) believe that garments have become a new medium for graphic artists. The usage of Various Design Software significantly improvised the assurance of the end outcome and thus helped in reducing the cost of physical prototyping, faster production, and scalable approach to obtaining solutions.

With the advent of software technologies Prototyping, Simulation, and even the entire Design Process became very rapid and dynamic. The main advantages of using Computer Aided Designs (CAD) for creating designs are the capability to quickly try out numerous design ideas and reduce lead time simultaneously. Thus CAD software packages help the designer in experimenting with several textures, colors, and patterns for producing the perfect design. They provide a variety of sketch backgrounds and tools for designing and repeating patterns and texture mapping.

An Institutional Digital repository of Basic Patterns or Sloppers can help the students to design something new instead of making basic patterns again & again. Students can grade the size according to the customer/model with a few mouse clicks.

Garment designs for the 3D human figure are usually presented as sketches or photographs, which are 2-dimensional. Flat or 2D sketches, even rendered very nicely, are hard to imagine how it will be in the other angles, like side view, back view, zoom-in details, etc. According to Watanabe (2009), fashion design drawings will be more convincing if the model and the garments are portrayed with a certain 3-dimensional effect. Miller (1997) believes that 3D computer visualization is a very effective way to communicate design ideas.

Need for Digital Technology

- **Explore More Virtual Materials:** With CAD software there are a lot of materials such as zippers, buttons, fabric texture, and customized prints available in the digital library. Characteristics/Properties of different fabrics such as drapeablity, stretchability, transparency, texture, and color are among a few great capabilities offered via a digital library. By accessing a comprehensive library of commonly used fabrics one can immediately explore the design ideas.
- **Test Fit in Minutes:** Learning digital pattern making not only reduces the consumption of drafting papers but also makes students skilled with digital pattern making and ready for the future. Industries are investing a lot in these digital pattern-making platforms considering that such platforms allow grading, last-minute changes, and analysis of a sample in just a few minutes. And the bonus point is, if you have a database of great patterns loaded you can select one to start working from.
- **Save on Fabric/Muslin:** Digital simulation of garment construction is also saving a lot of muslin fabric which is required for making test fit. With the digital assembly of patterns, it is possible to check the test fit on a specific size of the dress form. If the garment doesn't fit properly or requires any change, one can easily make the required alteration in the pattern with just a click. As soon as you make any alterations your changes are synced with the garment draped

on the dress form or virtual model to check the test fit & drape of the garment.
- **Virtual Draping:** Yes, It is also possible to make patterns with the draping technique. By just drawing the shapes with a 3D pen tool on the avatar/virtual 3D dress form directly and you will get a flat 2D pattern of that shape.

Literature Review

Reviewing the Literature is very important for any research to find out the answer to an existing problem. It is academic writing, demonstrating knowledge & understanding of academic literature on a specific topic placed in context. It helps the researcher to identify the gaps between the available Information.

The entire literature search is divided into Six major facets or themes. The literature is classified and rendered accordingly. The five facets are: Role of Digitization in Fashion Design Industry

- Digital Pattern Making
- Traditional Pattern Making
- 3D simulation/ Prototyping
- Repository of digital assets
- Digitalization as a Sustainable Approach

Techniques in Pattern Making

Pattern-making is related to creativity and thinking skills. It is a skill that requires the students to use their exploratory ideation process (Jerimiah and Harrah, 2016). In Malaysia, the pattern-making course still adapts to the model-driven technique, whereas in other parts of Asian countries, like Singapore, La Salle University has introduced independent student work, in a technique-driven approach to fashion, toward collaborative endeavors (Jeremiah and Harrah, 2016). Fashion graduates still have a limited understanding of the scope of what pattern cutting entails (Jeremiah and Harrah, 2016). This has to do with how educators introduce pattern-making skills through teaching and learning methods.

Knowledge and technical skills are two very important components of the fashion industry. To produce beautiful and quality designs requires precise pattern-making skills and intricate sewing, the lack of which has resulted in the stealing and fabricating of ideas from other designers. In the end, the market will be laden with designs and fashions of the same "look and feel" (Morris, 2016).

Kaplan (2004) feels that computers and computer-controlled equipment aid in many functions such as design, marking, and cutting. Also, many employees seek designers who know how to use computer-assisted design. According to Chatterjee et al. (2010), the advent of the technology of computers, the apparel industry is changing from a traditional, labor-intensive industry to a highly automated and computer-integrated industry. Dickerson (2003) opines that the fashion industry of the new millennium is quite different from the one that existed a decade or two ago. New computer technologies have been incorporated into virtually all aspects of the industry to increase efficiency, save time and produce value for the consumer.

3D Simulation/ Prototyping: In this book the author discussed, Which tasks belong to the human and which to the machine? What tasks a machine should do and what tasks a human should do is first to make a list of the strengths of machines versus the strengths of humans and then use that list when considering the requirements of the task at hand. For example, if the task calls for precision and repetition, machines are probably a better option than humans. If instead, the task demands intuition or moral judgment, humans

Repository of Digital Assets

Butterworth Jody (2018) discussed the preservation of documentary heritage, which is at risk of neglect, physical deterioration, and destruction where resources are limited. By creating digital copies of materials and saving them for local institutions or libraries. Photography method is used to digitize the data, and a workflow is designed to archive that data. Workflow includes creation of digital images, renaming and organizing the digital images, exporting the digital images, and backing up.

Surinder Kumar (2012) stated that there are many technical as well as social issues in establishing digital repositories. Technical issues

such as selection of hardware, loading software, uploading data, setup test server, manage process and maintenance. DSpace and Eprints are the two major software packages used in establishing institutional repositories. In Health Sciences, there are four digital repositories. It is recommended that institution should make it mandatory for their researchers to publish their papers in the institutional repository and register their open access policy at Registry of Open Access Repositories Mandatory Archiving Policies (ROARMAP).

Kim and Park (2008) reiterate that 3-dimensional computer-aided design has become one of the most indispensable elements in modern industries. It is very difficult to find any design process that is not aided by CAD systems in traditional manufacturing processes of machinery, aircraft, and watercraft and most engineers take it for granted nowadays. "My background is 3D design and technical fashion and my enduring passion is in merging technical and aesthetic considerations providing comfort - physiological and psychological - and ease of movement, through appropriate garment fit" opines Watkins(2005).

Optitex specializes in the development of innovative, easy-to-operate 3D CAD/CAD solutions for cut fabrics. It creates a virtual world when testing manufacturing designs. Optitex uses 3D visualization technology for garment fitting by synthetically reproducing a three-dimensional model of clothing items. Optitex develops innovative, user-friendly 2D and 3D solutions for all cut-fabrics and fashion related industries (Indian Textile Journal, 2010).

Bijan Kumar and Roy (2013) discussed the mechanism to organize valuable knowledge resources in the forms of theses, dissertations, project reports, courseware, pre-prints, etc through research and development activities. He is saying there is no mechanism for holding the such intellectual output of the university for scholarly communication. According to him, these resources are not available and inaccessible due to the absence of appropriate mechanisms. He solved this problem by implementing IDRs to make digital libraries where these resources are accessible and organized.

MerlinOne (2018) in this blog author discussed Copyright of any asset. They mentioned copyright is a major concern for photographers, and in the digital world, copyright infringement is rampant. To protect your work, add copyright information to every photo while importing it to DAM. Embedding copyright details into photo's metadata, ensure that that information travels along with artworks wherever they go. While it won't prevent copyright infringement, it can help to deter theft. If someone uses your artwork without your permission, copyright information embedded in the image's metadata proves ownership.

Graphic Designers are also in the same situation nowadays, they also need to mention copyright information about their creations.

Summary

The above review revealed that there were several studies conducted on digital Patterns making & traditional pattern-making methods. Lack of Access to digital data archives & Repository management. But studies showed that there was a lack of research on practicing digital pattern-making in design institutions in India.

There was a lack of studies on practicing digital Pattern making at Fashion Design Institutes in India. Therefore, the researcher felt the necessity to study & identify the bottlenecks, to understand the challenges faced by design institutes & students in practicing Digital Pattern making & Virtual Draping.

Problem Statement: Though the Digital method of pattern-making is more efficient, accurate & time-saving, still, industries and students are practicing the traditional method of pattern-making. The cause of this affinity towards traditional methods is to be understood.

Aims and Objectives

To find out the gaps/ reasons for not practicing Digital Pattern making & sampling instead of traditional methods.

To identify the solutions for making the digital method more easy and more accessible for students and the industry.

To develop a framework for integrating digitization in pattern-making & garment sampling.

Research Methodology

Descriptive or Case Study Research:

To study the effectiveness of the implementation of digital software (an in-depth case study of selected students in selected design colleges & then compare them to see if any general trends emerge.) Processes involve different methods of collecting data like questionnaire tools, unstructured interviews, expert Interviews, observations, and video recordings to understand the existing scenario of pattern making.

Qualitative Research Methodology

"Multi-method" qualitative research is the most appropriate one for present study. As it refers to using more than one data collection technique and applying multiple methods to analyze these data using non-numerical (qualitative) procedures to answer the research question.

To collect data via face-to-face interviews. A semi-structured interview protocol will be developed to guide the interview sessions.

Experimental Research:

To determine if the new teaching method can bring around a change in mindset towards better results.

1. Analyze/ compare the efficiency of students while working with digital v/s traditional method of PM & sampling.
2. Understand how a focus on digital design on curriculum will drive the change toward adoption.
3. Comparison and result validation.
   Virtual Sample/ prototype of the garment using Clo 3D software

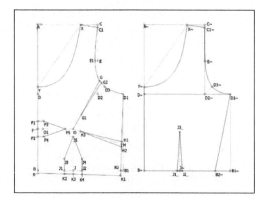

Fig. 1. Digital Pattern of Blouse by using Seamly 2D (Open Source) Software

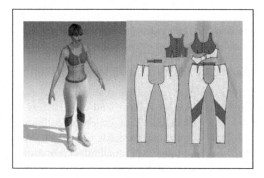

Fig. 2. Results (Observation and Inferences of Data Analysis)

Results for Pilot Study conducted with students of UG 5 & PG3 who had practiced digital Pattern making & 3D simulation of garment for approx One month. Data of 20 Respondents have been collected & Presented below.

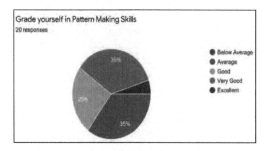

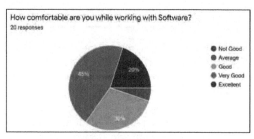

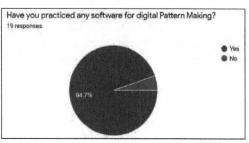

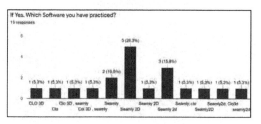

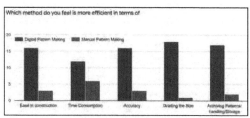

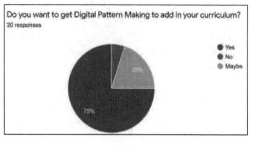

From the above Responses the Researcher analyses that students can practice Digital Pattern making After 10–15 Classes & they want to get it added to their curriculum to make the Sampling Process Virtual analysis of design more accurate & fast

Conclusion

Based on data collection & analysis researcher concludes that the digital method of Pattern making & Draping/ 3D simulation of Garments is more efficient, accurate, time-saving, sustainable & easy. Fashion Design Industries are looking for employees who can work on the digital medium of prototyping but institutes are not practicing it through open source & less expensive/affordable software are available nowadays. Fashion Design Institutes in India should add these practices in their curriculum.

References

Azman, Syed, and Maryam, Sharifah. (2019). Integrating Innovation in Pattern Making Teaching and Learning for Higher Education in Fashion Design.

Bailey, D. E., and Leonardi, P. M. (2015). What Product Designers Let Technology Do. In Technology Choices: Why Occupations Differ in Their Embrace of New Technology (pp. 69–102). The MIT Press.

Butterworth, J., Pearson, A., Sutherland, P., and Farquhar, A. (2018). A workflow for digitisation, Remote Capture: Digitising Documentary Heritage in Challenging Locations Open Book, 1st ed., vol. 1, Cambridge, UK, pp. 121–138.

Kumar, Surinder (2012). Setting up Digital Repositories: Challenges & Issues, National Informatics center, New Delhi, 2012.

Roy, Bijan, and Mukhopadhyay (2012). An Analytical Study of Institutional Digital Repositories in India. Library Philosophy and Practice, 2012, pp. 1–13.

MerlinOne: Digital Asset Management Tips for Photographers, October 2018.

Kaplan, N. S. (2004), Changing trends in apparel industry, Abhishek publications, Chandigarh, India, p. 2.

Kim, S., and Park, C. K. (2007), Basic garment pattern generation using geometric modelling method, International Journal of Clothing Science and Technology, 19(1), pp. 7–17.

Identifying the Pattern and Incidence of Significant Health Issues Among Garment Industry Employees

Khushi Jain,[a] Deven Patil,[b] Divya Parekh,[c] Bhubneshwar Goyary,[d] Parth Shikhare,[e] Atharva Malthankar,[f] and Arunachalam[g,*]

Karnavati University, Gandhinagar, India
E-mail: *arunachalam@karnavatiuniversity.edu.in

Abstract

Background: Unlike the Textile industry in India, which is the second-largest exporter of textiles with a share of 5% of global trade, The Garment industry also contributes to economic growth. It is the second largest sector to employ thousands of workers from across the country. Many unskilled laborers from rural areas work in this sector. Most of them work on the basis of hourly pay. Working continuously for longer durations of time without any rest, and in absence of any safety gear leads to major health-related issues among these workers. Materials and Methods: A survey was carried out using a search strategy, to collect the reasons behind this situation. The sole purpose of this study was to obtain information about any possible occupational origin of health-related hazards.

Results: The studies revealed that major health problems amongst garment workers were musculoskeletal, respiratory, and cardiovascular disorders. The most commonly affected regions were the neck and back followed by the shoulders.

Conclusion: This review on the health problems faced by the workers revealed that musculoskeletal disorder was the most prominent. Majority of the studies were carried out on women and it was found necessary that some specific changes must be made aiming to prevent musculoskeletal disorders among garment workers.

Keywords: Health Issues, Garment Industry, Rural Workers, Hazards

Introduction

India's economy is expanding at a rapid rate. The textile market is a major driver of economic expansion. The textile and clothing industry is widely recognized as one of the world's oldest and most ubiquitous. It's the country's second-biggest employer, behind the government. The Indian textile sector is dominated by mom-and-pop shops. The cotton and textile industries provide numerous job openings. A lot of low-skilled employment in rural regions is included in this. Prolonged sitting, standing, highly repetitive work, lifting of heavy objects, working with hands lifted to shoulder level or even higher, and working with a twisted or bowed back have all been linked to impaired work ability and enhanced long-term illness, all of which are commonplace in garment factories. Both the work and the hours put in by the employees are strenuous. The informal economy and small-scale enterprises pose unique risks to their employees. Numerous health issues, including those specific to garment workers, have been uncovered by international studies of the industrial workforce as a whole. The increased risk of reduced job performance, musculoskeletal problems, cardiovascular disease, long-term sickness absence, early retirement, and all-cause mortality among workers with high physical labor demands are well recognized. Physically demanding jobs can cause a variety of health issues, especially in people who perform the same tasks day in and day out. Problems with breathing, heart, intestines, muscles, and bones, as well as malnutrition, are the most typical outcomes. The cutting, stitching, and finishing work done in garment factories require extreme focus and concentration, resulting in frequent headaches and eye strain for the workers. The workers spend most of their waking hours in the plant and earn just enough to get by. It's hard for them to afford the groceries and time to prepare the food they need. That's why they risk food poisoning, diarrhea, gastric pain, malnutrition, abdominal pain, and other issues by opting for a less-than-sterile fare. Nutritional deficiencies, decreased appetite, diarrhea, hepatitis (jaundice), and food poisoning are only some of the issues linked to the food consumed by employees in the readymade garment sector, according to recent research.

Background of Current Research

We conducted this study to learn more about the workload and physical discomfort experienced by people who use sewing machines in various parts of India, as well as to identify and explain potential ergonomics issues in their workstations. Using a case study methodology, we conducted a small number of in-depth interviews to collect data and get insight into the issue at hand. The hope was that by pinpointing these issues, regulations could be drafted and preventative actions could be implemented to alleviate many of the musculoskeletal disorders that employees experience.

Literature Review

The study included 50 operators of sewing machines with short work cycles (30–60s) and 20 seamstresses as a control group, all of whom were selected at random. There was parity between the two groups when it came to age and tenure. Sewing machine workers had the most sedentary work environment, according to a survey of employees' postures. Sewing machine workers more commonly reported pain in the back, shoulders, and knees.

In Sweden, 75% of a sample of people who use sewing machines report having had neck and shoulder difficulties in the preceding 12 months, and 51% had had such problems during the last week. The prolonged hours spent in an awkward position at work were blamed for the discomfort. What's more, the research showed that many people had previously resigned from their jobs, with health concerns being a major factor in the decision of many. Since these operators found their work to be dull and repetitive, the authors suggested a change in schedule and assignment of responsibilities (Blader et al., 1991). Among Norwegian employees, those who operate sewing machines had a higher likelihood of reporting pain and discomfort than those who worked in offices (71% vs. 95%). Sewing machine operators also took more sick days than office workers (70 vs. 51 percent). Pain was most commonly reported in the upper back and shoulders, although it was also commonly experienced in the head, upper arms, and lower back. Sewing machine operators reported significant rates of physical and psychological symptoms, but a direct link could not be demonstrated (Westgard and Jansen, 1992). Working with a sewing machine was shown to be quite stressful for Norwegian workers. The job of using a sewing machine has been classified as "low" in Canada in terms of energy requirements. Operators were blamed for their physical issues since their jobs forced them to repeat up to 1500 repetitions of the same motion every day in a stationary position. Similar to the previously stated Norwegian study, the shoulder area was the most often reported source of discomfort in the Canadian study, with the hands coming in a close second. It's worth noting that the sample sizes in both pieces of research were rather low. Canadian researchers found that those working in the clothing industry had the greatest rates of clinical complaints. Twenty-nine employees had forearm/hand cancer, nine had tendinitis, and nine had carpal tunnel syndrome (Vezina et al., 1992). According to research conducted on American garment workers, 52% of employees reported experiencing upper back discomfort, 49% reported having neck pain and 48% reported having hand pain. Kelly said she believes bad posture, excessive repetitions, and uncomfortable seating are to blame for her health problems. Because Yu and Kesselring's chairs altered the angle of the seat pan and backrest, they were able to alleviate a lot of the aches and pains these workers were feeling (Halpern and Dawson, 1997). Lower back, upper back, and mid-back complaints predominated in the Corbett and Bishop body chart survey, followed by those about the shoulders, neck, and legs. Data are shown as a percentage of respondents who selected a certain degree of pain intensity from a scale of 0–10 for each body region (where 10 is the highest level of pain). Specifically, 32.5% of all p-a-r-t-i-c-ants reported severe pain, while another 37.7% reported considerable discomfort, in their upper backs. Only 7.9% said they didn't feel any discomfort in this area. The majority of patients (66.2%) reported moderate to severe lower back pain, 39% reported moderate to severe discomfort, and 10.1% reported no pain in the lower back. Of the study's participants, 26.4% reported severe pain in their mid-back, 34.5% reported moderate discomfort, and 8.8% reported no mid-back pain at all (Bandyopadhyay et al., 2012). At least five people reported severe discomfort, and no systematic interventions were taken.

Problem Statement

Our main focus is to reduce all the obstacles faced by Sewing Machine operators and make their working environment more efficient. In the present scenario due to the Absence of proper Ergonomic working space, Equipment, Guidance & Motivation, the Sewing Operators regardless of Age, Sex & Personal Health Status face a lot of Physical & Mental Health related issues. Some of the most common issues are Back, Neck, Shoulder & Thigh pain. The reason is the prolonged period of working time in addition to repetitive & pressure on muscles,

tendons, joints & bone. Despite that, the rise in psychological problems is also prevalent. Therefore, the changes are a must for the Sewing Machine Operators' well-being & to increase productivity.

Fig. 1. Bad postures while working on sewing machines

Aim

Our main aim is to identify the pattern and prevalence of major health problems among garment workers who are working on sewing machines for long hours.

Objective

Using this method, we searched many databases, including PubMed, Scopus, Science Direct, and the Cochrane Library. This evaluation includes all publicly accessible studies for which there is both a full-text article in the database and a comprehensive description of the research process.

Methodology

Over all approach is to solve the health issue among garment factory workers. What kind of problems do people face while working with sewing machines? There are a different kinds of people who work on sewing machines. So, we are redesigning the sewing machine table and trying to find out the perfect solution. Basically, we are focusing on the back pain issue which is caused by working for long hours.

We collected some data from scratch. Then we created a set of questions to collect the data needed. Created some set of questions so as to come up with problem areas so that we can figure out the solution for the same. What is their age? What are the working hours? Are they getting paid enough compared to the work they do? What kind of

health issues do they face? Do they get any kind of medical facilities? Are there any hurries in completing the work?

With the help of a literature review and interviews, we studied the details of health issues faced. Then we decided on the problem statement which we are going to solve. Finally, we came up with a particular solution and focused on solving the problem of back pain while working on sewing machine.

Results

This study has been carried out on sewing machine workers who regularly use sewing machines and are sources of their income. The research type was a case study and the data was collected using a primary method of data collection (interviews were taken of nearby local tailors, workers and people working in boutiques and shops). The target groups were divided according to the age group of the respective people. The two targeted groups were the middle age group (30–50 years) and the elder age group (50–70 years). The number of people interviewed in the age group of 30–50 years, was 27 and the 51–70 years age group people was 24. The study also showed a comparison between the two groups.

Table 1. Physical Issues

	Physcial issues	30–50 Years (27 people)		51–70 Years (23 people)	
		Male (16)	Female (11)	Male (12)	Female (11)
1	NECK	25.00%	27.20%	8.33%	0.00%
2	SPINAL CORD	12.50%	9.09%	33%	18.18%
3	ELBOWS	6.25%	9.09%	17%	9.09%
4	WRIST	6.25%	9.09%	0.00%	9.09%
5	SHOULDERS	12.50%	0.00%	8%	18.18%
6	LOWER BACK	0.00%	0.00%	25%	18.18%
7	ANKLE	6.25%	9.09%	8.33%	0.00%
8	CERVICAL	6.25%	9.09%	25%	9.09%
9	MATERNITY	12.50%	18.18%	8%	0.00%
10	HEADACHE	12.50%	9.09%	0.00%	9.09%

Table 2. Mental Issues

	Mental Issues	30–50 years (27 People)		51–70 years (23people)	
		Male (16)	Female (11)	Male (12)	Female (11)
1	WORK LOAD	12.50%	9.09%	16.66%	9.09%
2	SATISFACTION	12.50%	0.00%	8%	18.18%
3	RELIEF	6.25%	9.09%	16.66%	9.09%
4	FINANCE	25.00%	18.18%	25.00%	9.09%

5	RESTRICTIONS	6.25%		9.09%	0.00%	9.09%
6	ANXIETY	12.50%		0.00%	8.00%	18.18%
7	STRESS LEVEL	12.50%		9.09%	17%	0.00%
8	MENSTRUAL CYCLE	0.00%		18.18%	0%	9.09%
9	MATERNITY LEAVE	0.00%		18.18%	0%	9.09%
10	HOBBY	12.50%		9.09%	8%	9.09%

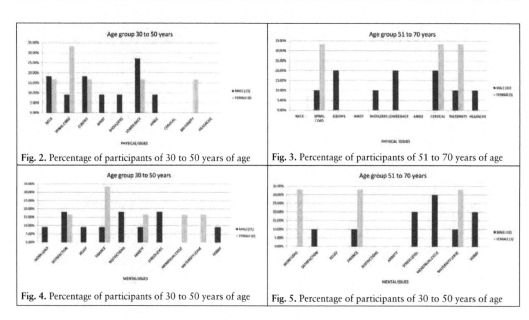

Fig. 2. Percentage of participants of 30 to 50 years of age

Fig. 3. Percentage of participants of 51 to 70 years of age

Fig. 4. Percentage of participants of 30 to 50 years of age

Fig. 5. Percentage of participants of 30 to 50 years of age

This was a study on workers operating sewing machines regularly and as a source of their income. This study was based on the collection of data, which was then categorized into two problem areas which are mental health issues and physical health issues. The data collected Then the classification of the data was done by gender differentiation.

The people in the middle age group faced more issues compared to that in the older age group people. The average working hours of the middle-aged group was 10 -12 hours and the average working hours of the older aged group were 6–8 hours. Most of them worked with unsuitable tools at sitting positions, on an unsuitable surface, and had to bend forward over the work. The area of pain for health issues experienced by middle-aged people was classified into the neck, spinal cord, elbows, wrist, shoulders, lower back, ankle, cervical pain, maternity issues, and headaches.

In the middle-aged group, the most common health issue seen was neck-related consisting of 25.00% in males and 27.20% in females. Apart from neck pain, the other issues occupying most percentages in males of the middle-aged group were spinal cord, shoulders, maternity, and headaches-related problems, consisting of 12.50%. The problems related to elbows, wrists, ankles, and cervical pain were lesser occupying 6.25% in males of the middle-aged group. There was no sign of lower back pain in the same.

Apart from neck pain, maternity leaves related problems were seen as the most occupying one in females of the middle-aged group, which is 18.18%. The problems related to spinal cord, elbows, wrists, ankles, cervical pain, and headaches were lesser occupying 9.09% in females in the middle-aged group. There was no sign of lower back pain in the same.

The health issues faced by the older age group people because of operating sewing machines were lesser compared to that of the middle-aged group people.

The most common and larger percentage in the older age group was occupied by spinal cord pain in males which is 33%. Apart from spinal cord-related pain, the highest percentage was occupied by lower back and cervical pain which was 25% in males of the same age group, followed by elbow pain which is 17%. The lesser occupying problems in health were neck pain occupying 8.33%, followed by shoulders-related problems occupying 8%. There was no sign of wrist pain or headaches in the same age group of males.

For females of the older age group, the most common problems were spinal cord, shoulders, and lower back, occupying 18%. Apart from the spinal cord, shoulders, and lower back-related health issues, the other problems faced by females of the same aged group were elbows, wrists, cervical pain, and headaches, occupying 9.09%, followed by maternity issues occupying 8%. There was no sign of ankle and neck pain-related issues.

The study also showed that the older aged group of people experience headaches more often compared to the middle-aged group people.

The classification of mental health-related problems and issues was done under ten different categories, which are workload, satisfaction, relief, finance, restrictions, anxiety, stress level, menstrual cycle, maternity leave, and hobby. In the middle-aged group, the most common issue was finance-related problems, consisting of 25% in males and 18.18% in females., finance-related longing, with finance, workload, satisfaction, anxiety, hobby, and stress level issues occupying the most amount which is 12.50% in males. The least issue in males related to mental health was restrictions and relief (the amount they get for relief) occupying 6.25% in total. There was no

sign of menstrual cycle or maternity leave-related issues.

The females face issues related to maternity leaves and the menstrual cycle the most along with finance-related issues, occupying 18.18%. The less percentage of issues in females were workload relief anxiety-related problems were not seen in females.

There was seen a good difference in the percentages of the issue of mental health in the older and middle-aged groups, except for the fact that the finance-related aged groups were the most common one consisting 25% in males of the older age group. Apart from finance, the data showed that the males of the older age group experienced more stress level-related issues, occupying 17%. The other problems which were workload and relief occupied 16.66% of males in the older age group. The less percentage of issues seen were satisfaction, anxiety, and hobby, occupying 8%. There was no sign of restrictions, menstrual cycle and maternity leave-related issues.

In the older-aged group of females, the most common issues were satisfaction and anxiety. Apart from satisfaction and anxiety, the data showed that the females of the older age group experienced workload, relief, finance, menstrual cycle, maternity leave, hobby and restrictions-related issues occupying 9.09%. There was no sign of stress level issues.

Conclusion

As we started our research, after taking some interviews and the data we received we came to the brief, that almost every worker is facing some kind of MSD (Musculoskeletal disorder), to prevent that we are planning to design a workplace setup for the workers, in which they will be able to work more efficiently with fewer health issues. On the other hand, we will be adding some safety features also to prevent them from getting any injury while working.

To begin with, we started improving the table where workers put all their things and where the actual machine is situated. We came up with a rough design solution as shown in Fig. 6.

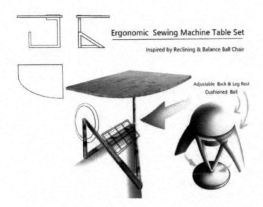

Fig. 6. Layout of the product

This table design is specifically made to look quite different from the other regular tables available in the market. It was seen that the table the workers work on was quite small and messed up, and the clothes lay everywhere, creating a huge mess. In such a mess prioritizing work was a big issue. Thus, we ideated that by increasing the size of the table this problem can be reduced almost by around 70 % now that workers can organize their work, making workplace area cleaner and more effective to use.

Also, the new table design that we are planning to implement, has a height and an angle adjuster, which can be customized according to the workers' needs. A slight change in the angle or even in the height can make a lot of difference! Although this does not completely solve the MSD disorder problem surely it does help to reduce it to a certain extent. As every worker is not the same everyone has their own opinions and thought processes thus, by giving a personalized choice, changing the height and angles of the table makes things work more efficiently. The product is an ergonomic table chair set which is designed keeping in mind to reduce the issues to the least.

Table – The rounded edge of the chair helps the worker to move and change their position while working. Also, the best part, the bottom area of the table gives space to move and position the leg comfortably.

Chair - Inspired by reclining and balance ball chair. The main part ball of the chair provides a good cushion. The adjustable leg, back, and neck rest allow more comfort. The rolling wheel at the bottom allows the user to move in different positions easily.

Fig. 7. Final product

References

Blader, P.S., Barck-Holst, P.S., Danielsson, P.S., Ferhm, P.S., Kalpamaa, M., Leijon, M., Lindh, M., and Markhede, M. (1991). Neck and shoulder complaints among sewing-machine operators. A study concerning frequency, symptomatology and dysfunction, Applied Ergonomics, 22, 251–257.

Westgard, N., and Jansen, N. (1992). Individual and work-related factors associated with symptoms of musculoskeletal complaints. II Different risk factors among sewing machine operators, British Journal of Industrial Medicine, 49, 154–162.

Vezina, N., Tierney, N., and Messing, N. (1992). when is light work heavy? Components of the physical workload of sewing machine operators working at piecework rates, Applied Ergonomics 23, 268–276.

Halpern, C. A., and Dawson, C. A. (1997). Design and implementation of a participatory ergonomics program for machine sewing tasks, International Journal of Industrial Ergonomics 20, 429–440.

Sealetsaa, O. J., and Thatcher, A. Department of Industrial Design and Technology, University of Botswana, Gaborone, Botswana psychology Department, University of the Witwatersrand, Johannesburg, South Africa

Bandyopadhyay, L., Bauru, B., Base, G., Holder, A. (2012). Musculoskeletal and other health problems in workers of small-scale garment industry – an experience from an urban Slum, Kolkata. IOSR Journal of Dental and Medical Sciences, 2(6), 23–8.

Graphic Trends and Developments

Harsha Gandrapu,[*,a] *Lolita Dutta,*[b] *and K. K. Singh*[c]

[a]Student, School of Visual Communication, United World Institute of Design
[b,c]School of Visual Communication, United World Institute of Design
E-mail: [*]kk@karnavatiuniversity.edu.in

Abstract

Graphic design has evolved both in terms of techniques and technology over the years. This is due to the inflow of ideas both from the people working in the field directly and those unrelated to this field, indirectly. Graphic design and graphics are constantly researched and modified according to the market requirements. Initially, the trends were not particularly followed during the graphic era but they were constantly developed over previous styles. Over time, increasing experiments and formulation of design principles along with graphic design theories increased the trends as a result of dictating style formation. Styles introduced by the designers during the past 5 decades or more are being followed by students and professionals.

Keywords: Typography, Europe, China, Alphabets, Art Movements, Bauhaus, De Stijl, Art Nouveau, Print, Printing Press, Gutenberg, Type Foundry, Movable Type, Propaganda, World Wars, Industrial Revolution

Introduction

Graphic Design is the process of visual problem-solving. It branches out across different themes, for example, branding, typography, illustrations, packaging, etc. It is also the method to arrange visual elements in a space. These days graphic design is also broadly included in visual communication, as it deals with communication through visuals. Visuals in the primitive era were based on Cave paintings, petroglyphs, pictograms, ideograms, and hieroglyphs. The Alphabet that we currently use was developed much later which currently slowly evolving into slang and emoticons and probably would lead to where we started, completing the circle.

The Introduction of Paper And Type

Paper was introduced to Europe by the Arabs. China is the center of paper manufacture, and the development of movable type printing appeared in China during the 11th century (Eskilson, S. (2019)). Gutenberg is credited for the introduction of movable type printing in Europe during the 15th century (Eskilson, S. (2019)). This allowed widespread printing across the continent. Around 1445, Gutenberg published his Bible. As these books were printed by movable type and were rubricated by hand, this increased the amount of time needed to complete each volume. Over time, the invention of two-color printing accelerated the printing process. The main purpose of rubrication and the two color printing methods was to create emphasis on particular letters as well as to decorate the book. Later this was replaced by Italics and Small Capitals (Eskilson, 2019).

In the 15th century, typography played a vital role in graphic culture in Europe. Letters were designed based on calligraphic forms

(Eskilson, S. (2019)). Gutenberg's Bible was elegant and straightforward, with the justified text divided into two columns, although there were no contemporary grid systems. The text was set in a type called Textura. Parallelly, printing of Arabic books in Europe led to development of Arabic typefaces during this period.

Johann Fust along with his assistant Peter Schoffer published the lavish Mainz Psalter in 1457. The book is a combination of woodcut illustrations and the type (Eskilson, S. (2019)). Then, Anton Komberge with a small team published The Nuremberg Chronicle, A high-quality book with text illustrations on the same page (Eskilson, S. (2019)). As a competition to the existing gothic style, a new style emerged. The Roman style was developed in Venice in the 1460s. This was developed due to the important role of printing during the renaissance period in Europe. As renaissance was totally influenced by scholars known as humanists, they adopted a type of handwriting called Carolingian minuscule based on the style of writing used for official documents. During the late fifteenth century, this style became known as Humanist minuscule.

Many Roman letterforms were developed because of the vast expansion of printing presses across Europe. Amidst this, there were many styles that were introduced to the world. Some of them are Old Style, Transitional and Modern. Typography in the renaissance was the most evolved field in terms of modern style. French printer and publisher Claude Garamond contributed to renaissance typography by designing the versatile typeface called Garamond, after his name. Garamond is one of the distinguished typefaces of the renaissance. Garamond also established the first type of foundry and initialized the modern style in Europe. The major change in construction type was in 1692. The invention of Romain du Roi led to the novel way to construct a typeface. Using a vertical and horizontal grid for creating fonts became the norm to create a font since then. A measuring system for the type was invented by Pierre Simon Fournier which used a scale based on inches which were further divided into 72 points (Eskilson, S. (2019).

Inspired by the works of Garamond, William Caslon, a British type designer Caslon set up a type foundry. Caslon's foundry designed a lot of typefaces that were much more modern compared to Garamond. In the 18th century, the fondness for national identities increased. These national identities were based on the official documents of the government. After 3 decades, the extreme contrast in the strokes of the letterforms was introduced by John Baskerville which was criticized as too thin strokes which were hard to read. Later, these typefaces were called Modern Typefaces. For printing these thin strokes, a new type of paper was introduced in France called Wove Paper by Ambroise Didot (Eskilson, 2019). He was also responsible for introducing a new type of measurement system which is based on foot or pied au roi in french. This was spread across Europe creating a standard for classifying types. Following the paths of Didot, Giambattista Bodoni further defined the Modern style.

Industrial Revolution and the Evolution of San Serif

The major developments in technology were during the industrial revolution. This was also the time when economies shifted from manual labor to the mass production of goods in the major cities. People started migrating to cities for better employment henceforth for better wages and salaries. There was a tremendous amount of increase in production houses which led to a lot of employment for people. The lifestyle of people changed. The thoughts of artisans and designers were ideological and strategic. Many companies started to market themselves for the wider reach of the audience. Print being the major medium of communication between companies and people, advertisements became popular for mass-produced goods to reach a wider audience. This also became the base for the expansion of mass literacy. As paper became cheaper and mass production began the volume of poster printing went up. Because of the urban mass culture and their demands, there was a need for a new style which was later labeled as San Serif. San Serif was mainly for decorative and display purposes during its inception. As there was uniformity in strokes compared to contemporary roman typefaces such as Didot, these kinds of typefaces were mainly found in the field of advertising and posters (Eskilson, 2019).

Wood type, easy, flexible, and economical was invented. To accommodate the ease of mass production for the wood type, Darius wells invented the lateral wood router. Later in the 19th century, Slab Serif was innovated. Letterforms were paired with heavy rectangular serifs (Eskilson, S. (2019). Later, newspapers were introduced and new technologies were invented namely, letterpress, lithography and chromolithography, rotary steam press, and photography which enhanced the range of possibilities for the mass production of printed materials. Photography was also invented during the Industrial Revolution. Photography studios became popular with the masses but never in the printing industry because of printing difficulties until the 1920s. Till then newspapers relied on illustrations for the visual representation of the text. Visuals in posters were still put in with lithographs or woodcuts.

Evolutions of Digital Products

For the most part, graphic design is based on hand skills. Layouts were visualized and ordered by hand, the type was carved before the industrial revolution and during the industrial revolution, the type was specified and ordered. However, this slowly changed when a lot of advances took place in the 1980's and 90's. Invention of digital computers altered the working process in design. A major change was observed in the amount of time required to correct errors, which was made easy compared to past methods.

The first major leg in the digital era for graphic design was when Apple programmed Macpaint for Macintosh computers. The Postscript page description language from Adobe Systems enabled pages of type and images to be assembled on the digital screen. Computers were still very new, people were getting used to them and the software was not well developed and was still experimental. The type was not the one that is regular for printing and layouts were very

different. When computers were first introduced, type and images were layered, type columns were overlapped, and the contrast in type sizes was very drastic. As this was happening, swift changes were parallelly happening hence introducing transparency, stretching, scaling, bending, and even layering visual elements on a digital screen.

Photoshop and Illustrator revolutionized design when they were introduced for Windows-based systems. As Windows-based systems were cheaper, they became available to more people. Digital revolution in graphic design expanded when the Internet became accessible to the public.

The Influence of Art Movements on Graphic Design

Photoshop and Illustrator revolutionized design when they were introduced for Windows-based systems. As Windows-based systems were cheaper, they became available to more people. Digital revolution in graphic design expanded when the Internet became accessible to the public.

Graphic design or any design for example is driven by ideas or concepts that drive the visuals. During the time when books were hand-printed and earlier, there were no defined ideologies. Art movements started contemporary to the industrial revolution. This led to the development of modern art and design. Ideologies were defined by a group of people to design and create things. Graphic Design was also influenced by these movements. A few most influential art movements are:

Art Nouveau or New Art (1890–1914) was the first art style to be considered a Modern style. Art Nouveau is the revamped version of Art Deco which was inspired by Gothic Style. Art Nouveau focused on natural forms. This representation was made simple and decorative due to the industrial revolution, Lithography was introduced to the world which allowed the mass production of color posters. Women were on posters that supposedly symbolized modernity. Due to a continuous change in political ideologies, posters were not only confined to museums but they could also be seen on the streets.

Bauhaus (1919–1933) was the first professional design school in the world. Set up in Germany, Bauhaus was based on the ideology "Forms follows Function." Contemporary design schools were and are built on Bauhaus ideologies. Bauhaus was one of the styles that promoted and adopted the change in the novel method of designing i.e., technology. Bauhaus made people realize how machines and technology could speed up the design process. The use of geometrical shapes and the function of the shape is the unique selling point of a design in Bauhaus style.

Swiss Style emerged in Switzerland in the 1950s. This style spread so quickly because of its cleanliness, use of geometric forms, and simplicity. Swiss style draws its inspiration from various styles namely constructivism for its simplified use of geometric forms, De Stijl for the usage of grids, and Bauhaus for its simplicity and ideology of less is more.

Propaganda and Graphic Design

Graphic design till the end of the 19th century was used as a medium to make a mark in society against the elite, to communicate a message, to communicate an ideology, to make people realize the reality. World Wars enabled propaganda in the graphic world. Politics entered the play and changed the way of communication.

Posters were a major means of communication to spread the news and information henceforth propaganda. Posters were the center of attraction to draw people to support the war by any means. The vital poster themes were recruitment, finance and home front issues. Although posters designed by both the forces (Central powers and Allied Powers) were different, the underlying purpose was the same. In one way the common purpose was to raise patriotism and nationalistic fervor in people. On the other hand, it was also clear that countries were raising money from various sources for keeping the economies going.

Not only during war, but propaganda was also around for a long time. Propaganda starts when there is a chance to promote something or an idea or for a cause. It is the bridge between the people and the party who wants to sell something. This leads to constant income for the seller.

Conclusions

The world that we live in has many products that play a significant role in our daily life. Design has an impact on those products, the graphics and the signs we observe, the news we read, etc.

Reiterating, graphic design is the process of visual problem solving. The current state of graphic design has been reached after a continuous evolution through time. Graphic Design has undergone many modifications due to the changes in society, art movements, trade between countries, the urge to earn money, politics, and of course the people themselves.

Understanding the history of graphic design is very important as looking back and analyzing the work of the past can bring a significant change in the workflow of today. Although the digital era facilities have greatly improved working conditions, the prerequisite of graphic design has not changed, yet i.e., the creativity of the designer.

References

Eskilson, S. (2019). Graphic design: A history. Laurence King Publishing.

Bibilography

Eskilson, S. (2019). Graphic design: A history. Laurence King Publishing.

Meggs, P. B., and Purvis, A. W. (2016). Meggs' history of Graphic Design. Wiley.

Propaganda graphic design: Propaganda Graphic Design: History and Inspiring Examples, https://creativemarket.com/blog/propaganda-graphic-design.

Pomroy, K. (2021, October 4). The evolution of graphic design styles, https://theartcareerproject.com/evolution-graphic-design-styles/.

Shehab, B., and Nawar, H. (2020). A history of arab graphic design. The American University in Cairo Press.